HEADHUNTING AND THE BODY IN IRON AGE EUROPE

Across Iron Age Europe, the human head carried symbolic associations with power, fertility, status, and gender. Evidence for the removal, curation, and display of heads ranges from classical literary references to iconography and skeletal remains. Traditionally, this material has been associated with a Europe-wide 'head cult' and used to support the idea of a unified Celtic culture in prehistory. This book demonstrates instead how headhunting and head veneration were practised across a range of diverse and fragmented Iron Age societies. Using case studies from France, Britain, and elsewhere, it explores the complex and subtle relationships between power, religion, warfare, and violence in Iron Age Europe.

Ian Armit is Professor of Archaeology at the University of Bradford. The author of more than eighty academic articles, he has also written numerous books, including *Anatomy of an Iron Age Roundhouse*, *Towers in the North: The Brochs of Scotland*, and *Celtic Scotland*.

Headhunting and the Body in Iron Age Europe

IAN ARMIT

University of Bradford

CAMBRIDGE
UNIVERSITY PRESS

CAMBRIDGE
UNIVERSITY PRESS

Shaftesbury Road, Cambridge CB2 8EA, United Kingdom

One Liberty Plaza, 20th Floor, New York, NY 10006, USA

477 Williamstown Road, Port Melbourne, VIC 3207, Australia

314–321, 3rd Floor, Plot 3, Splendor Forum, Jasola District Centre, New Delhi – 110025, India

103 Penang Road, #05–06/07, Visioncrest Commercial, Singapore 238467

Cambridge University Press is part of Cambridge University Press & Assessment, a department of the University of Cambridge.

We share the University's mission to contribute to society through the pursuit of education, learning and research at the highest international levels of excellence.

www.cambridge.org
Information on this title: www.cambridge.org/9780521877565

First published 2012

A catalogue record for this publication is available from the British Library

Library of Congress Cataloging-in-Publication data
Armit, Ian.
Headhunting and the body in Iron Age Europe / Ian Armit.
p. cm.
Includes bibliographical references and index.
ISBN 978-0-521-87756-5 (hardback)
1. Headhunters – Europe. 2. Human body – Symbolic aspects – Europe. 3. Human remains (Archaeology) – Europe. 4. Rites and ceremonies – Europe. 5. Violence – Europe. 6. Iron age – Europe. 7. Europe – Social life and customs. 8. Europe – Religious life and customs. I. Title.
GN575.A76 2012
306.4–dc23 2011032287

ISBN 978-0-521-87756-5 Hardback

For my son, James

Contents

Illustrations and tables

Illustrations

Tables

Acknowledgements

This book took a little longer to write than I had originally imagined, and a great many people and organisations have helped along the way. The project was funded by the Leverhulme Trust, whose award of a Research Fellowship enabled me to travel and conduct research in various parts of Europe. Queen's University Belfast granted me sabbatical leave to carry out my initial research in France in 2005–6, and generous support for the project was subsequently provided by the University of Bradford following my move there in 2006.

Much of my research was done in Aix-en-Provence, where I was based at the Centre Camille-Jullian at the Université de Provence for various periods from 2006 to 2010. I owe a great debt to Dominique Garcia for making this possible, and for many other kindnesses, including advice and information on recent Spanish and Catalan sources. Patrice Arcelin generously shared his expertise on a wide range of issues relating to the southern French Iron Age; the fieldwork I was able to conduct at his invitation at Entremont in 2010–11 has also helped deepen my understanding of that crucial and iconic site. Many long discussions with Loup Bernard have also been immensely helpful in my attempts to get to grips with the complexities of the Iron Age in the region. The intricacies of the Glanon sequence were explained to me by Anne Roth Congès, who also gave freely of her time in helping to track down relevant extracts from the late Henri Rolland's excavation diaries. I would also like to thank Thierry Janin at Lattes and Eric Gailledrat at Pech-Maho who not only discussed their own excavations in detail but also helped with access to what would otherwise have been inaccessible literature. Frédéric Marty, of the Musée Archéologique at Istres, also provided a wealth of advice and information during our collaborative fieldwork at Le Castellan, Istres, in 2007–8. Many of the observations relating to the topography and siting of the sanctuaries at Roquepertuse and Glanon, discussed in Chapter 4, were made during collaborative fieldwork with Mags McCartney in 2004–5.

Access to museum stores was crucial to the completion of my work in France and would not have been possible without the help of several individuals; especially Véronique Legrand, Sabrina Lamotte, and Claude Galdeano of the Hôtel de Sade, Saint Rémy de

Provence, who arranged access to the Glanon material during the renovation work on the museum. Staff at the Musée d'Apt kindly allowed me access to the sculptures from Rustrel despite the rather unexpected (and lengthy) closure of the museum.

Many other friends and colleagues helped ease the path to eventual completion in various ways, including Katharina Becker, Michelle Bonogofsky, Richard Brunning, John Collis, Tom Dawson, Gerard Fercoq du Leslay, David Field, Vicky Ginn, Alfredo Gonzalez Ruibal, Nick Haimendorf, Mark Hall, Delphine Isoardi, Rob Janaway, Jody Joy, Chris Knüsel, Ian Kuijt, Jim Mallory, Alan Macfarlane, Florence Mocci, Conor Newman, Ian Ralston, Katherina Rebay-Salisbury, Becky Rennell, Jim Rylatt, Rick Schulting, Fiona Shapland, Scott Speal, Janet Trythall, Tiffany Tung, and Kevin Walsh.

Some parts of the book build on papers written and presented during the period of my Leverhulme Fellowship. Parts of Chapter 5 have been reworked from a paper first presented at the European Association of Archaeologists' conference in Zadar in 2007 and subsequently published in the volume *Body Parts and Bodies Whole*. I am grateful to the editors, Marie-Louise Stig Sørensen, Katherina Rebay-Salisbury, and Jessica Hughes, for their permission to include this material here. Work on the human remains from the Sculptor's Cave also plays a big part in Chapter 5, and this has benefited hugely from collaborations with Rick Schulting, Chris Knüsel, and the late Ian Shepherd. Similarly, some parts of Chapter 6 originated in a paper first presented at the European Association of Archaeologists' conference in Krakow in 2006, subsequently published in *Atlantic Europe in the First Millennium BC: Crossing the Divide*, and I am grateful to the editors, Tom Moore and Lois Armada, for permission to use this material here. Certain sections dealing with La Tène art, in Chapters 2 and 4, had their origins in my contribution to *Relics of Old Decency: Archaeological Studies in Later Prehistory (Festschrift in Honour of Barry Raftery)* edited by Gabriel Cooney, Katharina Becker, John Coles, Michael Ryan, and Suzanne Sievers. My thanks again go to the editors for their permission to rework this material. Original illustrations were created by Dan Bashford, Rachael Kershaw, and Libby Mulqueeny. The bibliography and index were organised and formatted by Catriona Armit.

I am especially grateful to Chris Fowler and Fiona Shapland for reading the whole book in draft, and for their many useful and encouraging comments. I would also like to thank all at Cambridge University Press for their help and patience, especially Beatrice Rehl, Amanda Smith, James Dunn, and Brian MacDonald.

I never did manage to get back to Rum to check the photographs discussed in Chapter 3. Since I didn't have a book in mind (or a notebook in hand) when I first saw them, I hope my descriptions from memory bear at least some relationship to reality.

Finally, I would like to thank Catriona and our children, Rowan, James, and Theo, who have grown up with this book for rather longer than was ever intended.

Note: All AMS dates are quoted at two sigma. All translations are by the author except where otherwise indicated.

Bradford 2011

CHAPTER 1

Detached fragments of humanity

The Berawan of Long Teru lost their stock of skulls in a disastrous fire that consumed the entire longhouse. It was an incident that was often recalled, and the poor woman, now elderly, whose kitchen fire started the blaze has never been allowed to forget it. But oddly enough, the loss of skulls was not lamented. Though it was claimed that the heads had the potential to bring benefits to the community, the service of them was considered onerous. The heads had to be 'fed', so it was said, with small offerings, and kept warm with a fire that never went out. Women had to avoid that part of the longhouse veranda, or pass by in a crouched position. Any contact with heads was dreaded, so that only old men, weary of life, would dare move them, whenever the house needed rebuilding.

Metcalf 1996, 251

And so they met and there was a hard battle, and not long ere Melbricta fell and his followers, and Sigurd caused the heads to be fastened to his horses' cruppers as a glory for himself. And then they rode home, and boasted of their victory. And when they were come on the way, then Sigurd wished to spur the horse with his foot, and he struck his calf against the tooth which stuck out of Melbricta's head and grazed it; and in that wound sprung up pain and swelling, and that led him to his death.

Orkneyinga Saga, 5, trans. Dasent, 1894

Introduction

Some of the best-preserved prehistoric buildings in Europe are to be found in the Western Isles of Scotland. Along the coastal fringes of islands such as Lewis and North Uist are the buried remnants of extraordinary structures, long since engulfed by sand. One wet spring, in the late 1980s, I was busy excavating a settlement in the small township of Cnip, on the west coast of Lewis (Ill. 1.1). People living nearby had noticed the remains of some stone buildings eroding onto the beach, and a rescue excavation had been hastily arranged. The settlement at Cnip turned out to be a remarkable site, spanning the first century BC to the second century AD. The original building was a wheelhouse, a type of drystone round-house dug down into the sand. Soaring drystone piers (the 'spokes' of the wheel) formed a circle of small cells around a central, communal space. The walls still stood above head height in places.

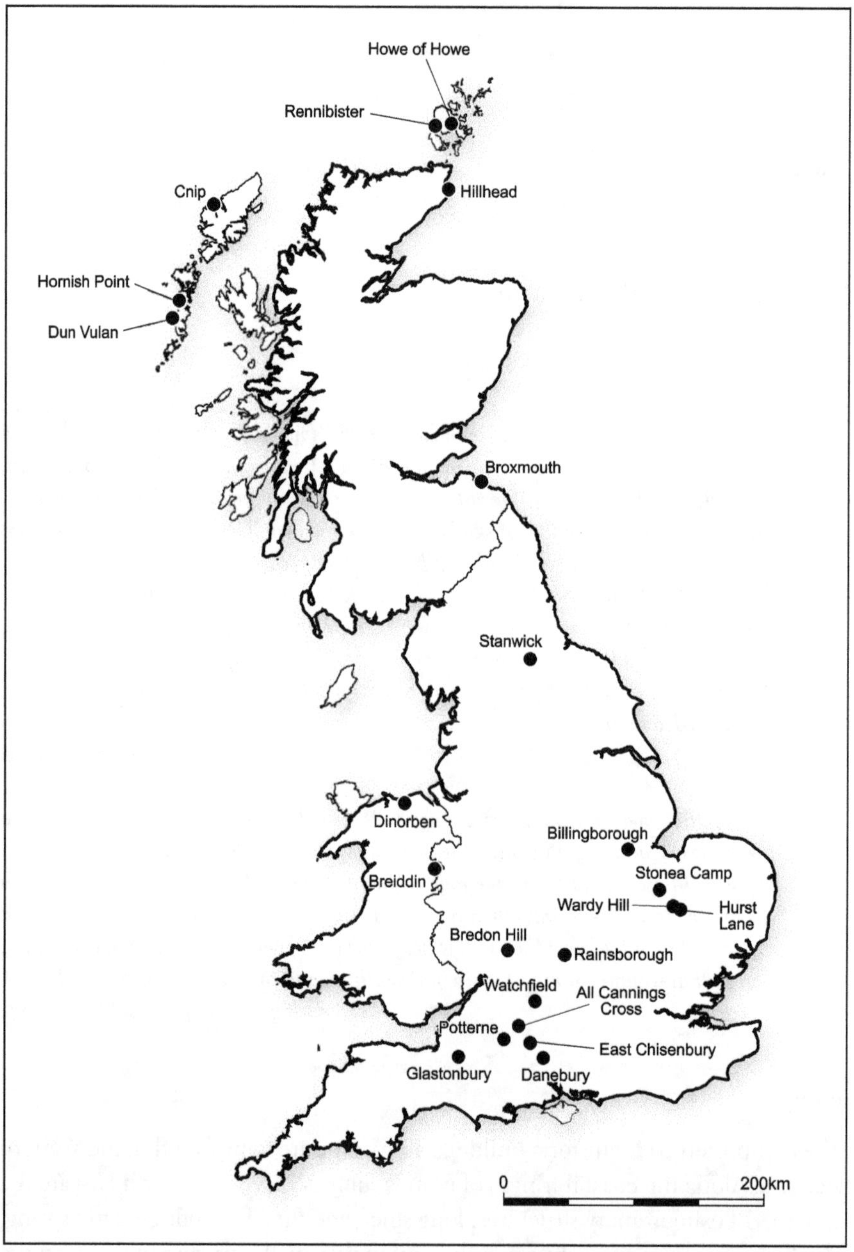

ILLUSTRATION 1.1. Main British sites discussed in Chapter 1. (Drawn by Rachael Kershaw)

This exceptional survival meant that the houses at Cnip could be dissected in great detail (Armit 2006a). When we began dismantling the walls, it soon became apparent that each stage in the life of the settlement had been marked in some way. In most cases, small deposits of carefully selected objects had been placed in pits, under thresholds, or behind the walls of the houses as they were built. Behind one wall, for example, were several

ILLUSTRATION 1.2. Votive deposit, comprising a cranium, a rounded stone, and some fragments of pottery and bone, placed in a pit below a small cell at the Cnip wheelhouse, Isle of Lewis, during the first century AD. (Photograph by author)

separate deposits: a complete pot, a length of articulated cattle vertebrae, and the head of a great auk, a now-extinct sea bird.

One of these deposits was particularly striking. Sometime during the first century AD the inhabitants of the wheelhouse had decided to create a small cell within the remains of a disused building. Before laying the foundations, someone scooped a hollow into the sand and placed in it a few objects, specially selected for the occasion (Ill. 1.2). Among these was the upper part of a human cranium, laid in the base of the hollow. Next to it was a smooth, rounded stone, which seemed to echo the shape of the cranial vault, and which was quite unlike the usual angular building stones found around the site. There was also a second cranial fragment, probably human but perhaps animal, and two scraps of pottery. Once placed in the hollow, this collection of broken and fragmentary remains was covered over with sand and buried beneath the slab floor of the new building (Armit 2006a, 244–8).

On one level this deposit fits rather neatly into the wider pattern of votive deposition at Cnip. Yet the deliberate incorporation of human remains marks it out and raises some important questions. Who was this person: perhaps an ancestor, or an enemy, or maybe a recently deceased inhabitant of the wheelhouse? Why was it judged appropriate to place human remains in this particular place and not in others? And why was only the head deposited?

The cranium itself yields few clues. It belonged to an adult, probably middle-aged. Its partial condition suggests that a significant period of time had elapsed between death and this ultimate burial. What had happened during that time? Was the head curated in some way by the inhabitants of Cnip, as a defleshed skull or cranium, or perhaps as a fleshed head, preserved by smoking or drying? Or had the cranium simply been retrieved from a grave, or some other location, with this votive deposit in mind? The smaller cranial fragment found next to it showed traces of gnaw marks, suggesting that it at least had been left exposed to animals, either within the settlement (where dogs and pigs were kept) or elsewhere.

The alkaline qualities of the Hebridean machair sands are ideal for the preservation of bone. Yet, although thousands of animal bones were recovered, the excavations at Cnip yielded only three other pieces of human bone. One was a fragment of tibia, or shin-bone, from a domestic midden. The others, found within the buildings, were, once again, pieces of cranium. One fragment from a young adult, found in wall packing, bore a series of cut marks, which were hard to interpret. Some of them could be the result of scalping, but this would not explain them all. Some may result from an unsuccessful attempt at trepanation (McSweeney 2006, 134–5), although they may equally have been made shortly after death. The third piece, from the cranium of a middle-aged adult, found at the entrance to a small cellular building, had unquestionably been modified after death, because a hole had been drilled from both inside and out (136). This may have been intended to allow the head, or cranium, to be suspended from a cord.

The small but intriguing assemblage of human bone from Cnip was plainly not randomly generated. Three of the four pieces (the only three found inside the houses) were cranial fragments, all from middle-aged or young adults. All had been modified and/or curated. For whatever reasons, the Iron Age inhabitants at Cnip had a strong interest in human heads. How might we interpret such material? With only four bones to work with, should we even bother to attempt interpretation at all?

The published literature reveals some intriguing parallels for the Cnip cranial fragments. One of the most dramatic was recovered by the Caithness landowner and antiquarian, Sir Francis Tress Barry, during his excavations at the complex Atlantic roundhouse of Hillhead (Anon. 1909). This calvarium, or skullcap, had been drilled through with three evenly spaced holes. These suggest that the skull, or fleshed head, like the drilled fragment from Cnip, might have been strung up for display (Ill. 1.3). Interestingly, this fragment was also found at an access point into a building, in this case on the floor of the entrance passage leading into the roundhouse (Tress Barry n.d., 7). Slightly further north, at Rennibister in Orkney, disarticulated bones representing six adults and at least 12 children were discovered in a rock-cut 'earth house', which probably dates to the early part of the Iron Age (Marwick 1927a; Armit and Ginn 2007). From the published records it appears that the

ILLUSTRATION 1.3. Perforated skullcap from Hillhead, Caithness. (Photograph by Fiona Shapland)

bodies had been carefully sorted after a considerable period of decomposition. Four of the skulls had been placed beside one of the pillars supporting the roof. In each case, the cranium had been inverted over a mandible (Marwick 1927b, 299).

Further instances come from the excavations at the complex roundhouse of Dun Vulan in South Uist, where the long sequence of occupation yielded numerous small fragments of human bone (Parker Pearson and Sharples 1999; Chamberlain 1999). These again included a wholly disproportionate number of cranial and mandible fragments, some of them in contexts suggestive of votive deposition. Radiocarbon dates suggest that some of these fragments were several centuries old at the time of deposition (Mulville et al. 2003). For instance, one piece of human mandible recovered from a stone-lined drain was radiocarbon dated to 110 BC – AD 130: yet the small rectangular building that this drain served was not itself built until some time during the third and fourth centuries AD. The excavators thus argue that this carefully curated fragment, some centuries old, was deliberately offered up as a foundation deposit when the later building was constructed (23–4). As at Cnip, human remains at Dun Vulan had apparently been curated for a significant period. Further skull remains from various contexts in the long-lived settlement of Howe of Howe in Orkney, although not as well-dated, hint at similar processes (Ballin Smith 1994; Armit and Ginn 2007). Indeed, the more closely one looks in the records of earlier excavations in Atlantic Scotland, the more of these apparently isolated and anomalous human bones one finds (Armit and Ginn 2007).

Midden bones

The situation is similar elsewhere in the British Iron Age, where fragments of human bone are often encountered in seemingly unlikely places. All Cannings Cross, in Wiltshire, for example, offers some parallels. This site, excavated between 1911 and 1922, dates to the transition from the Late Bronze Age to the Early Iron Age. It comprised an extensive area of dark soils, apparently unenclosed, incorporating occasional laid floors, hearths, pits, and even some post-built structures (Cunnington 1923). All Cannings Cross has been linked to other sites in and around the Vale of Pewsey, characterised by similar spreads of dark soil (Lawson 2000, 265). These have sometimes been described as middens, on the basis of their rich assemblages of discarded rubbish and their general lack of evidence for permanent occupation. They were, however, more than simple rubbish dumps. Instead, they seem to have been used for feasting and other activities associated with periodic gatherings of people – events that generated huge quantities of debris. David McOmish (1996, 75) has suggested that the term 'ceremonial feasting places' would be a better fit.

At All Cannings Cross, among the broken pottery, animal bone, and other debris, the excavators found more than 30 human cranial fragments, scattered and dispersed with no obvious pattern to their distribution. Aside from these, no other human remains were found. As in Atlantic Scotland, some of the All Cannings Cross fragments had been deliberately modified, apparently to be 'used for scraping or other purposes' (Cunnington 1923, 40). One had been worked into a small circular roundel, 'almost exactly the size of a penny-piece' and had a hole bored well off-centre (Keith 1923, 41, plate 26). Judging from the wear marks, it had apparently been carried or worn for some time, perhaps as a charm or amulet. A similar perforated roundel, made from the occipital bone 'of an old person', was found at Glastonbury lake village in Somerset (Bulleid and Gray 1917, 405). However, the Glastonbury piece is around 7cm in diameter with a central perforation, resembling a spindle whorl.

Another midden, at Potterne, some 10 km from All Cannings Cross, covers a larger area of around 3.5 hectares (Lawson 2000). Excavations of a small part of this site uncovered an enormous artefactual assemblage and around 134,000 animal bones (Locker 2000, 101). Compared to this, the human bone assemblage was tiny, with only 139 fragments recovered (J. McKinley 2000, 96). Yet, once again, these few bones raise many questions. Unlike the material from All Cannings Cross, the human bone at Potterne was not restricted to cranial fragments, though these did make up more than half the assemblage. Both sexes were represented as well as a range of age groups, including a child of around six (table 9). The absence of mandibles and cervical vertebrae suggests that it was defleshed crania, rather than heads, that were present on the site. Indeed, the virtual absence of facial bones may suggest that only calvaria, or cranial vaults, were present. The rest of the human bones had been equally selectively obtained. Most were long bones, with the femur predominating, while axial bones (such as vertebrae and ribs) were extremely rare. Small bones, such as those from the hands and feet were entirely missing. Selectivity was also marked in other ways. Among the leg bones, for example, there was a marked preference for right over left limbs; and, although numerous neonatal and foetal bones were present in the assemblage, there were no cranial remains from these individuals (99 and table 9).

None of the cranial fragments from Potterne seem to have been worked into objects, but at least two had a 'polished' or 'ivoried' appearance. Although the suggested causes of this condition included possible variations in the depositional environment, or 'some form of human manipulation' (J. McKinley 2000, 97), the same condition in the animal bone assemblage is interpreted, much less ambiguously, as being due to 'some cooking treatment, such as boiling' (Locker 2000, 103). Because there is no sign that any of the human bones were butchered, boiling may have formed part of the preparation of heads for display rather than consumption.

Human remains were also found in small-scale excavations at East Chisenbury, another of these midden sites, which extends over nearly 4 hectares (McOmish 1996). Although only a tiny area was examined, the excavators once again found human remains including four skull fragments (David Field pers. comm.). One of these seemed to have been deliberately placed, since it was surrounded by broken sherds from a single vessel and a small block of stone (McOmish 1996, 73). Seven centuries and nearly 600 miles may separate them, but the resonances with the cranial deposits from Cnip are strong.

Trophies and offerings

A different sort of head deposit was encountered by Sir Mortimer Wheeler during his excavations at Stanwick, in North Yorkshire in the early 1950s. Stanwick is an enormous complex, with enclosures covering more than 250 hectares, and it seems to have been built at the end of the Iron Age, in the middle of the first century AD. It appears to have been an important stronghold of the Brigantes, the most powerful northern English tribe at the time of the Roman invasion (Wheeler 1954). Although Wheeler's excavations were restricted to a few short stretches of ditch and a tiny fraction of the interior, he did encounter some striking deposits. Wheeler's Site B examined a length of ditch adjacent to one of the gateways and probably dating to the second half of the first century AD. In the waterlogged ditch terminal were various objects, including an iron sword in a bronze-bound wooden scabbard and the cranium, mandible, and cervical vertebrae of a middle-aged man. The cranium bore traces of at least three severe wounds inflicted by a sword or axe at the time of death (Osman Hill 1954). It appeared that this was the head of a warrior killed in battle and subsequently decapitated. The association of the cranium, mandible, and vertebrae shows that it was a fleshed head that was originally deposited in the ditch.

Wheeler (1954, 53) suggested that both the sword and the severed head may have formed parts of a trophy, displayed over the gateway, which subsequently tumbled into the ditch terminal when the site was destroyed. There is no apparent indication, however, in Osman Hill's (1954) osteological report as to how the head might have been fixed to the gate structure, and we should not exclude the possibility that the head, and perhaps the sword and scabbard, were simply placed into the ditch terminals. Interestingly, although this particular head deposit is often cited (e.g., L. Laing, 1981, 115), there were other cranial fragments found at Stanwick. Two other ditch cuttings, one of them more than a kilometre away at the far south of the enclosure system, yielded a further four pieces of human skull, which were not described in any detail in Wheeler's report (Osman Hill 1954, 56). Compared to the overall extent of the ditches at Stanwick, Wheeler's excavations were minimal, making it all the more surprising that so many cranial fragments were recovered. We must, presumably, expect that many more skull fragments remain within the several kilometres of unexcavated ditch.

Several other Iron Age British sites seem to support Wheeler's suggestion that human heads were put on display. At Bredon Hill, in Gloucestershire, for example, three crania and six mandibles were found during a 1930s excavation, under the debris from the collapsed gate. The excavator suggested that these were the remains of trophy heads, displayed above the gateway (Hencken 1938, 57). We should bear in mind, however, that many other human bones were found in the debris surrounding the entrance to this site, representing up to around 64 people, many of whom appear to have been subject to 'extensive mutilation … possibly on ritualistic grounds' (55). In the early Iron Age levels of the hillfort of Dinorben, in northeast Wales, fragments of human crania were found in the floors of three houses, as well as in one of the guard chambers at the main entrance. A further mandible fragment, recovered from the ditch next to the entrance, was interpreted by the excavators as a fallen part of a decayed trophy head, formerly displayed over the gateway (Gardner and Savory 1964, 221). As at Bredon Hill, however, it should be noted that other postcranial fragments found in the Early Iron Age levels did not attract detailed discussion in the excavation report. Elsewhere, heads seem to have been associated with the construction of ramparts, as at the Breiddin hillfort in Powys, where the maxilla (facial bone) of an adult had been incorporated within the Late Bronze Age rampart (Musson 1991, 23).

The phenomenon of display is not restricted to hillforts. Numerous skull deposits are recorded from the early excavations at the Glastonbury lake village (Bulleid and Gray 1917), where they dominated a human bone assemblage otherwise largely restricted to two partial burnt skeletons and around 10 neonatal burials. Most of the material was fragmentary, but in at least two cases, crania and mandibles were found together (Coles and Minnitt 1995, 170–4), suggesting the deposition of fleshed heads. Both showed evidence for unhealed sword injuries, suggesting a rather violent death. At least one had suffered significant damage to the foramen magnum, suggesting they it had been displayed on a spear or stake (ibid.; Boyd-Dawkins 1917).

I could, of course, go on. This brief survey of the British evidence omits a host of intriguing assemblages; skull fragments displaying signs of perimortem trauma from the interior of Broxmouth hillfort in East Lothian (Armit and McKenzie in prep.); a severed head in the 'guard-chamber' at Rainsborough hillfort in Northamptonshire (Banks 1967); evidence of decapitation and ritual deposition in the grain pits at Danebury (Craig et al. 2005); the structured deposition of sometimes modified skull fragments from a range of Iron Age sites in the eastern counties of England, including Wardy Lane, Billingborough, Hurst Lane, and Stonea Camp (Evans 2003); and innumerable other examples (Wilson 1981). Nonetheless, the basic point is clear – that Iron Age communities across Britain engaged in a range of practices relating to the removal, curation, and display of the human head.

Interpretations

How should we interpret these strange and highly selective assemblages of human remains, these 'detached pieces of humanity' (Cunnington 1919, 25)? Writing her report on the excavations at All Cannings Cross, Maud Cunnington had little doubt that she had found evidence for Celtic headhunting, as described in the writings of the classical authors (Cunnington 1923, 40). Indeed, wherever human skull fragments have been found on Iron Age settlements, the Celtic 'cult of the head' has been a popular explanation (e.g., Whimster 1981, 189; Wait 1985, 120), drawing on a powerful blend of classical and medieval literature, as well as archaeological and iconographic material from across much of Europe. The Celts, it was thought, saw the head as the seat of the soul, and thus severed heads could retain special powers after death. By cutting off and keeping the heads of their enemies, warriors could gain control over the spirits of the dead. By the same token, the heads of important individuals might be venerated after death. Identifications of 'head-hunting' or the 'veneration of the head' in the archaeological record have thus tended to be seen as indicating local adherence to a perceived pan-Celtic tradition, linking the islands of northern and western Scotland, and the chalklands of southern England, to the Mediterranean coasts of France and Spain.

Can we then understand the contents of the pit at Cnip, the cranial fragments from All Cannings Cross, and all the other instances discussed here as local expressions of some deep-rooted, distinctively Celtic, cosmological principle? This is doubtful. The concept of a culturally unified Celtic people in prehistory has been vigorously questioned in recent times (e.g., S. James 1999; Collis 2003), and there has been an increasing awareness of marked regional variations within the traditionally Celtic world of the European Iron Age. Iron Age communities of Atlantic Scotland were, in many ways, very different from their contemporaries in Wessex (e.g., Armit 2003; Cunliffe 2005), and neither need have had much in common with the continental communities described in the classical literary descriptions of the Celts. More fundamentally, however, special treatments of the head are present in chronologically disparate archaeological contexts in Europe, from Mesolithic 'skull nests', as at Ofnet (Orschiedt 2002, 2005) and Kaufertsberg in southwest Germany (Orschiedt 1998), to the relics of medieval Christian saints. In fact, similar practices can be documented from the earliest times through to recent centuries in most parts of the world (e.g., Chacon and Dye 2007). In such circumstances, it is difficult to explain headhunting, and head veneration, by reference to some Celtic cultural milieu.

It is also increasingly evident that, despite the high visibility of skull fragments at many Iron Age sites, other parts of the body were also subject to unusual treatments. We have already seen how the human bone assemblage at Potterne had a disproportionate representation of femurs, an under-representation of axial bones, and a bias towards right, as opposed to left, limbs. The Iron Age hillfort of Danebury, in Hampshire, has yielded considerable evidence for a special interest in heads (e.g., Walker 1984; Craig et al. 2005), yet, as Niall Sharples (1991, 81) has pointed out, numerous other body parts, including articulated limbs and pelvic girdles, were also deposited in pits around the interior of the hillfort. Nobody has yet argued the case for a 'Celtic cult of the pelvic girdle'. It is clear, however, that any attempt to understand the treatment of the human head in the Iron Age must also address wider attitudes to the human body.

Another issue, often overlooked, is the frequent association of human and animal bones. At Watchfield, in Oxfordshire, for example, a human cranium had been placed into a pit with a pig skull (Roberts and McKinley 2003), and at Hornish Point in South Uist, the remains of a young boy had been quartered and placed in four pits along with the butchered remains of young cattle, sheep, and pigs (Barber 2003). Neither of these instances is resonant of conventional rubbish disposal. Instead, the deliberate admixture of animal and human bodies seems to carry quite specific meanings. The killing, consumption, and deposition of animals may in some cases have substituted for the similar treatment of humans. In other cases, the deposition of particular animals, or animal parts, may have lent additional meaning to the human remains that accompanied them into the earth. Elsewhere we may be seeing a blurring of categories between animal and human (Hill 1995, 16).

In the following chapters, I want to re-examine the archaeological evidence for Iron Age headhunting, and associated practices, from a rather different perspective. Although the classical and early medieval literature dealing with the Celts remains essential, I try to develop new approaches using, primarily, anthropological studies of more recent headhunting societies and close contextual analyses of particular Iron Age communities. Anthropological literature on headhunting has been affected by the reluctance of recent generations of anthropologists to focus on aspects of non-Western cultures that might lend support to popular caricatures of the 'primitive'. Nonetheless, detailed case studies and synthetic works exist for certain areas, notably Southeast Asia (e.g., R. Rosaldo 1980; Hoskins 1996a) and Oceania (e.g., Aswani 2000a), as well as an enormous body of early ethnographic literature of highly variable quality. The insights that such studies might generate have been largely ignored in past studies of Iron Age Europe. In part, this is because the existing literature on the Celts has sometimes seemed to provide all the information we need. I do not suggest that the anthropological literature can provide ready-made templates that can simply be applied to the communities of Iron Age Europe. Yet it can potentially furnish us with fresh ideas and approaches, opening up new avenues for interpretation of the Iron Age material and revealing connections and meanings that were previously dormant.

Anthropological studies, as we will see in Chapter 3, reveal the centrality of headhunting within many non-Western cultures and the wide range of cultural and material expressions associated with the treatment of the human head. Headhunting, clearly, is far more than the collection of battle trophies. Heads may be taken from the out-group, through acts of group-sanctioned violence, or from the in-group, as a product of funerary practice. The head can be a potent symbol, with associations relating to power, fertility, coming of age, the acquisition of status, and gender. The control and deployment of this symbol can potentially tell us a great deal about wider social relationships and practices.

Ideas and practices associated with headhunting manifest themselves quite differently, even in neighbouring communities, and this idea of 'difference' will be important when we come to look again at the societies of Iron Age Europe. As will become clear in Chapter 2, I am not setting out to define a distinctively 'Celtic' belief system, applicable across the length and breadth of Iron Age Europe. I want instead to explore the nature of specific Iron Age communities, in particular historical circumstances. This requires a close analysis of the archaeological material within particular regions, rather than recourse to pan-Celtic explanations. Variations on the theme of headhunting, curation and display, can help us to reassess the social and cultural relationships between Iron Age societies without resorting to blanket attributions of shared Celtic ethnicity.

Some definitions

Headhunting

The anthropologist Janet Hoskins (1996b, 2) defines headhunting as 'an organised, coherent form of violence in which the severed head is given a specific ritual meaning and the act of head-taking is consecrated and commemorated in some form'. This is a useful starting point, encompassing the taking of head trophies in war, as well as decapitation as a component of sacrifice. In either case, it could cover the removal of heads from members of either the in-group or the out-group.

Yet, Hoskins's definition does not necessarily encompass all of the categories of behaviour that I want to consider. The emphasis on consecration is probably valid in the specific region to which Hoskins's formula applies (Southeast Asia), but it implies a clear connection between headhunting and religion that we might want to question. It could be argued that a good deal of human trophy-taking, at other times and places, has had less to do with spiritual beliefs than with the brute demonstration of power and the dehumanisation of the enemy. The widespread decapitation of Japanese war dead to provide souvenirs for Allied soldiers during World War II could, for example, be interpreted in this way (S. Harrison 2006).

For the Berawan of Borneo, the act of decapitation was merely the first stage in an ongoing relationship between head-taker and victim. This relationship was one that could pass down the generations and involved long-term curation and 'nurturing' of the head, by feeding, sheltering, warming, and the showing of due deference (Metcalf 1996, 251). For the medieval Earls of Orkney, the taking of a head was a more straightforward business (*Orkneyinga Saga*, 5, trans. Dasent, 1894). Quite what Sigurd's intentions were with regard to the heads of Melbricta and his unfortunate followers is not entirely clear, but most likely they would have been discarded after a rather briefer and less reverential period of display than practised by the Berawan. Sigurd's motives, as related by a cultural insider, seem to have been primarily based around the display of power and the denigration of a despised enemy. The element of 'consecration' here is much less obvious, if it exists at all. Similarly, the frequency with which beheadings in combat were recorded in the medieval Irish annals, in a Christianised society, might suggest that the spiritual element was unnecessary for the maintenance of a headhunting culture. Individuals and groups within nominally Christian societies, however, retain all sorts of beliefs that might seem incompatible with modern scriptural interpretations, and it would be unwise to assume that these decapitations lacked any 'specific ritual meaning' in Hoskins's sense. Nonetheless, I want to keep the possibility open, at least for now, that headhunting, in its widest sense, need not have an overtly religious dimension.

There are other analogous practices relating to the treatment of the human head that do not fall strictly within Hoskins's definition, yet which are still clearly related to headhunting as she describes it. The early Irish literature, once again, describes how victorious warriors would remove the brains of their defeated opponents and mix them with lime to create a hard and durable 'brain ball', which could be shown off on suitable occasions (K. Jackson 1964, 20). Other forms of human trophy-taking, such as scalping, or the removal of ears or noses, could be taken as a shorthand for headhunting and may invite similar sorts of interpretation. We might also want to consider judicial beheadings in state societies like those of medieval Europe or modern-day Saudi Arabia, where severed heads may be displayed, in a ritualised fashion, after the event.

For present purposes, then, I define headhunting as a form of group-sanctioned, ritualised violence, in which the removal of the human head plays a central role. It commonly involves the curation, display, and representation of the head, often within a religious context, but all of these elements need not be present in every case. The more specific term, 'predatory headhunting', is also used to describe the violent targeting of outsiders as a source of head trophies.

Head veneration

The pervasive interest in the human head, however, encompasses a still wider territory than even this expanded definition can accommodate. In a number of ethnographically documented contexts, certain communities retain, display, and venerate the heads of their own kin (e.g., Barth 1987). These are obtained not through violence but in the course of conventional funerary practices, even if only a small proportion of heads are retained. These may belong to particularly important, influential, or unusual individuals, whose particular life histories made them appropriate foci for communal ritual and memory, but equally they may be chosen quite arbitrarily. Among the Mountain Ok peoples of Inner New Guinea, for example, skulls were displayed in central 'ancestor houses' where they were painted, on certain occasions, in the colours of particular clans and played a central role in communal ceremonies (Barth 1987). Individual skulls might be used to symbolise the category of all ancestors, and the specific identity of the skull's original owner need not be especially important to its posthumous ritual life. Ideas of a similar kind underlie that importance attached to the relics of Christian saints in medieval Europe (Woodward 1993), although here the specific identity (real or imagined) of the original owner of the body part in question was, of course, crucial.

Headhunting and head veneration are by no means mutually exclusive. The Berawan obtained their collection of skulls through violence directed against outsiders, yet subsequently venerated them over many generations (Metcalf 1996). Elsewhere, for example, both the New Zealand Maori (Vayda 1960) and the North Indian Nagas (Jacobs 1990) curated the heads of both enemies and ancestors, though they were differently treated in each case. The practices and beliefs involved, as we shall see throughout the following chapters, are so intertwined that the attempt to define headhunting and head veneration as separate categories serves little useful purpose. For present purposes, the term 'head veneration' is restricted to contexts where no element of interpersonal violence is involved, for example, where heads are obtained exclusively as part of secondary funerary practices. This does not, of course, mean that an element of veneration is not present in many other contexts.

Cosmology, religion, ideology

Interpretations of headhunting in prehistoric societies, as in the ethnographic record, inevitably involve discussions of cosmology, religion, and ideology. Yet archaeologists have markedly divergent views over the relationships between these concepts. Insoll (2004, 23), for example, sees religion as an overarching structural principle that determines the ways in which other aspects of human life are organised. Although I agree with his view that the role of human mental life has been underplayed in archaeology, at the expense of (*inter alia*) technology, trade, and subsistence, I think that religion is too narrow a term in this context. Instead I would conceptualise religion, cosmology, and ideology as an interlinked group of mental domains. None of these need have primacy, but each clearly has the capacity to influence the others, especially over a period of several generations or centuries. As these terms are used throughout the book, it is worth setting out what I mean by them.

Cosmology, at least in the context of premodern societies, relates to the ways in which people understand their world, its origins, and the place of humans within it. It includes 'mental geographies' such as the common understanding, in ancient times, of the world as a great circle centred on the Mediterranean, as, for example, in the maps of Anaximander and Hecataeus. This inner circle of civilisation was girdled by concentric zones of progressively less civilised, stranger, and more fantastical inhabitants (Piggott 1968, 80). Such ideas may sometimes be closely related to the concept of religion, but they could be argued to be rather more fundamental, in the sense that similar cosmologies, for example, might coexist with quite different religions. Cosmological understandings may operate at quite unconscious levels, governing the ways in which people categorise objects in the world around them, including various human (and what may be perceived to be subhuman or near-human) groups, animals, plants, and so on.

By religion, I mean a body of commonly held beliefs relating to the nature and workings of the supernatural. This body of belief need not be codified in any formal way, nor need the boundaries between individual 'religions' be clearly definable. I assume that the communities of Iron Age Europe all held beliefs that we would (if we had access to their long-vanished minds) characterise as religious. I also assume that beliefs in these nonliterate societies were highly fluid across both time and space. In other words, no single 'Celtic religion' ever existed.

The concept of ideology is clearly related to both cosmology and religion. For Marx and his followers, ideology was a means by which the dominant class legitimated its position through the promulgation of a system of ideas that enshrined a 'natural' social order. Such ideologies might be inextricably intertwined with religious ideas, in which the position of the ruling class is presented as divinely sanctioned (if the ruling class is not itself divine). Such ideologies are often envisaged as a form of concealment, in which the dominant elite pulls the wool over the eyes of a subservient populace. From an external perspective, this may be a valid reading of the situation, but it perhaps underplays the extent to which members of the dominant elite are themselves bound by their own ideological constructs. Ideology may indeed represent 'false consciousness' – in other words, a distorted perspective obscuring the real operations and exploitations within a society. Yet it need not be seen as a conscious trick, knowingly played by one privileged group over the lower orders.

The boundaries between these three domains are necessarily fluid (Ill. 1.4). Those with a Marxist bent, for example, would probably see religion as no more than a component of ideology, whereas others (e.g., Insoll 2004) would prefer to see both ideology and cosmology more or less subsumed within religion. I am not interested in establishing firm boundaries between what will always remain fuzzy concepts. What is important here is that the relationships between these conceptual domains are open to change over time and space. Equally, changes in one are very likely to lead to changes in the others, particularly in a preliterate context. For instance, a new elite group, emerging through

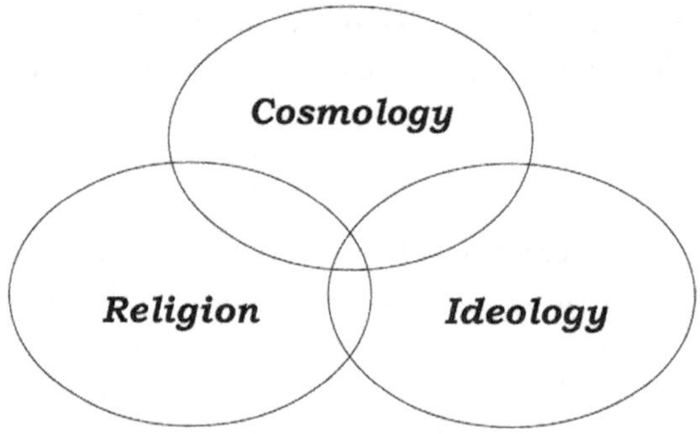

ILLUSTRATION 1.4. Religion, cosmology, and ideology.

economic or military preeminence, may rather quickly establish new ideologies to legitimate its privileged position (e.g., Earle 1997). It may well exert power or influence over religious structures that might shift or distort the content, or at least emphasis, of religious understandings. In time, this may affect the cosmological basis on which understandings of the world are founded, embedding a largely arbitrary social order within a stable 'natural' system. Cosmology, religion, and ideology are thus used here as heuristic devices to enable us to understand some of the different manifestations of the human head in Iron Age practice and belief.

Ritual violence

> Ritualisation is both a way of acting which reveals some of the dominant concerns of society, and a process by which certain parts of life are selected and provided with an added emphasis. (Bradley 2005, 34)

If headhunting is to be understood as a form of ritualised violence, then we need to consider what is meant by ritualisation. Richard Bradley (2005) has defined this as the process by which particular acts are given special emphasis, through the formality of their performance. Ritualised actions might simply elaborate on routine, day-to-day activities, or else they might be performances specifically created to achieve a particular purpose or express a particular idea or set of relationships. The kind of complex liturgical ritual that comes quickly to mind for modern westerners fits into this latter category, where the purpose is mediation with the supernatural. Ritualised actions might be repeated many times and may be seen as 'traditional' in character. Equally, they may represent creative variations on a set of commonly understood themes, such as we might see in the individualistic ritual performances of shamans (Jordan 2001, 100). In small-scale societies, lacking any central religious authority, we might expect a good deal of innovation and variety in ritual performances, especially where these are infrequent and initiated by part-time, rather than fully specialised, ritualists (e.g., Hill 1995, 116).

Perhaps the most important aspect of Bradley's concept of ritualisation is that it firmly separates ritual from religion; two concepts that are frequently confused. Ritual acts need not be religious in character. Though they may involve interactions with the supernatural, rituals may also be essentially secular, centring on communication primarily between performers and the audience. But ritual acts do, I would suggest, necessarily relate to one or more of the triad of religion, cosmology, and ideology.

Decapitation, like other acts of violence, may be ritualised as a means of accommodating it within a religious, cosmological, or ideological framework. In this way, ritualisation can act to legitimize violence that might, from another perspective, be seen as motivated by a simple desire for personal power, slaves, land, or tribute. Yet, in other contexts, heads are taken primarily to fulfil religious or cosmological obligations. The ritual element here is not a post hoc rationalisation of violence: it is the prime motivating factor. These ideas will become clearer in Chapter 3 when we consider some actual case studies drawn from the anthropological literature. For now, what is important is that the ritualisation of violence can take many forms, and have many motivations.

For present purposes then, when I refer to ritualised violence, I mean violence as formalised action, or conscious performance, understood in relation to prevailing ideologies, religious beliefs, or cosmological structures.

From here on ...

In the following chapters, I examine the evidence for headhunting and related practices in Iron Age Europe in relation to these ideas. I do not set out to provide a new interpretation of the 'Celtic cult of the head'. In fact, I do not attempt to set up any pan-European explanatory scheme at all. Nor do I provide any sort of comprehensive survey of the relevant evidence, which is extremely common and diverse across much of Europe (e.g., Hartl 2005). Instead, the focus is on the role of headhunting, and ideas about the head, in specific Iron Age communities. I focus throughout on a single detailed case study; the Iron Age of southern France, where we have an unparalleled conjunction of literary, iconographic, palaeopathological, and archaeological evidence for the development of beliefs and practices associated with the human head. Within each chapter, however, I also draw on shorter case studies, dealing with particular sites or objects from elsewhere in Europe, especially northern Scotland, Ireland, Belgium, Iberia, and central and northern France, to explore related aspects of the treatment of the head.

In Chapter 2, I examine the origins and development of the conventional belief in a 'Celtic cult of the head'. There I discuss the various strands of evidence that have been combined, over many years, to define this phenomenon and examine the basis of its longevity as a Celtic trope. In particular, this involves a detailed consideration of the literary sources, predominantly from the classical world and from early medieval Ireland and Wales, which first laid the seeds of this idea. I also discuss the ways in which the value of this concept has been eroded in the face of new interpretations of Europe's prehistoric past. Then, in Chapter 3, I briefly examine the development of anthropological writing on headhunting. Here I consider, in particular, the potential of past ethnographic work to inform and focus our interpretations of prehistoric societies.

In Chapters 4–6, I consider the role of the human head in Iron Age cosmology, religion, and ideology. Chapter 4 introduces the major regional case study that forms the backbone of the rest of the book; the Iron Age of southern France. The wealth of evidence for special treatment of the human head in this region takes the form of elaborate stone sculptures, skeletal remains suggesting the curation, display, and deposition of human heads on both domestic and religious sites, and a range of textual references. The core of Chapter 4 comprises a discussion of the earliest manifestations of this Iron Age interest in the head, through a close study of several important stone carvings that depict human heads grouped into elaborate and regular compositions. I argue that these carvings are analogous to other monuments concerned with the severed head found at other times and in other places, within societies that believed in an association between the human head and the fertility of people, animals, and crops. Against this background I then reconsider some other depictions of heads from Iron Age Europe, in stone, wood, and metal, and suggest that a widespread cosmological relationship may have existed here too between heads and fertility.

The focus shifts, in Chapter 5, towards the role of severed heads as liminal objects, caught between the mundane world of the living and the otherworld of the dead. Here the discussion moves from the realm of broad-scale cosmology to the more specific and localised religious beliefs and practices through which Iron Age communities engaged with the supernatural. Whereas the discussion in Chapter 4 considers Iron Age iconography with little reference to the locations in which it was displayed, Chapter 5 brings these special places to the fore. Again I concentrate on the evidence from the southern French Iron Age, where activity at sites such as Roquepertuse suggests the emergence of increasingly formalised religious practices, but I also consider a series of cave sites that highlight the persistence of related ideas in the later prehistory of northern Europe.

In Chapter 6, religion gives way to ideology, as I examine the distinctive and individualistic sculptures that emerge in the third century BC in southern France, and the human remains that testify to the reality of the practices they depict. These images graphically portray the importance of headhunting to the emerging elites of the region and place art, violence, and religion at the centre of the Iron Age political scene. Headhunting, I argue, plays a key role in the establishment and legitimisation of these new militaristic authorities, just as it did in a range of ethnographically documented societies of more recent times. The evidence is not restricted to the classic southern French oppida like Entremont but can be paralleled far to the north at central Gaulish sanctuaries such as Ribemont-sur-Ancre. Here, although we lack the contemporary iconography, the evidence from human bones themselves provides stark evidence of decapitation and display.

The focus in Chapter 7 shifts away from the head to consider other aspects of body treatment and deposition. In particular, I explore the deposits that combine human and animal body parts and the potential cosmological links between the two. Evidence from the more fragmented, small-scale societies of Atlantic Scotland is used to contrast with the more hierarchical social forms that emerged in southern France. Some of the actions and beliefs behind the deposition of human remains in various states and combinations are considered in relation to ideas of individuality, the self, and personhood in prehistory. These further dimensions to ritual violence and the body help place the head deposits within a broader context.

Finally, in Chapter 8, I draw together some of these themes and consider their broader implications. How might we interpret the nexus of ideas that seems to emerge in Iron

Age Europe surrounding the role and importance of the human head in the realms of cosmology, religion, and ideology? To what extent can the special treatment accorded to the human head be seen as a pan-European phenomenon? And how does the treatment of heads relate to wider attitudes to the human body? Last of all, I return to where I started, at the Cnip wheelhouse in Lewis, and explain whose head it was in that pit.

CHAPTER 2

A remarkable spiritual continuity?

If any single belief can be claimed to pervade Celtic superstition it is the cult of the severed head.

L. Laing 1981, 113

The severed head was probably to Celtic religion what the cross is to Christianity.

Newman 1993, 22

Heads are ... the most powerful symbol of political dominance in the Celtic world. Here the dead influence the living more than in any other culture of Europe.

Enright 2006, 55

The cult of the head

The ancient Celts have long been regarded as headhunters. Early archaeologists were quick to link new discoveries, like the carved heads from Entremont in Provence, with the classical writings of Poseidonius, who described headhunting among the Gauls (Déchelette 1914, 1536). The French antiquarian Camille Jullian (1908, 201) even devoted a section to the practice in his influential *Histoire de la Gaule*, describing severed heads as a 'talisman of the first order'. As early as 1911, the Saluvii of the southern French Iron Age were even included in wide-ranging discussions of headhunting focused primarily on the emerging ethnographic record (e.g., Sébillot 1911, 135).

By 1913 the French archaeologist Adolphe Reinach (1913) had drawn together many of the textual sources describing Celtic headhunting, relating these to the accumulating archaeological and iconographic evidence. This idea of a spiritually infused, distinctively Celtic cult of the head was developed further during the second half of the twentieth century. Particularly important were the writings of Celtic scholars such as Pierre Lambrechts (1954) and Anne Ross (1967), who elaborated on the connections between headhunting, head veneration, and pan-Celtic belief systems stretching from prehistory into the Middle Ages. For these authors, the veneration of the human head constituted 'a remarkable spiritual continuity' (Lambrechts 1954, 22) spanning two millennia or more. These ideas have been extremely influential in establishing the cult of the head as one of a package of cultural traits thought to define Celtic identity. Quotations spanning the last three decades that assume the existence of a cult of the head can be found repeatedly across the

ILLUSTRATION 2.1. La Tène art is full of examples of human heads embedded within apparently abstract vegetal patterns. This gold ring from Rodenbach in the German Rhineland bears two opposed heads, one of which is clearly visible in this drawing. The heads are highly stylised, composed from similar elements to those which make up the remainder of the decoration. It probably dates to the late fifth century BC and is around 2 cm in diameter. (Drawn by Dan Bashford, after Jacobsthal 1944, plate 52, no. 72)

vast body of general and synthetic works dealing with the ancient Celts. Yet do they really reflect a prehistoric reality?

This belief in the existence of an ancient Celtic head cult was based on several strands of evidence. Most fundamental were classical descriptions of the Celts, which give frequent mention to practices associated with decapitation. Also important was a second collection of literary material, this time deriving primarily from Ireland and Wales in the Early Middle Ages. Although first written down by Christian monks, these sources are often thought to preserve elements of pre-Christian oral tradition (e.g., K. Jackson 1964). A third strand of evidence came from the art and iconography of Iron Age Europe, where the disembodied human head, in a range of guises, is a frequent and persistent motif (e.g., Almagro-Gorbea and Lorrio 1993; R. Megaw and Megaw 1993; Ill. 2.1). Finally, and perhaps giving most force to the idea, was the discovery on numerous archaeological sites of human skulls, sometimes pierced with heavy nails and frequently deposited in apparently significant locations.

In this chapter I first examine the literary sources, and the ways in which they have been combined with archaeological and iconographic evidence to resurrect this lost prehistoric cult. I then look at the ways in which this initially powerful and dramatic theory began to break down as modern scholarship has changed its thinking on the nature of the ancient Celts.

The classical literary tradition

The ethnonyms *Celtae/Keltoi* and *Galli/Galatai*, in Latin and Greek respectively, were used by classical authors to characterize the barbarian groups resident on the northern and western fringes of the Mediterranean world. The term *Keltoi* has the longer pedigree, its earliest surviving usage being in the writings of Herodotus, dating to the fifth century BC. It was probably used earlier, by the Carthaginian Himilco in the sixth century, and by Hecataeus in the early fifth, although these references survive only in the writings of later authors. Polybius, a Greek historian writing in the second century BC, seems to have understood the terms *Galatai* and *Keltoi* as synonyms, and indeed the same linkage had probably been made by earlier Greek writers since the mid-fourth century BC (Williams 2001, 19). This does not, of course, mean that the usage of these terms in antiquity was either precise or consistent; indeed, the relationships between them can be read in several, mutually contradictory ways (Collis 2003, 98–103). Nonetheless, it is these descriptions of the *Celtae/Keltoi* and *Galli/Galatai* that have formed our image of the 'ancient Celts'.

Headhunting is a recurrent motif in the classical writings on the Celts. Perhaps its earliest surviving occurrence is the *Histories* of Polybius, written in the mid-second century BC. Polybius would have encountered Celtic communities at first hand, as he retraced Hannibal's route through northern Italy, Gaul, and Spain, and he seems also to have fought against the Galatians as a young man in Asia Minor (P. Freeman 1994, 96). In his account of the Battle of Telamon, fought between Rome and a confederation of Celtic tribes in Etruria in 225 BC, Polybius mentions the decapitation of the Roman consul Gaius Atilius Regularis:

> In this action Gaius the consul fell in the mellay fighting with desperate courage, and his head was brought to the Celtic kings. (Polybius, *Histories*, 2.28.10, trans. Paton, 1922)

Although written several generations after the event, Polybius's account appears to draw on the earlier writings of Fabius Pictor, who took part in the campaign (P. Freeman 1994, 99). A similar account, which also purports to describe events of the third century BC, comes from the work of the much later Roman author Marcus Junianus Justinus, writing during the third century AD (though drawing on the lost work of the first-century BC author Pompeius Trogus). In his *Epitoma Historiarum Philippicarum*, Justinus describes the death of the Macedonian king, Ptolemy Ceraunus, at the hands of Celtic invaders in

279 BC (24.5). In this account, the king's head, spiked on a lance, was paraded around to strike fear into the hearts of his army. Although there is no particular reason to doubt their historical veracity, these accounts present decapitation as an essentially political gesture, serving as proof of the death of enemy leaders. Neither implies any specifically Celtic practice of headhunting.

A second reference from Polybius is more suggestive. This comes from his account of the Second Punic War and describes the defection of a group of Celts attached to the Roman army during Hannibal's advance into northern Italy in 218 BC:

> The Celts … fell upon the Romans who were encamped nearest to them. They killed or wounded many, and finally, cutting off the heads of the slain, went over to the Carthaginians. (Polybius, *Histories*, 3.67, trans. Paton, 1922)

Although too brief to carry much interpretive weight, this passage is interesting because, unlike the death of Gaius in the previous extract, the taking of these heads apparently adds nothing to the narrative. Instead, it reads more as a casual allusion to a form of barbaric behaviour with which Polybius expects his readership to be familiar. It may, in fact, suggest the existence, as early as the second century BC, of the treacherous, headhunting Celt as a literary trope.

Later in his *Histories*, Polybius provides a third and more detailed account of a separate case of Celtic head-taking. The relevant passage describes a Roman punitive campaign of 190 BC against the Tolistobogii, a Galatian tribe, and their king Ortiagon:

> By chance, one of the prisoners captured when the Romans defeated the Asian Galatae at Olympus was Chiomara, wife of Ortiagon. The centurion in charge of her took advantage of his soldierly opportunity and raped her. He was indeed a slave to both lust and money, but eventually his love of money won out. With a large amount of gold being agreed on, he led her away to be ransomed. There was a river between the two camps, and the Galatae crossed it, paid the ransom, and received the woman. When this was accomplished, she ordered one of them with a nod to kill the Roman as he was making a polite and affectionate farewell. The man obeyed and cut off the centurion's head. She picked it up and rode off with it wrapped in the folds of her dress. When she reached her husband, she threw the head at his feet. He was astonished and said, 'Wife, faithfulness is a good thing'. 'Yes', she said, 'but it is better that only one man alive would have lain with me'. (Polybius, *Histories*, 21.38.1–6, trans. Freeman, 1994)

According to Plutarch's later account, 'Polybius says that he had a conversation with this woman in Sardis, and that he admired her good sense and intelligence' (*De Mulierum Virtutibus*, 3.22, trans. Babbitt, 1931). As with the first extract, the act of decapitation is central to the narrative; the severed head acting as proof of the death of a specific individual. There is nothing in the story as related by Polybius to suggest any ritualisation of the act. Nonetheless, a pattern of 'Celtic' behaviour is apparently beginning to form.

Although written much later, during the Augustan period, Livy's *History of Rome* (*Ab Urbe Condita*) also covers the events of the third and second centuries BC and provides us with a reference to head-taking some 70 years earlier than the first incident mentioned by Polybius. This relates to the Roman defeat at the Battle of Clusium in 295 BC, during the Third Samnite War:

> Some authors say that the entire legion was wiped out there, not a man being left to carry the tidings, and that though the consuls were not far from Clusium at the time, no report of the disaster reached them until Gaulish horsemen appeared with the heads of the slain hanging from their horses' chests and fixed on the points of their spears, whilst they chanted war-songs after their manner. (Livy, *History of Rome*, 10.26, trans. Roberts, 1912)

As with Polybius's description of the Celtic mutiny in 218 BC, the implication here is that head-taking is a generic Celtic practice – part of 'their manner', like the chanting of war songs. A similar allusion to customary trophy taking comes from Diodorus Siculus, writing in the mid-first century BC about the aftermath of the Gaulish defeat of the Roman army at the Battle of Allia, more than 300 years earlier. This disastrous encounter occurred in the lead-up to the sack of Rome in 387 BC, and, according to Diodorus Siculus's account, the surviving Romans had time to fall back and fortify the Capitol:

> for the Celts spent the first day cutting off, according to the custom, the heads of the dead. (Diodorus Siculus, *Bibliotheca Historica*, 14.115.5, trans. Oldfather,1954)

Although the episode is intrinsic to the narrative, and may thus come directly from an earlier source, there are problems in any attempt to use it as evidence for institutionalised trophy taking at such an early date. We know, for example, that Diodorus Siculus borrowed much of his descriptive material on the Gauls, including a passage on their propensity for headhunting, from Poseidonius, who was writing in the earlier part of the first century BC. It is not impossible that this passage may have coloured Diodorus Siculus's rendition of much earlier events.

Returning to Livy we find a second relevant passage that forms one of the most-quoted references to head-taking from the classical literature. Taken from his account of the Second Punic War, it relates the death, in 216 BC, of the consul elect, Lucius Postumius, in Cisalpine Gaul:

> Postumius died fighting with all his strength trying to avoid capture. The spoils stripped from his corpse and the severed head of the commander were taken by the Boii to their holiest temple. Then, after they removed the flesh from the head, they adorned the skull with gold according to their custom. They used it as a sacred vessel to give libations on holy days, and their priests and custodians of their temple used it as a goblet. (Livy, *History of Rome*, 23.24, trans. Roberts, 1912)

The historical veracity of this passage has been questioned by Jane Webster (1996, 117–18), largely on the basis of its omission from Polybius's earlier account of the same campaign. John Collis (2003, 21–2) has also attacked Livy more generally, his criticisms again based largely on discrepancies between his work and that of Polybius. Polybius, however, was by no means the only source used by Livy. Indeed, they may well have drawn on common sources, including Fabius Pictor, and it is quite possible that the material they selected simply reflects the different emphases of their respective narratives. Postumius had been sent northwards to create a diversion for Hannibal's Gaulish allies (Polybius, *Histories*, 3.106.6), while the main focus of events lay far to the south, in Apulia. Whichever way one looks at it, this ill-fated northern mission was little more than a footnote to the much more significant Roman defeat at the Battle of Cannae, only a few days earlier. Polybius dismisses this minor action in a sentence, regarding it perhaps as something of an anticlimax, in narrative terms. He says only that Postumius and his force were attacked and 'utterly destroyed' (*Histories*, 3.118.6). Livy, however, provides greater detail, using the episode to paint a picture of the ever-increasing woes visited on Rome during this dire chapter of the Second Punic War. Livy may thus have drawn more detail from his sources on this particular campaign than Polybius had done.

Writing rather later than Livy, towards the end of the first century AD, but again dealing with the events of the Second Punic War, was the Latin poet Silius Italicus, whose major work, the epic *Punica*, is the longest surviving Latin poem. Silius Italicus has been widely derided as 'a dull man who wrote a bad poem' (Duff 1934, xii), but he does contribute one brief episode, from his description of the Battle of Ticinus in 218, which is relevant here. It describes the death of the Roman Quirinius at the hands of the Gallic Boii and Senones:

> Then Vosegus cut off his head from behind, and carried off the helmet hanging by its plume with the dead man's head inside it, and hailed his gods with the war-cry of his nation. (Silius Italicus, *Punica*, 4.215, trans. Duff, 1934)

Given the generally florid style of Silius Italicus's writing, there seems little reason to give credence to this extract as representing a genuine historical incident. The names of many of those mentioned at Ticinus, for example, are plainly invented (Duff 1934, 180). It may be that Livy's description of the death of Postumius was the inspiration for this passage, although it may simply have relied on the general background awareness of his audience, drawn from Livy and others, that headhunting was a Celtic habit. Like Livy before him, however, Silius Italicus also mentions the use of human skulls as drinking vessels (*Punica*, 13.482–3). Indeed, this practice was one that Roman writers came to associate particularly with the Gauls. In his *Epitome*, which drew largely on Livy's histories, the second-century AD Latin writer Florus describes the depredations of the Scordisci, a Gallic tribe raiding territories in northern Greece and the north Adriatic coast in 114 BC. Among numerous other atrocities, he mentions that they used their victim's skulls as drinking cups (*Epitome*, 1.39.1–3). This description of the Scordisci is echoed by Ammianus Marcellinus in the late fourth century AD (*Res Gestae*, 27.4.4), and again in a particularly graphic passage by the early Christian polemicist Paulus Orosius, written in the early fifth century AD, and describing events from the Macedonian War of the early first century BC:

> When there was need of a cup, they made use of the bones of the human head, still bloody and covered with hair and the brain badly scooped out through the inner caverns and poorly bedaubed; they eagerly and without any feeling of horror made use of these as real cups. (Paulus Orosius, *Historia Adversus Paganos*, 5.23.18, trans. Deferrari, 1964)

TABLE 2.1. *The principal classical sources for Celtic headhunting, excluding some of the more obviously derivative sources (Marcellinus, Paulus Orosius, etc.)*

Event	Location	Principal source
Fourth century BC		
387 Battle of Allia Decapitation of the Roman dead	Central Italy	Diodorus Siculus (mid-first century BC)
Third century BC		
295 Battle of Clusium Collection of trophy heads	Northern Italy	Livy (late first century BC)
279 Celtic invasion of Macedonia Decapitation of Ptolemy Ceraunus	Macedonia	Justinus (third century AD)
225 Battle of Telamon Decapitation of Gaius	Northern Italy	Polybius (mid-second century BC)
218 Defection of Celts to Hannibal Decapitation of former Roman allies	Northern Italy	Polybius (mid-second century BC)
218 Battle of Ticinus Decapitation of Quirinius	Northern Italy	Silius Italicus (late first century AD)
216 Ambush by Boii Decapitation of Postumius	Northern Italy	Livy (late first century BC)
Second century BC		
190 Campaign against Tolistobogii Decapitation ordered by Chiomara	Asia Minor	Polybius (mid second century BC)
114 Thracian War Skulls used as cups by the Scordisci	Northern Greece and Croatia	Florus (mid-second century AD)
First century BC		
90s Poseidonius in Gaul Display of heads by Gauls	Southern France	Diodorus Siculus (mid-first century BC) and Strabo (early first century AD)

Whatever truth was contained in the earliest versions of these tales, by the early centuries AD the motif of the Celtic skull-cup had clearly become a well-established literary convention.

The accounts discussed so far represent the only substantive literary evidence for headhunting by the Celts in the period before the first century BC. Most of the writers lived long after the events they describe, which are concentrated in the century or so between 295 and 190 BC. With the exception of Pompeius Trogus, who claimed descent from the Vocontii of southern Gaul (Collis 2003, 22), all of the writers were non-Celts, writing about people they saw as culturally alien. All of the events took place around the northern fringes of the Greek and Roman world (mostly in northern Italy, northern Greece, and Asia Minor: Table 2.1), and many have a dramatic or literary logic, which need not relate to any strictly historical truth. None of the references need suggest any institutionalised practice of headhunting or veneration, although we might note Livy's use of the phrase 'according to their custom' when describing the treatment of Postumius's skull (*History of*

TABLE 2.2. *Schematic representation of the chronology of the various headhunting episodes described in the classical texts (top row), and the broad periods during which the texts themselves were written (bottom row)*

4th BC	3rd BC	2nd BC	1st BC	1st AD	2nd AD	3rd AD
■ - - - -	■■■■■■	■ - - -■-	■			
		—	——	——	——	——

Rome, 23.24). At the very least, however, they do indicate, that the collection of human heads as battle trophies was regarded as a characteristic Celtic practice by Greek and Roman writers from at least the mid-second century BC (in the writings of Polybius), and probably rather earlier (in his lost sources) (Table 2.2).

Perhaps the most important reference to Celtic headhunting, however, comes from the writings of Poseidonius, a Greek philosopher and historian born in Apamea, in Syria, around 135 BC. Although resident for much of his life in Rhodes, Poseidonius was unusually well travelled for his time and was one of the few classical authors to have visited Gaul. The date of his trip is disputed, but a combination of personal and political circumstances suggests that it most likely took place during the 90s BC, when Poseidonius would have been in his late 30s or early 40s (Kidd 1988, 16).

Poseidonius's life spanned a period of considerable sociopolitical upheaval in Gaul. At the time of his birth, the Gaulish tribes retained their power and autonomy. The Greek colony of Massalia (modern Marseille) and its satellite colonies along the southern French coast were politically, culturally, and commercially important, but exerted little, if any, political domination over their Gaulish neighbours. The annexation of southern Gaul by Rome in 121 BC, however, and the establishment of the province of Gallia Narbonensis marked the beginnings of Roman political domination over the southern tribes, and persistent interference in the affairs of those to the north. By the time of Poseidonius's death around 50 BC, Caesar had completed the conquest of the whole of Gaul.

The relevant sections of Poseidonius's work were contained in book 23 of his *History*, which, although it does not itself survive, was heavily plagiarised by later authors. The material on the Gauls seems to have been conceived as an introduction to his historical account of the wars between Rome and the southern Gaulish tribes that culminated in the foundation of Gallia Narbonensis. From passages in the work of later writers (principally Diodorus Siculus, Strabo, and Athenaeus), it is possible to reconstruct substantial sections of the lost original (Tierney 1960), including the passage referring to the treatment of heads. Material deriving from Poseidonius is also contained in Caesar's *Gallic War*, although the extent of these borrowings is disputed (Nash 1976). Caesar has nothing to say on the subject of Celtic headhunting, although, somewhat ironically, the last book of the *Gallic War*, written by Caesar's general Hirtius, does describe the execution by the Roman army of Gutuator, the 'principal war-monger' of the Carnutes who 'was flogged to death and his head cut off' (8.40.3).

As a youth, Poseidonius had been a pupil of the Stoic philosopher Panaetius, and Poseidonius himself became the leading Stoic of his day. As such, he believed that his own

Mediterranean world was in a state of degeneracy and decline. The barbarian nations to the north still possessed, in his view, some trace of the moral and religious virtues lost by Poseidonius's own society. For Poseidonius, the Druids, the priests of the Celtic tribes, 'preserved something of the golden age' (Momigliano 1975, 70), and it appears that he was keen to present Celtic customs in a positive light where he could (Piggott 1968, 83–5). It is this, perhaps, that gives added weight to his accounts of headhunting, which he clearly found repugnant. The crucial passage is rendered by Diodorus Siculus as follows:

> They cut off the heads of enemies slain in battle and attach them to the necks of their horses. The blood-stained spoils they hand over to their attendants and carry off as booty, while striking up a paean and singing a song of victory, and they nail up these first fruits upon their houses just as do those who lay low wild animals in certain kinds of hunting. They embalm in cedar-oil the heads of the most distinguished enemies and preserve them carefully in a chest, and display them with pride to strangers, saying that for this head, one of their ancestors, or his father, or the man himself, refused the offer of a large sum of money. They say that some of them boast that they refused the weight of the head in gold; thus displaying what is only a barbarous kind of magnanimity; for it is not a sign of nobility to refrain from selling the fruits of one's valour, it is rather true that it is bestial to continue one's hostility against a slain fellow-man. (Diodorus Siculus, *Bibliotheca Historica*, 5.29.5, trans. Tierney, 1960)

A second version of the same passage from Poseidonius turns up in Strabo's *Geographia*, with certain changes of emphasis:

> In addition to their witlessness they [the Celts] possess a trait of barbarous savagery which is especially peculiar to the northern peoples, for when they are leaving the battlefield they fasten to the necks of their horses the heads of their enemies, and on arriving home they nail up this spectacle at the entrances to their houses. Poseidonius says that he saw this sight in many places, and was at first disgusted by it, but afterwards, becoming used to it, could bear it with equanimity. But they embalmed the heads of distinguished enemies with cedar-oil, and used to make a display of them to strangers, and were unwilling to let them be redeemed even for their weight in gold. The Romans have put an end to these customs and also to their sacrificial and divinatory practices opposed to our customs. (Strabo, *Geographia*, 4.4.5, trans. Tierney, 1960)

Both Diodorus Siculus and Strabo take considerable liberties with Poseidonius's work, as we can see by comparing their sometimes crudely condensed summaries with equivalent passages in Athenaeus, writing in the second and third centuries AD. Athenaeus appears to quote his source largely verbatim, but sadly does not include the passage on headhunting among his Poseidonian borrowings. Nonetheless, the two surviving renderings of this influential passage are sufficiently close to give a reasonably reliable picture of what Poseidonius might have said.

The first part of the Poseidonian account, particularly as rendered by Diodorus Siculus, is strikingly similar to Livy's description of the Battle of Clusium (*History of Rome*, 10.26). In both cases, victorious Gallic warriors chant war songs and hang the heads of their

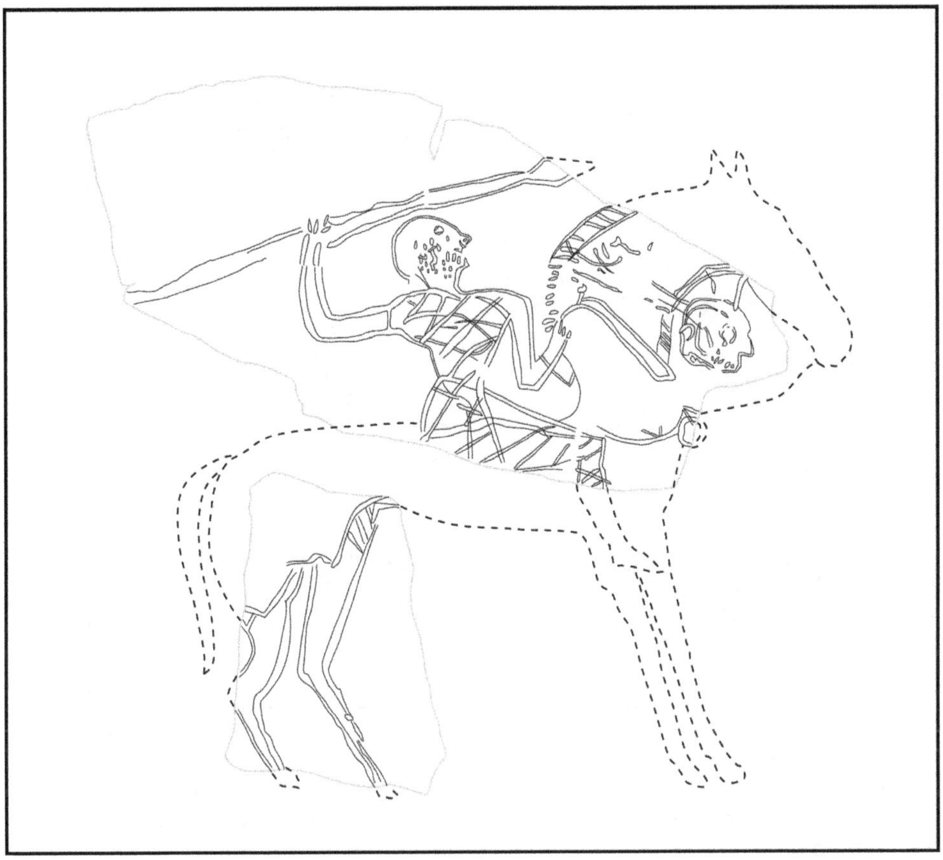

ILLUSTRATION 2.2. Pottery fragments from the Iron Age settlement of Aulnat, near Clermont Ferrand in the Auvergne, bear graffiti showing a mounted warrior with a severed head tied around his horse's neck. (Drawn by Dan Bashford, based on Périchon 1987, plates 1, 3, 8, and 9)

enemies around the necks of their horses. Although Livy purports to describe much earlier events (dating to 295 BC), we should bear in mind that he would have had access to Poseidonius's writings. We must be wary, therefore, of the possibility that Livy adapted certain useful details, such as the suspension of severed heads from the necks of horses, and transposed them to the earlier period. Certainly, this particular Celtic habit finds independent support from the Iron Age iconography of both France and Spain around the time when Poseidonius was writing. Closest in date is a piece of graffiti on a pottery vessel from Aulnat, near Clermont-Ferrand, showing a mounted spearman with a human head suspended from his horse's neck (Ill. 2.2). Found in deposits dating around 120–80 BC (Périchon 1987, 678), this fits remarkably well with the Poseidonian account. Other finds, however, extend the currency of this motif back closer to the period with which Livy was concerned. Within southern France itself, a warrior with a severed head dangling from his horse's neck is depicted on a pillar from Entremont, probably dating to the third or second century BC (Salviat 1993, 214). Earlier still, perhaps, are several zoomorphic fibulae from Spain, probably dating to the third or second century BC, which depict horses, with or

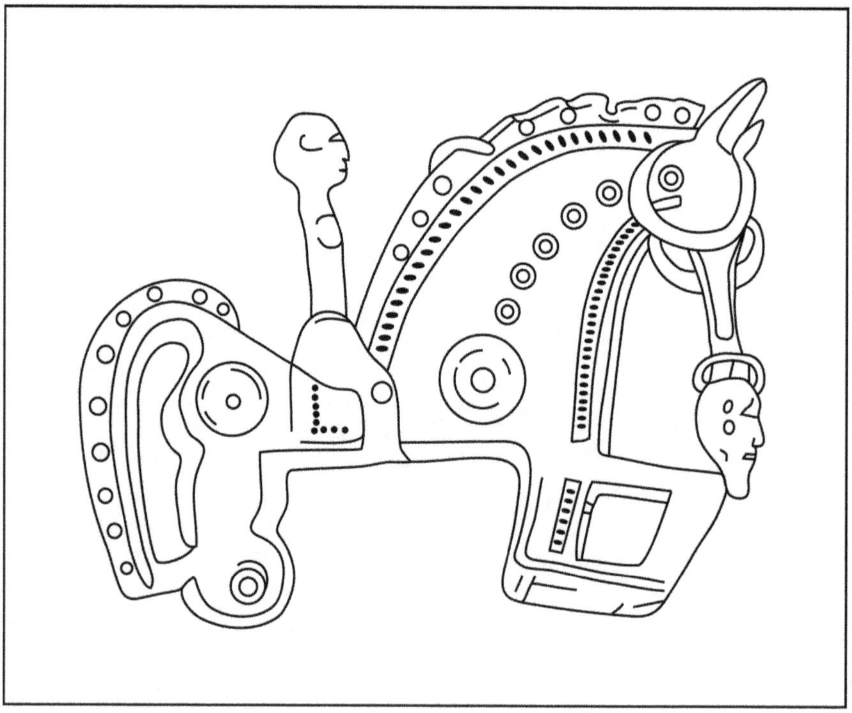

ILLUSTRATION 2.3. Small bronze fibula from an unknown Spanish Iron Age context showing a mounted warrior with a severed human head hung from his horse's neck (drawn by Dan Bashford, after Almagro-Gorbea and Lorrio 2004, fig. 3). This appears to be a recurrent motif in Iberia, with similar examples occurring in the second-century BC tombs of Numantia and elsewhere. (Lorrio and Ruiz Zapatero 2005, figs. 22 and 23)

without riders, once again carrying severed human heads (Almagro-Gorbea and Lorrio 1993, 231–3; Ill. 2.3).

The remainder of the Poseidonian account describes the subsequent treatment of these trophy heads. According to Diodorus Siculus, treatment is based on the status of the victim. Most heads were apparently displayed in the open 'at the entrances to their houses', while those of high-status individuals were embalmed in cedar oil and carefully curated over several generations. Both Diodorus Siculus and Strabo make specific reference to the enduring value of these curated heads, referring to the refusal of the owners to sell them even for enormous amounts. As Kenneth Jackson notes (1964, 37), the gold on offer was presumably an attempt at posthumous ransom by the victim's kin, since we have no reason to believe in a prehistoric collector's market in secondhand (or, strictly, thirdhand) heads (*contra* Brunaux 1988, 110). The power of heads in this society evidently had a long life, if descendants over three generations were still concerned with their recovery.

Although venturesome for a Greek writer, we should not assume that Poseidonius carried out any extensive explorations among the Gauls. Indeed, Kidd has suggested that most of his material could have been collected from the relative comfort of Massalia (Kidd 1988, 16). The most adventurous excursions for which we have evidence were a brief exploration of the stony plain of La Crau, near the mouth of the Rhone, and a visit to at least one Gaulish settlement, where Poseidonius would have seen the heads described in the excerpts from Strabo and Diodorus Siculus. As Nash points out (1976, 119–20), curated heads from more or less the right period have been found on indigenous Gaulish sites within 15 kilometres of Massalia, for example, at the oppidum of La Cloche, near Les Pennes Mirabeau (Salviat 1972). There is nothing, therefore, to suggest that Poseidonius travelled beyond the area of direct Roman control. Indeed, his observations need relate only to a localised set of practices in the region around Massalia itself.

Whatever detail and nuance may have been lost with the disappearance of the original Poseidonian account, we are patently left with some problems of interpretation concerning this evocative passage. Who, for instance, were the headhunters? Did they belong to a specific tribe or group of tribes, and, if so, where were they located? Did they compose the whole adult male warrior group, or was the practice restricted to an elite element as Diodorus Siculus's reference to 'attendants' might be taken to imply? Were the heads displayed at the entrances of individual houses (implying that they were the trophies of individual warriors) or the entrances to villages or fortifications (implying a value to the community as a whole)? What was the significance of the display of these heads? Was it simply a means to show off one's military prowess, or was there a religious connotation, which may have been obscure to Poseidonius, or even suppressed by him in his efforts to portray the higher qualities of Celtic, and specifically druidic, religious thought. Or had the Roman suppression of indigenous religious practice by the 90s BC removed much of the overt ritualisation from the practice of headhunting?

The Irish and Welsh literary traditions

I broke a hundred forts
I slew a hundred stewards
I bestowed a hundred mantles
I cut off a hundred heads

(From 'The death-song of Uther Pendragon', *Book of Taliesin*, XLVIII, trans. Skene, 1858)

The occurrence of head-taking in the medieval written versions of earlier Irish oral tales has been compared explicitly to the accounts of Poseidonius and others (e.g., Le Roux 1958; K. Jackson 1964, 35–7). Although the earliest surviving manuscripts only rarely predate the twelfth century AD, the origins of these stories are often thought to go back many centuries, to a pre-Christian, pagan Celtic milieu. As such, they have been seen as preserving vestiges of the same Celtic head cult glimpsed through the classical accounts.

Jim Mallory (1981, 108) records more than 30 episodes of head-taking in the Ulster Cycle of tales alone, including the following exchange, between the Ulster hero Conall and his Connaught counterpart Cet:

> 'I accept your challenge to single combat, Cet', said Conall. 'I swear what my tribe swears, that since I took a spear in my hand I have not often slept without the head of a Connaughtman under my knee, and without having wounded a man every single day and

> every single night'. 'It is true', said Cet. 'You are a better hero than I am. If Anlúan were in the house he would offer you yet another contest. It is a pity for us that he is not in the house'. 'He is though', said Conall, taking the head of Anlúan from his belt, and throwing it at Cet's breast with such force that a gush of blood burst over his lips. (*Scéla Mucce Meic Dathó*, 16, trans. Chadwick, 1927)

Elsewhere, Conall avenges the death of the Ulster hero Cú Chulainn by taking the heads of his killers and presenting them to the grieving widow (Enright 2006, 19). In yet another tale, Fothadh Cannaine, a war leader described in one eighth-century text, proclaims that he 'may not drink ale unless it be drunk in the company of white faces'; his war band wastes no time in procuring a freshly severed head to meet this demand (54).

Most of the references to headhunting in these Insular tales depict the practice as a form of martial display, reflecting the prowess of individual heroic warriors. Heads are taken generally for revenge, to denigrate the enemy, to prove a kill, or in a burst of uncontrollable ferocity. The paucity of references to the veneration of severed heads is hardly surprising when we consider that these tales were being transcribed by Christian monks, unwilling to give credence to pagan belief. Nonetheless, there are a few instances where, even through the distorting mirror of later transcription, it is clear that certain heads had supernatural powers. Following the Battle of Belach Mugna in AD 908, for example, the head of the defeated Irish king and cleric Cormac mac Cuilennàin was venerated by his victorious enemy, Flann, and went on to generate a series of prophecies and miraculous occurrences (Radner 1978). Another such head belonged to the high king of Ireland, Fergal mac Máele Dúin, who was killed at the 'Battle of Allen', according to a tale for which the earliest extant manuscript dates to the late fourteenth century:

> Then Fergal's head was washed and plaited and combed smooth by Cathal, and a cloth of velvet was put round it, and seven oxen, seven wethers and seven bacon-pigs (all of them cooked) were brought before the head. Then the head blushed in the presence of all the men of Munster, and it opened its eyes to God to render thanks for the respect and great honour that had been shewn to it. (*Cath Almaine*, 26, trans. Stokes, 1903)

The treatment afforded to Fergal's head is all the more remarkable when we consider that it was his killers who were providing the hospitality. This apparent veneration of a 'trophy head', echoing the ethnographic account of the Berawan described in Chapter 1, demonstrates yet again how hard it will be archaeologically to distinguish between the heads of ancestors and enemies.

A further casualty of the Battle of Allen was the young Donn-bó, a renowned minstrel, whose head was plundered from the battlefield by one of the victorious warriors (*Cath Almaine*, 20, trans. Stokes, 1903). Placed on a pillar at the victory feast, Donn-bó's head 'turned his face to the wall of the house so that it might be dark to him' and sang so that 'all the host were weeping and sad at the piteousness and misery of the music' (21). Whilst couched in Christian terminology (Donn-bó's head is asked to make music 'for the sake of God's Son'), the piece appears to preserve ideas quite alien to conventional Christian sentiment. The story may have been inspired by a genuine battle recorded in the Annals of Tigernach in the eighth century AD (Stokes 1903, 43), but the central motif, of the severed head endowed with supernatural powers, has been linked to much earlier, perhaps even prehistoric traditions (e.g., Enright 2006).

Similar tales of head-taking and curation occur in the Early Medieval Welsh literature, for example in the *Mabinogion* tale of Branwen, daughter of Llŷr. Suffering from a poisonous spear wound, Bran the Blessed, king of Britain, commands his followers to cut off his head. After more than eighty years, during which the head remained free from decay and able to communicate with its retainers, it was finally buried on a hill in London (Gantz 1976, 79–81). Bran's posthumous pronouncements recall those of the mythical Norse figure Mimir, who appears for example in the *Heimskringla*, committed to writing by Snorri Sturluson in the thirteenth century AD (S. Laing 1907). After decapitation, in this case by his enemies, Mimir's head continued to furnish Odin with wise counsel. The similarities between Bran and Mimir perhaps suggest echoes of a broadly similar myth across much of northern Europe in the medieval period and perhaps earlier.

It used to be thought that the Insular tales provided, in Kenneth Jackson's memorable phrase, a 'window on the Iron Age' (K. Jackson 1964). More recently, however, this idea has fallen from favour in the light of the accumulating archaeological evidence. Mallory (1981, 109), for example, has suggested that much of the material culture described in the texts may relate more to the period in which they were first committed to writing than to the prehistoric milieu in which they are often thought to have originated. The sword forms described in the Ulster Cycle, for example, 'seem to be no earlier than the Early Christian period' (108).

By implication, much the same suspicion might be levelled at many of the literary motifs contained within the written versions. For instance, as Mallory points out (1992, 139–41), the Irish annals for the seventh to twelfth century AD, which chronicle contemporary events, are littered with references to beheadings in combat. This is precisely the period during which many of the oral tales were first being transcribed. So can we be sure that the frequent decapitations that occur in the written versions of the tales actually derive from the oral versions? Or were they, like so much else, inserted here and there to accord with medieval Irish expectations?

At best, the beheadings in the tales present a rather heroised picture. Decapitation, using a typical short sword of the Irish Iron Age, would have been possible, but it would have required a certain amount of time and patience. Mallory (1981, 108) points out that 'a survey of the Early Iron Age swords from Ireland leaves us hard-pressed to find a weapon capable of cutting off a hand much less separate a stiff-necked Irishman from his head'. This is not to say that heads were not taken; the discovery of decapitated corpses in two first-century AD graves at Knowth, Co. Meath, would suggest that they were (O'Brien 1990, 38). So too would a spectacular recent find of a decapitated bog body from Old Croghan in County Offaly, which provides further evidence, this time from the period between the fourth and second centuries BC (Kelly 2006). However, the swift and elegant swipe with the hero's sword, as described in the tales, would simply not have been possible with the swords available during the Irish Iron Age.

The pan-Celtic head cult

The first systematic study of the Celtic head cult was published in 1913, by the young French archaeologist Adolphe Reinach, chiefly remembered today for his work on Greek and Egyptian antiquities. Reinach based his analysis primarily on the sorts of textual sources that we have already discussed, relating them to the iconography of Gaulish coinage and the Iron Age stone statuary of southern France, as well as to later, Gallo-Roman imagery. Interestingly, he also considered possible parallels with the beliefs and practices of contemporary headhunting peoples, known through the ethnographic accounts that were by then becoming available. Reinach's death in action a year later, during the First World War, prevented the further development of these ideas, and it was not until the middle years of the twentieth century that the notion of a Celtic head cult assumed its more fully developed form, in the work of the Belgian scholar, Pierre Lambrechts.

Lambrecht's full-length treatment, *L'exaltation de la tête dans la pensée et dans l'art des Celtes* (1954), although drawing heavily on Reinach's earlier work, introduced a major new strand of evidence to the debate; the role of the disembodied human head in Celtic, or La Tène, art. This art style, which is found widely throughout temperate Europe during the Iron Age, takes its name from the waterlogged votive site at Lake Neuchâtel in Switzerland, first excavated during the mid-nineteenth century. It has long been regarded as the art of the ancient Celts, and the terms La Tène art and Celtic art have (for better or worse) usually been regarded as synonymous (V. Megaw 1970; Harding 2007). Despite the manifest problems of isolating and defining an art 'style' in the fluid social conditions of prehistoric Europe (e.g., Taylor 1991), many modern archaeologists continue to find the broad equation of La Tène and Celtic art helpful (e.g., Fitzpatrick 2007, 341).

In contrast to the classical arts of the Mediterranean, La Tène art is overwhelmingly abstract in character, dominated by complex curvilinear motifs, which find their fullest expression in high-status metal objects. Although many of its central motifs, including the distinctive plant forms and the human heads themselves, can be traced back to Mediterranean models, these become thoroughly transformed in La Tène art, creating radically distinctive designs. In contrast to the art of the Mediterranean, La Tène art contains very few depictions of the complete human form. Indeed, Paul Jacobsthal (1944, 161), whose work underpins all modern studies of the subject, called it 'an art of ornament, masks and beasts, without the image of Man'. Yet, for Lambrechts, it was the human heads or 'masks' that were the crucial element.

While some scholars, such as Albert Grenier (1922), argued that Celtic art was essentially decorative and devoid of deeper meanings, Lambrechts followed Jacobsthal in his belief that Celtic art contained hidden narratives, 'where the tale of mythology was told through zoomorphic disguise' (Jacobsthal 1944, 162). Yet, for all his belief in the meaningful nature of La Tène art, Jacobsthal himself did not attribute great importance to the frequent depiction of human heads, which he saw as simple derivations of Greek or Etruscan prototypes, used as essentially decorative devices (12–24). They did not, therefore, compromise his view of Celtic art as being essentially devoid of people. For Lambrechts, however, these Mediterranean prototypes were adapted to embody wholly indigenous ideas regarding the special status of the human head. Indeed, it was this focus on the head

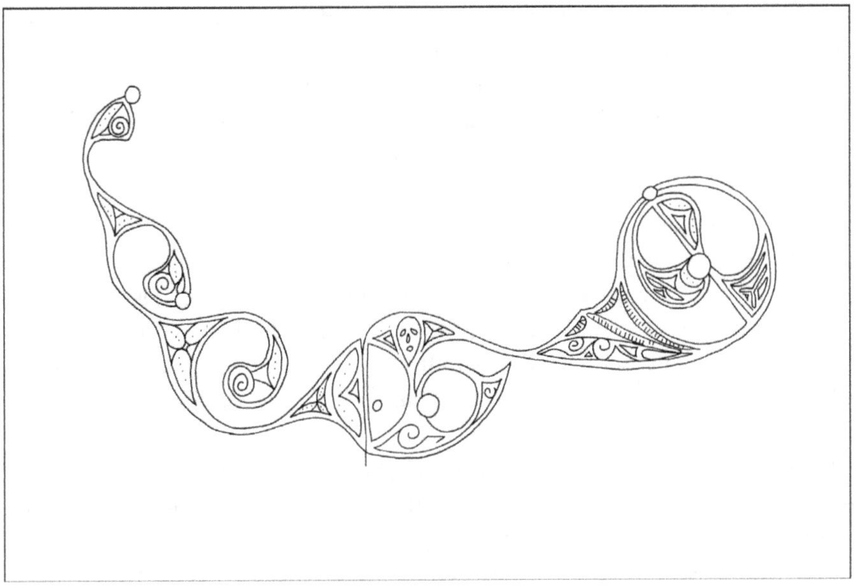

ILLUSTRATION 2.4. Drawing depicting the incised decoration from one of two probable drinking horns associated with the Torrs pony cap, from a hoard in southwest Scotland. A disembodied human head can be seen within the central part of the decorative scheme. The horn was probably made sometime around the third century BC. (Drawn by Dan Bashford, after R. Megaw and Megaw 1989, 197, and Atkinson and Piggott 1955)

that, for Lambrechts, defined Celtic art as a distinct style within the wider Eurasian artistic tradition of which it formed one regional expression. The head was, for Lambrechts (1954, 20), the central religious symbol and *motif national* of the Celts.

Lambrechts (1954, 19) outlines a number of characteristic traits of these Celtic heads: the face is seen from the front; it often lacks ears, or has animal rather than human ones; the nose is characteristically triangular; mouths are highly varied; eyes are usually horizontal, often seeming to be closed, or else are oversized and bulbous; beards are unusual but moustaches common; the forehead is often decorated in some way; and many heads, particularly in the earliest phase of the art style during the fifth century BC, bear a distinctive 'leaf crown' (see Ill. 4.14). Complete or partial heads can often be discerned within complex abstract or vegetal patterns, with elements of the heads merging with other, seemingly abstract elements of the design (e.g., Szabó 1993 and Ill. 2.4). For Jacobsthal (1944, 19), they represent 'the mechanism of dreams, where things have floating contours and pass into other things'.

The depiction of disembodied heads in Celtic art need not of course have any connection to headhunting. In his study of South American Moche art, Christopher Moser (1974, 30) recognised as evidence of decapitation only those heads that showed blood flowing from the neck, were held by the hair or suspended by a cord, or were associated with decapitated bodies. Certainly the heads depicted in La Tène art would fail his test. Celtic art, however, lacks the obvious narrative content of Moche art, and we cannot expect such unambiguous signals. What is clear is that these heads are both frequent and important.

Often they appear to be shown as sleeping, dreaming, or dead. Frequently they appear in pairs, or in groups, and often they are semiabstract. Even in the rare instances where the whole body is represented, such as the bronze sculpture from Bouray, or the small stone figure from Euffigneix in northeastern France, the head is often disproportionately large (R. Megaw and Megaw 1989). There is little doubt that these heads would have had meaning, for at least some of those who viewed them.

Undoubtedly the most important figure connected with the idea of the Celtic head cult in the anglophone literature was the British Celtic scholar Anne Ross. In a series of papers dealing specifically with the ritual importance of the human head (e.g., 1958, 1962), and culminating in her *Pagan Celtic Britain* (1967), Ross developed the first major synthesis of the Insular evidence. Although dealing with the whole range of pagan Celtic religious practice, the longest chapter of *Pagan Celtic Britain* was devoted entirely to 'the cult of the head'.

Given the relative paucity of pre-Roman Iron Age representations of the human head in Britain, Ross leant heavily on Romano-British iconography in her attempts to delineate the main strands of pagan Celtic religion. In particular, she focused on a large number of crudely carved stone heads that she believed, on stylistic grounds, to date to the Roman Iron Age or earlier. The belief that some of these heads were genuinely ancient was reasonable enough. Lambrechts (1954, 88) had already considered continental stone heads like that found buried in a pit just outside the square enclosure or *viereckschanze* at Mšecké Žehrovice in Bohemia, and the Janus head from Roquepertuse in Provence. To these we could add numerous other examples, such as the Janus head from Badaczony-Lábdi in Hungary (Szabó 1965) or the tricephalic bust from Beaucaire in Languedoc (Py 1990, 819–21). There was no doubting the Iron Age credentials of these sculptures. The British series, however, is more problematic.

Human faces certainly occur on Roman-period artefacts in Britain, many of which were probably made by indigenous people. Notable examples include decorative rooftiles (*antefixa*) and pots, such as one from a votive deposit at Coventina's Well at Carrawburgh Fort on Hadrian's Wall (Allason-Jones and McKay 1985, no. 144). Coventina's Well also produced part of an adult female skull as well as three small bronze masks, which probably served as decorative mounts, suggesting something more than a casual interest in the human head (Ross 1967, 141). Finding unequivocal evidence of carved stone heads that can be securely dated to the Roman period is more difficult, but there are nonetheless some good candidates, including one from the Welsh Roman town of Caerwent, found within what appears to have been a fourth-century temple (Boon 1976).

More usually, however, such heads have been found devoid of any archaeological context and dated on stylistic grounds alone (Ill. 2.5) – a tricky exercise given the intrinsic simplicity of the carving. A three-faced stone (or tricephalos), discovered near the source of the river Clyde in Scotland, for example, was dated by Ross (1974) to the Roman period although it was devoid of any archaeological context and located some 6 km from the nearest Roman installation. Elsewhere in Scotland, a stone block with four simple heads carved in relief has been suggested as representing a pagan religious site predating the Pictish monastery at Abernethy (Proudfoot 1997; Hall 1997). Another Scottish tricephalos, this time apparently from Sutherland (though it actually turned up in a Glasgow antique

ILLUSTRATION 2.5. Crude stone head, from Coupar Angus in Perthshire. This head is typical of many hundreds of similar examples found in various parts of Britain and Ireland. Although some may date to the Iron Age, the great majority are likely to be far more recent. (Courtesy of Perth Museum and Art Gallery, Perth and Kinross Council)

shop and is made of non-Scottish granite), has also been compared to continental carvings. Combining its stylistic affiliations, its alleged region of discovery, and its nonindigenous material, Ross (1967, 108) felt able to suggest that 'the head can probably be dated to the turn of the Christian era, and may have been brought into Britain by Celts escaping from Roman troops in Gaul'. This, however, seems an overly optimistic reading of an intrinsically difficult and patently ambiguous object.

In Ross's view, the apparent floruit of carved stone heads in the Romano-British period was the result of the prohibition by the Roman army of actual headhunting (Ross and Feachem 1984, 342). These surrogate heads were thus thought to 'stand for' the unavailable real heads in ritual acts that had their origins in the pre-Roman period. Thus the apparent concentration around Hadrian's Wall was thought to indicate a particular propensity for headhunting in that region prior to Roman pacification of the hill tribes (Ross 1967, 106).

Ross also believed that it was possible to trace a direct continuity between the head veneration of the pagan Celts and the persistent appearance of severed heads in the folklore of more recent times. One corpus of tales, for example, demonstrates an association between severed heads and wells (Ross 1962). One among many concerns the naming of two wells on the small Hebridean island of Vatersay. According to the tale, three brothers were murdered and decapitated at Tobar nan Ceann (the Well of the Heads). Their heads, having been retrieved by their father, were taken home. As they passed a standing stone, one of the heads pronounced that its own unborn son would avenge all three deaths. Years later, the dead man's son encountered the murderer and decapitated him as he drank from yet another well. The boy cast the murderer's head into this well, which became known as Tobar a'Chinn (the Well of the Head). For Ross, a connecting cord links such relatively recent folkloric accounts, through the pagan Celtic tales, to the Celts of prehistory. If this were true, these stories might provide an insight, however imperfect, into the lost cosmologies of the ancient Celts.

For Lambrechts and Ross, the Celts were clearly no mere headhunters. Rather, they practiced the 'exaltation' or 'veneration' of the head, which manifested itself in the display of enemy heads, the depiction of gods and of the dead, and a range of other practices, where the disembodied head stood for the whole being. Heads were apotropaic and prophetic and contained the individual's spiritual force; they were associated with regenerative powers and fertility and could mediate between this world and the next. They lay at the core of Celtic cosmology and religious practice. This was powerful and compelling stuff, and the influence of these two scholars is clear in some of the most important works on European prehistory from the second half of the twentieth century. Stuart Piggott (1965, 230), for example, refers to both in his discussion of headhunting, which he regards as an important element of Celtic religious practice, at least for the elite 'warrior-caste'. Terence Powell (1958, 108), in his influential synthesis, *The Celts*, describes headhunting as a 'horrifying Celtic custom'. Elsewhere, Nora Chadwick (1970, 49) writes that 'headhunting was universal, not only in Gaul but among the insular Celts also'. More-recent general works on the Celts have followed a similar line (e.g., S. James 1993, 82; Green 1997, 76–7), while specialist studies have developed the idea of a Celtic head cult in more detail (e.g., Sterckx 2005). Human heads, whether real or carved, came to be seen as an archaeological signature of the Celts. Recently, for example, the presence of headless and mutilated corpses has been taken as archaeological confirmation of literary evidence for a Celtic presence at Gordion in Anatolia (Dandoy et al. 2002). Headhunting and the idea of a head cult became, in other words, an inalienable part of the 'Celtic package'.

Heads without Celts, Celts without heads?

It would be misleading to suggest that the views of Ross, Lambrechts, and others were universally accepted (e.g., Le Roux 1964). The influential historian of pagan religions, Ronald Hutton (1991, 194–5), has dismissed the prevalence of the human head in Celtic art, much as Jacobsthal might have done, as 'nothing more than … a favorite decorative motif', and questioned the wisdom of generalizing from stories, like that of Bran the Blessed, where specific heads are endowed with mystical properties. Miranda Aldhouse Green (1986,

216), while accepting the importance of the head in Celtic spiritual life and art, contests the idea of a generic 'head cult'. Instead, she believes that heads were used simply as an iconographic shorthand for the whole being, especially for deities.

The insular stone heads are peculiarly problematic. The writings of Anne Ross in particular generated a flurry of studies of supposedly 'Celtic' stone heads in Britain and Ireland during the 1960s and 1970s. Numerous stone heads from West Yorkshire, for example, were meticulously catalogued and published by Sidney Jackson, in his *Celtic and Other Stone Heads* (1973), while Etienne Rynne (1972) did much the same for Ireland, in his study of 'Celtic stone idols'. Yet it is increasingly clear that many of these simple stone heads were actually carved much later. Indeed, there is increasing evidence for major regional folk traditions of head carving as late as the seventeenth and eighteenth centuries (e.g., Billingsley 1998). Heads more or less identical in their simplicity to the supposed Celtic ones can be found carved into wooden furniture and dated gravestones of the early modern period, and many of the stone heads built into the gables and door frames of rural cottages in the north of England were probably carved as symbols of luck or protection in relatively recent times. Numerous stone heads can now be identified as the work of individuals, such as the stonemason and poet John Castillo who was active in North Yorkshire in the early nineteenth century (Brears 1989, 32). Indeed, the great majority of Sidney Jackson's West Yorkshire heads now appear to be of early modern date and are of considerable interest in their own right as 'a confident cultural gesture of the upper peasantry of the seventeenth century' (Billingsley 1998, 2). They have nothing whatever to do with the Celts.

Similar problems exist for many of the Irish stone heads, which again may have been carved until quite recently (Hickey 1985). John Billingsley, for example, challenges Ross's (1967) assertion that the three-faced Corleck head from County Cavan can be dated to the Iron Age 'on stylistic grounds alone'. In her view, the three faces on the Corleck stone relate to Celtic notions of sacred triplism, and she suggests that the hole at the bottom of the head would enable it to be 'attached to a stone pillar, or base, on which it would be exhibited and venerated'. In the light of the emerging evidence for the early modern carving of stone heads, Billingsley (1998, 90) suggests that the head may actually have been set up on a wooden pole as an 'Aunt Sally' at a travelling fair. Because the Corleck head was, like almost all such heads, an unassociated find, it is hard to know how to judge such divergent hypotheses. Yet the growing realization that few stone heads can firmly be dated to the prehistoric or Roman period has eroded much of the bedrock underlying the 'cult of the head' hypothesis, in the British Isles at least.

Cypriot heroes and African giants

Even the most dedicated Celtic head cult enthusiasts have recognised that headhunting was practiced elsewhere in the ancient world. Lambrechts (1954, 23), for example, mentions the Germans, Iberians, and Scythians in this regard, as well as occasional instances among the Romans, and Ross (1967, 162) mentions briefly the pre-Celtic origins of head veneration in Europe. Lambrechts makes special mention of the archaic Greeks, whose mythology seemed to preserve ideas analogous to those of the Celts. The birth of Athena from the head of Zeus, for example, can be interpreted, according to Lambrechts, in relation to an archaic notion of the human head as the centre of human power (25). Occasional pieces of Scythian art also seem to depict disembodied human heads either sleeping or dead (e.g., Reeder 1999, 224). For Lambrechts and others, however, Celtic practices differed in both quantity and quality. Nowhere else, in his view, was the human head displayed with such frequency, subtlety, and complexity as among the Celts. Kenneth Jackson (1964, 36) reflected this view when, in relation to head taking among 'the northern Germanic peoples', he concludes that 'there is nothing really to suggest that the special refinements known among the Celts were familiar to them'.

Was there anything truly 'Celtic' about this interest in heads, or did the societies that we group together in that category simply form part of a wider continuum of beliefs and social practices of which an interest in heads was one recurrent part? I have already alluded to headhunting among some near neighbours of the Celts, but there were many other such cases. Describing the aftermath of the German victory at the Teutoburg forest in AD 9, for example, Tacitus describes 'human heads, prominently nailed to trunks of trees' (*Annals*, I. 68). In some cases the specific parallels can be striking. The ritualised skull cup of the fallen Postumius, described by Livy in relation to the events of 216 BC, is almost directly mirrored in Herodotus's account of Scythian practice, written in the fifth century BC:

> They have a special way of dealing with the actual skulls – not with all of them, but only those of their worst enemies: they saw off the part below the eyebrows, and after cleaning out what remains stretch a piece of rawhide round it on the outside. If a man is poor, he is content with that, but a rich man goes further and gilds the inside of the skull as well. In either case the skull is then used to drink from. (Herodotus, *Historia*, 4.65, trans. de Sélincourt, 1952)

John Koch has also pointed to analogous practices documented as far east as China, where annals of the Han dynasty record the warring between the nomadic Yueji and Xiongu during the third and second centuries BC (Koch 1995, 30–1). Having killed the Yueji king, the victorious Xiongu reputedly made a drinking cup of his skull. Koch suggests that these tribal groups were Iranian or Tocharian speakers, placing them in the same broad linguistic milieu as both the Scythians and the Celts. There is also, however, much earlier archaeological evidence from China that would be impossible to link with European practices. From the Longshan site of Jiangou, for example, archaeologists have found drinking vessels made from human calvaria dating to as early as 2600–1900 BC (Underhill 2006, 261). Lest we are still tempted to invoke some shared Eurasian heritage behind such practices, it is worth noting that similar skull cups have also been documented as far afield as Fiji (Carneiro 1990, 204), the pre-Columbian Andes (Arkush 2006, 295), and the prehistoric southeast United States (Jacobi 2007, 321–4). In the eastern midcontinent of America, 'skull bowls' were manufactured through the Middle and Late Archaic and Early Woodland periods, spanning from around 4000 to 900 BC (Speal 2006)

Returning to the Scythians, it is clear that they did more than simply fashion goblets from their enemies' skulls. Skulls pierced, seemingly for suspension, have been found in the Touva tombs, and skeletal evidence of the related practice of scalping, also described

ILLUSTRATION 2.6. Scythian gold ornament dating to the late fourth century BC. The original object is around 5.5 cm tall. (Drawn by Libby Mulqueeny, after Schlitz 1994, fig. 365)

by Herodotus, comes from one of the Pazyrk kurgans (Schlitz 1994, 431–3). Although iconographic material is limited, a warrior holding a spear in his right hand and a severed head in his left, is depicted on a bell-shaped gold ornament from the end of the fourth century BC (Ill. 2.6). The severed head is rather larger than that of the warrior, which suggests perhaps that it had belonged to a giant or hero of some kind. The Scythians, it would seem, shared the Celtic interest in heads to a significant degree.

The Tauri, neighbours of the Scythians, inhabiting the Crimean peninsula, are also described by Herodotus in a passage strikingly reminiscent of Poseidonius describing the Celts:

> Any one of them who takes a prisoner in war, cuts off his head and carries it home, where he sets it up high over the house on a long pole, generally above the chimney. The heads are supposed to act as guardians of the whole house over which they hang. (Herodotus, *Historia*, 4.105, trans. de Sélincourt, 1952)

Such descriptions of headhunting among the Tauri may have underlain some later representations in the art of the Mediterranean world. Two severed heads, for example, appear on a late Etruscan funerary urn, probably from Chiusi, where they form part of a scene from Euripides' fifth-century BC tragedy, *Iphigenia in Tauris* (Bonfante 1984). But many other Mediterranean headhunting tales have no such obvious links back into the barbarian world. Hermary (2003, 527), for example, notes the frequent descriptions of head-taking and display as a trope in Greek literature, and the *Iliad* contains numerous references to decapitation. Although these Homeric accounts contain little to overtly link decapitation with religious belief, such links are clear elsewhere in the Mediterranean. Herodotus, once again, recounts a tale about a head displayed in Cyprus at the beginning of the fifth century BC:

> The people of Amathus, in revenge for his having laid siege to their town, severed the head from the dead body of Onesilus and hung it up above their gates. In time it became hollow, and was occupied by a swarm of bees, who filled it with honeycomb. In consequence of this the townspeople consulted an oracle, and were advised to take the head down and bury it, and, if they wished to prosper, to regard Onesilus henceforth as a hero, and to honour him with an annual sacrifice. This was done, and the ceremony was still observed in my own day. (Herodotus, *Historia*, 5.113–15, trans. de Sélincourt, 1952)

Interestingly, Herodotus does not pass judgement on the townsfolk, nor does he explicitly link the taking of heads to barbarian practice, although the residents of Amathus were fighting on the barbarian (Persian) side of that particular conflict. Herodotus's account does, nonetheless, show the displeasure of the gods at the displaying of a head (or at least the heroic head of Onesilus).

Other examples derive from Greek myths written down in the sixth and fifth centuries BC. The poet Stesichorus, a native of Sicily writing in early sixth century, provides the following story of Cycnus, later to be killed by Heracles:

> Cycnus, son of Ares, lived in the pass of Thessaly and beheaded strangers who came along in order to build a temple to Phobos from the skulls. (Stesichorus, *Greek Lyric III*, frag. 207, trans. D. Campbell, 1991)

And the same theme, of head display in a votive or religious context, is taken up by the fifth-century Theban poet Pindar, who recounts the activities of the African giant Antaeus, son of Poseidon:

> Yet, once on a time, from Thebes, the city of Cadmus, there went a hero, short in stature, but in soul unflinching, even to the home of Antaeus, in corn-bearing Libya, to wrestle with him and to stay him from roofing Poseidon's temple with the skulls of strangers. (Pindar, *Isthmian Odes*, 4.52–4, trans. Sandys, 1956)

Roman monumental art provides another rich source of iconographic evidence for head-taking in various military and political contexts. Numerous relevant scenes appear, for example, on Trajan's Column in Rome, which was erected in the early second century AD to commemorate Trajan's victories over the Dacians. One shows a Roman soldier fighting with sword and shield, while holding the decapitated head of a Dacian warrior in his teeth (scene 24); in another, the head and right hand of the Dacian leader Decebalus are presented to the emperor himself (Speidel 1970). The taking of trophy heads was thus seemingly appropriate material for state-sanctioned monumental art at the heart of the empire. Coin issues of the time even show Trajan standing with his foot on the severed head of the Dacian leader (Ferris 2000, 77). Similar motifs appear later in the second century AD, on the Column of Marcus Aurelius, where scene 66 depicts the severed head of a barbarian being presented to the emperor (Ferris 2000, 91), and again on the Arch of Constantine, erected in Rome during the early fourth century AD. The taking of heads is also far from uncommon in the literature of the Roman world. Silius Italicus, for example, describes the decapitation of Hannibal's brother, Hasdrubal, by the Roman general Nero. The severed head, impaled on pike, is taken to Hannibal's camp to taunt him (*Punica*, 15.806–23). Given the prevalence of such evidence, it would be difficult to build a case from the classical accounts alone for a specifically Celtic interest in heads.

The 'Celto-sceptic' backlash

In recent years there has been a tendency among Iron Age archaeologists, at least in the anglophone tradition, to distance themselves from traditional interpretations of Celtic headhunting (e.g., Merriman 1987; S. James 1999). Perhaps the most vocal has been English archaeologist John Collis, who, in his important recent study on the Celts, has summarily dismissed the 'cult of heads' phenomenon:

> One supposed aspect of Celtic religion which has attracted a wide-ranging literature is the supposed head cult. It is a mish-mash of information taken from various times and places, and much of the quoted evidence in fact comes from secular contexts with no religious connotations. Using the skull-cap of an enemy as a drinking bowl, and the display of a human head outside one's house, something which initially disgusted Poseidonius, is connected rather with social display and insulting one's enemies, and it is something which receives confirmation in the archaeological record, like the mounted warrior scratched on to a pot from Aulnat near Clermont-Ferrand, with a human head dangling from in front of the horse. (Collis 2003, 215–16)

Patently, in Collis's view, the notion of a ritualistic 'Celtic' interest in heads is without merit. On one level this argument is perfectly understandable. There is no doubt that the nebulous head cult has formed one recurrent element in the bulging bag of cultural traits conventionally used to characterise the ancient Celts. Like other components of the Celtic package, this head cult has been assumed to operate on a pan-European level, and little if any thought has been given to its geographical, chronological, or cultural variation within the Celtic world. Rather, it forms part of a timeless Celtic continuum seemingly immune to historical change. In this sense, Collis is right to dismiss it.

Yet it seems to me that there are two problems with this blanket condemnation. Firstly, although he accepts a ritual context for certain children's skulls buried in pits around Clermont-Ferrand, Collis (2003, 216) relegates the supposed Celtic interest in heads primarily to the secular realm. Within the terms of reference I am adopting in this book, he attributes headhunting to the realm of ideology but denies its rootedness in cosmology or religion. More than this, Collis explicitly situates headhunting in the context of 'insulting one's enemies'. Even the elaborately carved stone heads and curated human heads at the southern French sanctuaries of Roquepertuse and Entremont (which Collis characterises as 'not Celtic, but Ligurian', of which more later) are interpreted as 'enemies who are being insulted'. For Collis, the Celts take heads to terrify, to impress, to insult. They do it out of badness, as a simple by-product of war, to further assault the enemy through the despoilation, mutilation, and depersonalisation of the dead. They do it because it is calculated to offend. But can we truly support an interpretation of Celtic headhunting and display that sees it as so casual? Was it really so removed from wider beliefs and social practice? Collis is right that there was no pan-European 'cult of skulls', but the taking, curation, and display of human heads are no less widespread or important for that. Rather than dismiss the phenomenon, we must examine it far more critically than has been done up to now.

The second issue arising from Collis's formulation concerns his observation that the head cult derives from a 'mish-mash' of divergent sources. By this I take him to mean the conflation of the historically and geographically localised observations of Poseidonius with the diverse 'head-related' archaeological finds from the La Tène–using regions of Iron Age Europe; a head here, a statue there, and the odd image scratched on a pot. In this regard, the modern construct of a Celtic head cult reflects, in miniature, the larger creation of a prehistoric Celtic people from a piecemeal assemblage of cultural traits held together by a traditional reading of the fragmentary classical literature. In addressing the inconsistencies inherent in the archaeological evidence for a head cult, Collis (2003, 216) makes the observation that the northern French site of Ribemont-sur-Ancre has produced hundreds of headless corpses 'in contrast' to the profusion of heads at southern sites such as Entremont. This is quite true, and it is indeed a contrast. And yet surely the painstaking decapitation of hundreds of corpses at Ribemont displays a concern with heads just as clearly as their collection and display at Entremont. The specific practices may be reflections or reversals of one another, but the ritual grammar is patently linked. Rather than assume a ready, 'one-size-fits-all' explanation drawn from Poseidonius, we must problematise the whole issue of Celtic headhunting and display and explicitly examine its changing form through time and space. Indeed, it is precisely the 'mish-mash' that is important.

Using the literature

In a recent study, Michael Enright has considered potential symbolic and ideological links between the Iron Age Pfalzfeld pillar from southwest Germany and the Sutton Hoo sceptre from a royal grave in Anglo-Saxon England. Both have complex decoration involving disembodied human heads that, Enright argues, express certain enduring ideas relating to the nature of Celtic kingship, which continue to influence kingly ideology even in the Anglo-Saxon world (Enright 2006). This is a highly sophisticated analysis, which contains many important ideas. Yet it jars in certain ways with archaeological perceptions of Iron Age society. Perhaps most importantly, Enright's study underestimates the diversity of

Iron Age societies in Europe and the social disjunctions that separate the sculptors of the fifth-century BC Pfalzfeld pillar from those of the sixth-century AD Sutton Hoo sceptre. It assumes a common Celticity that homogenizes European prehistoric and early medieval cultures across more than a thousand years and as many miles. Yet, if that Celticity is simply an illusion, how can we hope to draw meaning from the vast array of literary, iconographic, and archaeological evidence? For many modern archaeologists, the answer has been to narrow the focus to the detailed analysis of smaller regions and shorter time periods and avoid anything that smacks of pan-Celtic generalization. The challenge, however, must be to address the innumerable linkages, in art, belief, and material culture, that cut across the regional archaeologies of Iron Age Europe, without recourse to simplistic notions of a Celtic continuum.

Certain recurring motifs in La Tène art hint at shared concepts across wide geographical areas, and these can be useful for giving us insights into the broad cosmological background. They tell us that the human head was important, and shared points of detail can suggest that more specific ideas were held in common at different times and places. The motif of 'opposed heads', for example, is very widespread (Ill. 2.1; also see Chapter 4) and hints at some commonality of symbolism, though we cannot be sure that its meaning was constant over time and space. But if we want to understand the role and significance of the head in more depth, then we have to tack between the details of specific regional and historical contexts and the broader European picture.

So where does this leave the literary sources? If it cannot support the hypothesis of a discrete Celtic 'cult of the head', what value does this material have? We have seen that the classical sources are of very variable quality and were mostly written long after the events they describe. In many cases, the act of head-taking serves a specific narrative purpose and need have no wider significance. Although the classical authors, certainly by the last century BC, regarded headhunting as something of a Celtic habit, the texts themselves attribute similar practices to a wide range of contemporary peoples. Indeed, as we can see from the state-sponsored monumental art in Rome, the imperial army itself not only took heads but curated and displayed them.

Despite this rather pessimistic assessment, however, there are occasional glimpses of something deeper. The Poseidonian account, filtered as it is through later copyists, describes a firsthand encounter with trophy heads, embalmed in cedar oil and displayed above the entrances to Gallic houses, close to first-century BC Massalia. It describes a society where heads were curated and could be ransomed, retaining their value over generations. Poseidonius presents these practices as essentially secular, implicitly linking them to the broader picture he paints of the Celts as violent and boastful, with the preserved heads representing nothing more than mementos of victories in combat. Any religious or cosmological basis to headhunting may simply have escaped him, given the brevity of his visit. Or perhaps it simply did not accord with his somewhat idealised view of druidic religion to allow that these people venerated severed heads of their enemies or ancestors. Even if we trust his observations, however, there is no reason to follow his interpretations of such practices. Like the earliest accounts of European writers describing headhunting in Southeast Asia, the classical accounts can often tell us what happened but rarely why. As such, they form just one strand of evidence, and not necessarily the primary one.

The early medieval texts are, if anything, even more problematic. It would be easy to dismiss them out of hand as either fanciful literary fictions or anachronistic interpolations based on the fluid political situation at the time they were committed to writing. Yet, even if their prehistoric credentials are in doubt, there are elements in these tales that seem unlikely to have their origins in a Christian context. The attention lavished on the severed heads of important individuals, like the Irish high king Fergal, and the posthumous musical abilities of Donn-bó's head, seem to carry echoes of deeper-rooted beliefs, even if the original context is gone. It seems reasonable, therefore, to consider these medieval writings as a potential source of interpretative ideas, in much the same way as we might treat the ethnographic material. Why, after all, should analogies drawn from nineteenth- or twentieth-century headhunters in Borneo or South America have any more relevance to the peoples of Iron Age Europe than the tales and traditions of their early medieval descendants?

In Chapters 4–6 I return to the literary evidence now and again where it seems to have particular relevance to the picture derived from the archaeology. Before embarking on the detailed study of the archaeological evidence, however, I want to step back from Iron Age Europe and consider the evidence for headhunting on a much wider scale.

CHAPTER 3

Shamans on the march

If we are quite frank, we shall have to admit that, even though the worst accounts of Kayan cruelty were substantially true, such behaviour would not in the least justify the belief that the Kayans are innately more cruel than ourselves. If we are tempted to take this view, let us remember that, after our own race had professed Christianity for many generations, the authority of Church and State publicly decreed and systematically inflicted in cold blood tortures far more hideous and atrocious than any the Kayan imagination has ever conceived.

Hose and McDougall 1912, 192

These people, like all cowards, are absolutely devoid of pity.

Up de Graff 1923, 274

From Rum to Riyadh

Some years ago I attended an archaeological conference on the Hebridean island of Rum. One of the main attractions was the opportunity to stay in Kinloch Castle, an opulent shooting lodge built by Sir George Bullough, a wealthy Edwardian industrialist. The Bullough family had decamped to the south many decades earlier, but the contents of the Castle had remained virtually untouched. There were letters in drawers, children's pictures shoved into loose-leaf folders, and piles of books on tables, as if the family were just about to return from a summer picnic. In the library were several leather-bound volumes of photographs, taken on Bullough's numerous foreign voyages. Leafing through them I came across two extraordinary photographs taken in China. The first shows a line of prisoners, kneeling by a roadside in a featureless landscape. They stare blankly at the camera. The second shows the same scene a little later. A neat row of freshly severed heads sits along the same roadside, the crumpled bodies lying behind.

Once the initial shock passes, a closer look reveals that these are not simple 'before and after' photographs, but were in fact carefully choreographed. The heads do not lie where they fell. Between the moments of stillness depicted, a scene of extraordinary violence unfolded. The guards, paid one imagines by Bullough, collected the heads and set them carefully in a line facing the camera. Perhaps a few had to be moved once or twice, to get the spacing just so.

In the past few years, the taking of heads has acquired a new and unexpected prominence. Political instability in Iraq, following the U.S.-led invasion of 2003, led to a renewed spate of hostage taking in the Middle East. In 2004, for example, Islamist militants abducted an American working in Saudi Arabia for a company supplying the U.S. military. A few days later he was beheaded and his body dumped in a remote part of Riyadh. Photographs of the execution and its aftermath were posted on the Internet. Yet before this recent conflict, beheading had not been a common method of execution among Middle Eastern militant groups. Its use here seems calculated to deliver the maximum symbolic effect. The Riyadh case demonstrates the contempt of the abductors for an individual seen as part of the Western military establishment, and thus a legitimate victim, but it also embodies the ritualisation deemed appropriate for this particular killing.

Perhaps ironically, in view of the extreme public outrage felt over this and other hostage murders, beheadings are not in themselves especially unusual in Riyadh. At the time of writing, the Saudi regime continues to employ public beheading as a routine form of capital punishment. The degree to which beheading is sanctioned by Islamic law is disputed, but the Saudi government justifies its approach to capital punishment through appeal to the rather wider concept of Islamic tradition. As such, it is deemed appropriate for a variety of transgressions ranging from murder to homosexuality: 52 men and one woman were beheaded during 2003 alone (Anon. 2004). These decapitations, as well as other judicial amputations, are highly visible public performances, taking place at prescribed times in places where the public might gather to watch. The beheading of the American hostage, and others similarly executed around that time, explicitly imbued the killers' actions with notions of justice and religious authority, which cannot be divorced from this wider context. The display of the severed head over the Internet is arguably little different in principle from the public display of amputated heads and limbs that, at least until quite recently, followed Saudi judicial executions. As with Bullough's Chinese images, the dehumanisation of the victims, the choreography of the killings, the careful arrangement of the heads, and the importance of display and curation (in this case through the modern medium of the photograph) have resonances with other times, places, and events.

Western reaction to hostage beheadings was predictably extreme. The deaths of these individuals commanded far more media attention than the many others who died more mundane but nonetheless violent deaths throughout the Middle East over the same period. Simply in terms of the reaction provoked, these beheadings were patently more effective than alternative forms of killing might have been. Yet why should we be so disturbed by this particular mode of execution? Repeatedly, press reports stressed the 'barbaric', 'grotesque', and 'medieval' nature of these acts. Beheading, clearly, was not something to be practised by any civilised society. Decapitation is somehow set apart from methods of execution more commonly practised in the West; electrocution, lethal injection, hanging, and, less commonly now, shooting. It is hard, however, to determine any rational basis for this civilised-barbaric distinction. It certainly does not reflect the technical efficiency of the various forms of killing or the levels of suffering endured by the victims. Perhaps it is the loss of the integrity of the corpse that causes such offence. Modern execution methods tend to keep the body whole. Indeed, one might track a trajectory in which Western sensibilities have increasingly favoured forms of capital punishment that involve progressively less visible marking of the corpse. A premium is placed on the appearance, though seldom the reality, of a sanitised, humane, and efficient dispatch. This, perhaps, accounts for the falling from favour of firing squads and hanging and the ascendancy of the lethal injection.

It would be wrong to portray judicial decapitation as entirely alien to the modern West, since the guillotine was deployed in France as late as 1977 (although latterly only on very rare occasions). The use of decapitation for 'special' executions – for example, of high status individuals and/or those convicted of treason – is particularly characteristic of Western nation-states from the Middle Ages onwards. This may reflect the desire to make a particular example of those individuals who are perceived to threaten the stability of the state most directly. A classic case involves the exhumation of Oliver Cromwell following the restoration of Charles II. Despite having been dead for some time, Cromwell was nonetheless 'executed' for regicide, an event that culminated in the decapitation of the corpse and subsequent display of the head (Tarlow 2008).

In medieval and Renaissance England, beheading came to be seen as the death of choice for those faced with a range of options. Traditionally this is explained as being due to the relative speed and painlessness of decapitation compared to the alternatives on offer, though this does not always square with the evidence, at least in the days before the introduction of the guillotine and its variants. The execution of Mary I of Scotland for example was a botched and prolonged affair that caused a good deal of suffering for the victim. It is possible, however, that part of the reason for this perception was the 'high status' nature of decapitation; a desire to keep up appearances right to the end. Despite this, however, decapitation has hardly been a characteristic method of execution in modern Western societies.

No killing of a defenceless and immobilised victim is ever likely to be without its ritualised dimension, and death by lethal injection is as structured and formalised an event as the most spectacular public executions of the medieval period. Nonetheless, the ideology underpinning present-day capital punishment in the West has moved away from death as spectacle. Audiences, where admitted, are strictly limited and kept at a distance. Beheading operates on rather different principles. Often public, always extremely bloody, it is hard to imagine anything more physically or symbolically invasive. The instantaneous removal of the head, the seat of the soul, the basis of our individuality, the part of the body that most defines who we are, remains perhaps the most appallingly brutal act that can be perpetrated against the human body.

Much of this is, of course, culturally specific. In societies that reify the individual, the destruction of physical integrity is especially shocking. Western funerary rites either leave the body intact (in appearance if not reality) or destroy it entirely through cremation and pulverisation of the cremated bone. There is little or no involvement of surviving kin with the physicality of the corpse. In societies where funerary rites involve the active dismemberment of the dead, as was often the case in prehistory, the act of beheading and the retention of heads, although no less powerful, may have had a quite different set of associations. Yet, when we consider the ways in which academics, ethnographers, and colonial officials have attempted to understand headhunting and associated practices, it is this Western disengagement with the body that we must confront.

For early European colonialists, encountering previously unknown peoples and attempting to come to terms with alien cultural practices and values, headhunting was the epitome of savagery. Headhunting raids seldom obeyed the proper rules of military engagement. They were often ambushes or surprise attacks aimed at overpowering individuals or small groups with little hope of defending themselves. Victims were often women, children, and the elderly, suggesting a rather ungallant approach to battle out of keeping with the ethos (if not the reality) of Western modes of war. In the absence of any real understanding of the cosmological background that so often structured headhunting practices, most saw the removal and display of heads as simple trophy taking – an abominable denigration of the dead. For nineteenth-century writers on the Nagas of northern India, for example, headhunting was a 'savage blot . . . on an otherwise splendidly warlike and virile character' (Jacobs 1990, 119). The remainder of this chapter examines the nature and content of ethnographic and anthropological writing on headhunting societies of the past few centuries. In particular, I hope to show how ideas drawn from this extensive, though uneven, literature might help structure our interpretations of ritual violence in the more distant past.

The savage blot

> In writing about head-hunting we risk evoking images of the savage other. (Aswani 2000b, 4)

In the nineteenth and early twentieth centuries, images of headhunting were an inextricable part of popular ideas of tribal peoples around the fringes of the European colonial world. More recently, however, headhunting has been effectively written out of the ethnographies of many non-Western peoples. Not unnaturally, anthropologists wishing to avoid stigmatising their subjects have downplayed or simply omitted those behaviours that seem to align too easily with the popular stereotype of the 'savage'. Phenomena such as headhunting, cannibalism, and human sacrifice, which were staples of nineteenth- and early twentieth-century ethnographic writing, and which helped form a persistent and pervasive popular caricature of the 'primitive', have thus almost disappeared from view.

This sensitivity has not been restricted to overly squeamish cultural outsiders. The same process can be seen where indigenous peoples and colonial states have collaborated to represent the past. The Musée Canadien des Civilisations in Quebec, for example, although located in the town of Hull, on the Québecois side of the Ottawa River, is actually less than a kilometre distant from the Canadian Parliament Building in Ottawa, on the opposite bank. As a celebration of indigenous culture it could hardly have a higher political or cultural profile. The building itself is lavish and modernistic, designed by Native Canadian architect Douglas Cardinal, whose own high-profile success encapsulates the overall message of the museum. For this lavish celebration of First Nations' culture, so close to the heart of the modern capital, is intended to demonstrate how far the relationship between the indigenous peoples and colonising Canadians has come since the bad old days when the 'Indians' were regarded simply as savages in the path of progress (Laforet 1992).

The spectacular displays in Musée Canadien des Civilisations have been assembled with the close and active involvement of the various First Nations communities and represent Canada's primary showcase of indigenous culture. The highlights include full-scale reconstructions of house fronts from the Canadian Northwest coast, complete with monumental carved house poles. There is also a good deal of material relating to the institution of the potlatch, where high-ranking individuals established and enhanced their social status by the redistribution and destruction of their own property.

Yet something is missing. When I visited, I was surprised by the lack of information of indigenous ritual practice, social inequality, or conflict. Leland Donald's book *Aboriginal Slavery on the Northwest Coast of North America* (1997) provided part of the missing dimension. In a painstakingly detailed account of the institution of slavery in the region (absent from the museum displays at Hull), Donald provides a vivid picture of vibrant and dynamic cultures in which the undoubted cultural and artistic achievements, now presented as the totality of Northwest culture, are balanced by the brutality of institutionalised slavery, headhunting, and ritual human sacrifice. It gives one a rather different perspective on the monumental carved Nuxalk house poles to realise that their erection was often accompanied by the ritual sacrifice of a slave (e.g., McIlraith 1922–4, cited in Donald 1997, 171–2). Similarly, the 'wealth' destroyed at potlatch ceremonies included human beings, slaves whose ritual murder constituted perhaps the most powerful manifestation of property destruction and the culmination of the ritual theatre of the potlatch. Slave killings at potlatch ceremonies are reported as late as 1880, and unrecorded instances may have occurred until much later (Donald 1997). Perhaps these aspects of Northwest culture are thought to threaten the rehabilitation of indigenous cultures within the modern Canadian state. Certainly they do not figure in the displays at Hull.

Leland Donald's main concern was slavery, but the indigenous peoples of the Pacific Northwest also took heads. One such group was the Kwakiutl, who were studied intensively by Franz Boas, one of the founders of modern anthropology. Yet, as Donald (1997, 65) writes, 'although Boas collected literally thousands of pages on a vast array of topics, there is no indication that he sought texts to explain or describe this culture trait [headhunting] from a Kwakwaka'wakw [Kwakiutl] point of view'. The same group was the subject of a silent documentary film, made in 1914 by the pioneer film maker and photographer Edward Curtis, and originally released under the title *In the Land of the Head-Hunters*. When re-released in 1974, however, in line with the changing sensibilities of the times, it was renamed *In the Land of the War Canoes* (ibid.). It was under the latter title that it was subsequently deemed culturally significant by the United States Library of Congress, ensuring its preservation in the National Film Registry. The film had not changed much during the interim period, but anthropology clearly had.

Along with cannibalism, slavery, and warfare, headhunting forms part of a 'savage package' that has been treated effectively as taboo by recent generations of anthropologists. Aside from the obvious modern distaste for such activities, there have been a number of other contributory factors to this neglect. Pacification by early colonial authorities led to the relatively speedy prohibition of undesirable cultural practices such as cannibalism and headhunting in many areas. Thus, anthropologists have seldom been able to study such behaviours at first hand, and they have rarely formed part of the living cultural systems, with which modern anthropology has been primarily concerned. Instead, the majority of our accounts come from the earliest (and frequently least reliable) observers. As the discipline has become increasingly politically engaged over the past few decades, any concerns with long-vanished, ill-documented, and politically unhelpful practices such as headhunting have come to seem even more remote from the mainstream. The problem has been exacerbated by what Tim Taylor (2002) has characterised as the 'visceral insulation' of the modern West – a distancing from the physical realities of death and bodily decay that make such behaviours as headhunting, human sacrifice, and cannibalism seem inherently improbable and alien. All this has created a vicious cycle since, with so little academic attention, the various components of the savage package have been left undertheorised – and even less attractive subjects of study for new researchers.

There have even been attempts to erase elements of the savage package entirely. William Arens's influential book, *The Man-Eating Myth* (1979), comprised a sustained critique of ethnographic and other records relating to cannibalism, ultimately suggesting that institutionalised cannibalism had never existed. Many anthropologists would probably now accept that such wholesale dismissals of innumerable firsthand observations go too far, and the archaeologist Tim Taylor (2002, 58–75) has provided a comprehensive rebuttal of Arens's claims. Yet Arens's work remains symptomatic of a wider neglect of the 'undesirable' in studies of non-Western cultures. A similar situation applies, for example, to the study of war in non-Western societies. Although nobody has yet taken the Arens line that the whole phenomenon is illusory, American archaeologist Lawrence Keeley (1996) has argued forcefully that anthropologists and archaeologists have effectively 'pacified the past'. By this he means that they have grossly underestimated the degree to which warfare was practised in nonstate societies, and the effects that it has had on the social and cultural development of human societies.

Headhunting, itself often considered as a by-product of warfare, has seen a similar neglect. Even today there are extraordinarily few academic studies dealing specifically with headhunting and related practices. Indeed, despite some important work on specific social groups, such as the Iban of Borneo (R. Rosaldo 1980) and the North Indian Nagas (Jacobs 1990), it was not until 1996, with the publication of Janet Hoskins's edited volume *Headhunting and the Social Imagination in Southeast Asia* (1996a), that the first serious comparative study the subject appeared. Despite the subsequent publication of important individual studies (e.g., Aswani 2000a; D. Arnold and Hastorf 2008), comparable syntheses for other regions have yet to emerge.

There is no question that headhunting was practiced by a great many societies widely separated in time and space. Yet it clearly did not spread, viruslike, among all these groups. Rather it developed time and again, in quite distinct cultural contexts. Richard Chacon and David Dye (2007) have recently presented a global survey of headhunting and other forms of violent trophy taking, listing instances from all parts of the world, from the Palaeolithic through to the modern times. Their grim catalogue provides a baseline for any potential study of the historiography of headhunting. There is no space, however, to embark on any such venture here. Instead, I will simply introduce some of the ways in which westerners have encountered, documented, and tried to come to terms with headhunting in some key historical contexts. From these, we can begin to sketch out some recurrent concepts and ideas that commonly underlie the practice of headhunting at various times and places.

Encounters with headhunters

> It is logical to suppose that unless some sort of government be started which will prevent head-hunting…., the races inhabiting New Georgia will gradually be exterminated. Except from a scientific point of view, I think one might be almost reconciled to this dispensation. The natives have their good points, certainly, but their bad are so much more conspicuous that the elimination of the race would be no great loss to the world. (Somerville 1897, 411)

> In unadministered territory, where head–hunting still flourishes, the inhabitants are kept braced and alert. This cannot but redound to their good. (Mills 1926, 209)

European societies were in contact with headhunting groups from at least the sixteenth century, with the establishment of Spanish colonies in the Philippines and Central America (Ills. 3.1, 3.2). While these early encounters provide us with some valuable evidence, such as Spanish accounts describing the monumental Mesoamerican skull racks known as *tzompantli* (Ill. 3.3), there were few attempts to move beyond basic reporting of exotic and barbarous practices. This changed as accounts of headhunting began to multiply in the travel and adventure writing of the eighteenth and nineteenth centuries. Detailed firsthand descriptions, such as John Rodgers Jewitt's (1985) account of his kidnap by the Nootka on Vancouver Island in 1803, fed a growing popular appetite for tales of treacherous savages and their outlandish practices. Some writers, like the adventurer Fritz Up de Graff (1923), who lived from 1894 to 1901 among the Shuar peoples of the Upper Amazon basin, even participated in headhunting raids and were unafraid to share the grisly details with their readers. Typically, such accounts play to the expectations of a Western audience in turn fascinated and repelled by such alien cultural practices. By the end of the nineteenth century the headhunter was a familiar figure in the European imagination, an exotic bogeyman from the furthest reaches of the known world.

Although the early sources naturally vary hugely in quality and reliability, they do have the considerable advantage of presenting firsthand observations of living headhunting traditions. Captain James Cook's encounters with the indigenous Maori populations at Queen Charlotte Sound in 1770 gave rise to several accounts. Sydney Parkinson (1773, 116), for example, a Scottish artist who sailed with Cook, reported seeing enemy skulls used as water containers and canoe balers. A generation or so later, the English-born missionary, Samuel Marsden recounted that the war canoes of the Maori war chief Hongi returned from one raid laden with up to 70 severed heads apiece (Walsh 1894, 613–14). These trophy heads seem to have been roughly treated, forming a focus for verbal and physical abuse. The mid-nineteenth-century missionary, John Alexander described one particularly horrific scene:

> Among these spectacles I was arrested by the ghastly appearance of a once human head. In mere derision it had been boiled, stripped of the skin and hair, and put on a post with a raw kumara [sweet potato] placed in the mouth. (Cited in Vayda 1960, 95)

ILLUSTRATION 3.1. Image of a 'reptilian being' clutching a severed human head from a Moche III bottle dating to the mid-first millennium AD. Such images demonstrate the time depth of beliefs associated with the severed human head in Central America in the centuries before the establishment of European contacts. (Drawn by Dan Bashford after Moser 1974, 31)

Other accounts describe the heads of enemy war leaders being impaled on posts and left exposed at the victor's enclosure (Vayda 1960, 95.). The primary purpose of headhunting, according to these descriptions, seems to have been to degrade and insult the enemy.

Yet, despite the emphasis in most of these accounts on violence and debasement, it is clear that the Maori also preserved the heads of certain ancestors. Heads of both enemies and ancestors were treated using a range of methods, including steaming, boiling, and smoking, to preserve their facial features and distinctive tattoos (Walsh 1894, 615). In contrast to the prominent display of trophy heads, however, ancestral heads were kept away from public gaze (611). Such heads could, for example, be stored in elaborately carved wooden boxes, and taken out for use in important ceremonies (Davidson 1984, 177).

ILLUSTRATION 3.2. Mictlantecuhtli, Aztec god of death, carrying a severed human head, from the sixteenth-century *Codex Laud*. (Drawn by Dan Bashford after Moser 1973, 44)

Indeed, according to Samuel Marsden (1932, 244), 'if they worship any idol it is the head of their chiefs'.

Although Maori headhunting was a long-lived indigenous practice, there can be little doubt that it was exacerbated by European contacts. The spectacular success of Hongi, the war leader mentioned by Marsden, was facilitated in large measure by the availability of firearms supplied by the British. Not only that, but European sailors provided a ready market for preserved trophy heads, which they took either as souvenirs or to sell as exotic curiosities or scientific specimens (Fulford et al. 2004, 131). The trade was sufficiently commonplace to appear in Melville's *Moby Dick* (1851), where the 'cannibal' harpooner Queequeg hawks preserved Maori heads around the Massachusetts whaling port of New Bedford. Heads with elaborate facial tattoos commanded a particular premium among collectors. European demand was such that a vicious cycle quickly emerged by which trophy heads were traded for muskets, which in turn enabled warriors to obtain yet more heads. In some cases, Maori chiefs reportedly killed their slaves to increase the stock of heads for sale (Vayda 1960, 96). This inflationary trade did not cease until laws were introduced in 1831 to make possession of trophy heads a criminal offence (Jørgensen 2005).

Like the Maori, the Nagas of northern India also retained the heads of both enemies and ancestors, treating each very differently. Accounts of Naga headhunting begin in the 1840s (e.g., Robinson 1975 [1841]; Butler 1847) and, despite attempts at pacification by successive colonial governments, continued well into the twentieth century, when some of our most detailed accounts emerge. Indeed, Jacobs (1990, 119) reports that, in certain parts of the region, Naga warriors continued to decapitate slaves on specific ceremonial

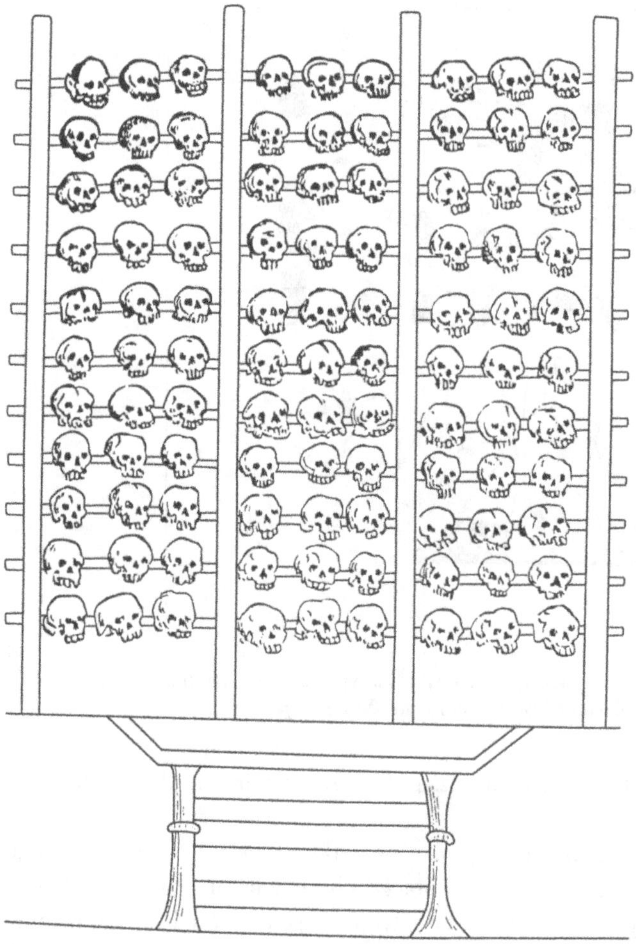

ILLUSTRATION 3.3. This image of an Aztec *tzompantli* is based on a sixteenth-century illustration by the Franciscan friar Diego Durán, but numerous indigenous representations are also known. Although the images usually, as here, depict crania, they would originally have been displayed as fleshed heads (Moser 1973, 28). (Drawn by Rachael Kershaw after Durán 1951)

occasions throughout the period of British rule, and headhunting raids are documented as late as 1958 (Elwin 1961, 12). While reports of Maori headhunting came to us mainly through the testimony of Christian missionaries, explorers, and soldiers, Naga headhunting persisted long enough to be studied by the first generations of professional anthropologists. The written accounts thus take on a much sharper focus. Although the rites associated with headhunting vary markedly, it is perhaps useful to give at least one detailed description of the treatment of captured heads, if only to give a sense of the complexity that such rites could entail. The following description is based on the firsthand observations of the anthropologist Christoph von Fürer-Haimendorf (1969, 97–9), who lived with the Konyak Nagas in 1936–7.

Following the welcome of the returning headhunters by the village elders and other men of the community, the captured head was taken to an area outside the village, close to where the heads of ancestors were kept in their stone 'cists' or pillars. There the trophy was pelted with raw eggs to 'blind the kinsmen of the dead foe'. Rice beer was then poured into its mouth, accompanied by invocations to the victim's kin to come and surrender themselves to the victorious headhunters. After this initial ceremony, the head was taken into the village itself where there was period of dancing in front of the cult house or 'morung', and the headhunters were ritually washed. The head itself was tied to a log gong, which was beaten during the festivities. Later the head was placed on a post in front of the morung, accompanied by a further period of dancing. On the day after its arrival, the head was taken in procession by the men of the village, all bedecked in body paint and ceremonial costume, to a ritual area specified for the purpose, where a 'priest' (usually the senior descendant of the village founder) cut off parts of the ears and tongue and once again called on the victim's kinsmen. This part of the ceremony was accompanied by the sacrifice of a chicken and reading of its entrails. The head was then hung from a tree and, for a period of a day, no work was done and the village was given over to dancing and feasting. After a month had passed, a further feast was held, during which the head was taken down from its tree, given rice beer to 'drink', and deposited in the morung.

In other Naga villages there were variations on this complex sequence of rituals. At Siong village, for example, the heads were taunted and speared in the eyes by widows who had lost their husbands to earlier headhunting raids; in other cases, the head was physically broken up and divided between members of the headhunting party (von Fürer-Haimendorf 1969, 99). It is seldom clear from the ethnographic accounts what exactly happened to the head at the end of its ritual use. In some cases they appear to have been displayed in the morung until they eventually fell apart, although there are other accounts of heads being exposed with the newly dead on corpse platforms, where they would have simply decayed until removed by weather, time, and scavengers.

As well as the Maori and the Nagas, much early discussion surrounded the headhunting communities of Southeast Asia, where ethnographic work began to focus from the late nineteenth century onwards (e.g., Hose and McDougall 1912). By now, headhunting had ceased to be simply a barbarous curiosity and was increasingly seen as an intrinsic part of the complex social lives of peoples worthy of study in their own right. A major marker point came in 1898, with the Cambridge Anthropological Expedition to the Torres Strait, between Australia and Papua New Guinea. Led by Cambridge anthropologist Alfred Haddon, the multidisciplinary team contained several individuals who were to become major figures in the development of British anthropology, including W. H. Rivers. Although headhunting was hardly the central concern of Haddon's team, the practice did now form an inextricable part of wider anthropological interpretations of life in the Torres Strait.

The search for 'soul-substance'

By the early twentieth century, European ethnographers had come to realize that headhunting had deep roots in indigenous religions and cosmologies. Removing the head of a victim, however repellent to Western sensibilities, was not simply an act of gloating triumphalism but was regarded as transferring certain spiritual or supernatural benefits to the head-taker. The English administrator Major Arthur Tremearne (1912, 152–3), for example, reported that the ghosts of headhunting victims in northern Nigeria were condemned to serve their killers in the spirit world, while Norwegian ethnographer, Carl Lumholtz (2004 [1920], 155) attributed similar beliefs to the Dayaks of Borneo. More influential, however, were the ideas of Albert Kruyt, a Dutch missionary and ethnographer working in Indonesia from the 1890s. Kruyt developed the idea that indigenous communities in that region saw the natural world as animated by a 'life-fluid' or 'soul-substance' (Needham 1976, 73). If the appropriate rites were followed, this mysterious 'soul-substance' not only could be appropriated but could be moved around and stored. In humans, 'soul-substance' was thought to be concentrated in the head. By taking the heads of enemies, therefore, headhunters could accrue benefits for their community; crops would grow more plentifully, humans and their domestic animals would multiply, and illness and misfortune would decline. By contrast, failure to gather heads would see a reversal of these benefits and an erosion of communal prosperity. Essentially the same logic underlay the retention of ancestral skulls.

Kruyt's theories became extremely influential and the perceived relationship between heads and fertility came increasingly to be recognised as central to the cosmologies of headhunting societies. Perhaps the clearest indication of how this link was seen to operate is given by John Henry Hutton, whose academic career as professor of social anthropology at Cambridge University had been preceded by 25 years working as an administrator for the Indian Civil Service. This lengthy posting had given Hutton unparalleled opportunities for travel among the Nagas, enabling him to observe, at first hand, societies whose contact with westerners was otherwise minimal. J. H. Hutton (1938, 12) explained how the Naga headhunter sought 'to lay in a supply of life-fertilizer for the benefit of his community, bringing it in the heads of his enemies, from which it exudes into the sacred stones of his village, to pass into the cycle of life of its crops, its livestock, and its human population'. Hutton's 'life-fertilizer', like Kruyt's 'soul-substance', acted in these interpretations as a mediating agent, between the taking of enemy heads and the benefits that the practice was thought to bring.

It was not until the 1970s that problems with the concept of 'soul-substance' were exposed, in an analysis by anthropologist Rodney Needham (1976). While accepting that some causal link was widely perceived between the taking of heads, fertility, and general prosperity, Needham pointed out that terms such as 'soul-substance' have no equivalents in the indigenous languages of the groups concerned. Indeed, when the Dutch ethnographer Jacob Elshout asked his contacts among the Kenyah of Borneo about this enigmatic 'fluid', he was 'surprised to find that [they] can say practically nothing about it' (1926, 211; translation in Needham 1976, 75). The notion of a 'soul-substance' was, in Needham's view, simply an interpolation that enabled Western ethnographers, steeped in an intellectual tradition of scientific materialism, to conceptualise the otherwise unintelligible causal links between severed heads and fertility. For Needham, there was no need for human skulls to act as containers of 'life-fluids' or 'soul-substance'. Taking heads simply brought fertility. If Western ethnographers could not understand why this should be so, and thus felt the need to explain it in terms of the transference of some invisible essence, then that that was simply a shortcoming of Western scientific modes of explanation. As Julian Jacobs (1990, 124) put it, with reference to the Nagas:

Nagas:	heads → good crops
Anthropologists:	heads → soul substance → good crops

Nonetheless, Kruyt's ideas capture some sense of the value of heads in certain cases. Among the Amazonian Shuar, for example, a concept approximating to 'soul-substance' may well have existed. Shuar *tsantas* (shrunken heads) remain ritually charged for the duration of their transportation and ceremonial assimilation into the headhunter's home village (Ill. 3.4). A series of rituals then serves to transfer spiritual energies (embodied in its *miusak*, one of the three souls that Shuar might acquire) from the head to the community. When these rites are complete, the tsanta is effectively used up and is subsequently retained only as a token, devoid of any power (Rubenstein 2006, 365). The Shuar miusak thus bears at least some resemblance to Kruyt's 'soul-substance'.

Heads and fertility

Needham's arguments highlight the near-impossible task of interpreting one society's cosmological understandings in the conceptual language of another – an observation with implications well beyond the realm of headhunting. From a strictly archaeological point of view, however, this may not matter too much, since we have little hope in any case of establishing the subtleties of prehistoric causalities. What remains clear is that a crucial link was thought to exist, in indigenous communities spread very widely across Southeast Asia and beyond, between headhunting and fertility. Indeed, the explanation most commonly offered by practitioners themselves, was that headhunting was necessary to ensure and enhance the fertility of crops, people, and animals, as in the following quotation attributed to a Penihing Dayak informant from Borneo in the early years of the twentieth century:

> If no heads are brought in there will be much illness, poor harvest, little fruit, fish will not come up the river as far as our kampong [village], and the dogs will not care to pursue pigs. (Lumholtz 2004 [1920], 155)

Similar sentiments are encountered even in more recent ethnographic literature; an Ilongot informant told Renato Rosaldo (1977, 168) in the early 1970s, for example, that 'when we took heads all the time, there was no illness'.

The connection between heads and fertility was explored in some depth by the anthropologist Derek Freeman (1979) during his fieldwork among the Iban of Sarawak (Borneo) in the 1940s and 1950s. During World War II, Iban warriors had fought alongside British soldiers against the Japanese military. Despite the colonial government's prewar prohibition of headhunting, wartime decapitation of the Japanese dead was commonplace and widely tolerated (e.g., Heimann 1999, 203). Trophy skulls collected

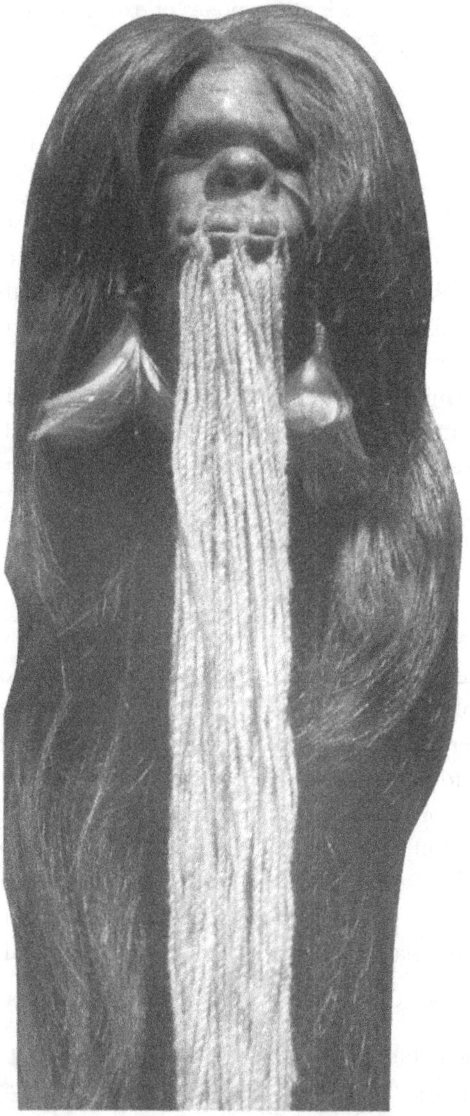

ILLUSTRATION 3.4. Shuar *tsanta*. (Photograph by author)

during this period remained very much in evidence at the time of Freeman's fieldwork, and the cosmologies surrounding the human head still formed a vibrant tradition. In one Iban myth, the war god Lang splits open a human trophy head with his sword, releasing seed that subsequently yields a human crop (D. Freeman 1979, 234). This association between heads and agricultural success is further reinforced by the Iban belief that the rice crop is possessed of a soul that shares its name (*semengat*) with the soul that resides in the human head. In Iban cosmology, the human head is also associated with the phallus, both being seen as containers of seed. This cosmological link was apparently widespread in Borneo, as illustrated in the practice, observed by anthropologist George Appell, of treating infertile women by placing a trophy head between their thighs (ibid., 237). In New Guinea, the Asmat made similar metaphorical associations between the severed human head and the fruit of the sago (Zegwaard 1959, 1039).

Although these connections emerge most clearly in the ethnographies of Southeast Asian headhunting societies (where they are almost universal), they also apply in other parts of the world – as, for example, among the Amazonian Shuar (e.g., Rubenstein 2006). Indeed, they can even be identified in the pre-Columbian societies of Central and South America, where the human head was a dominant and persistent motif in indigenous art. Moche art, from the first half of the first millennium AD, contains many images of severed heads often associated with supernatural beings or masked human figures (Moser 1974, 37). In virtually all cases these can be plausibly identified as embodiments of a fertility deity associated with agricultural production. Nasca painted pottery similarly shows trophy heads with plants sprouting from their mouths (Proulx 2006, 109–10). Echoes of similar beliefs survived long after the formal adoption of Christianity. In one Guatemalan community, for example, three 'ancestral' skulls continued to be paraded around the village at specific points in the agricultural cycle well into the 1950s: the skulls were housed for the rest of the year in a side altar of the local Catholic church (Wright 1988, 55). As in Southeast Asia, it seems that the taking or offering of heads was regarded as beneficial for the fertility of the crop.

With very few exceptions (e.g., van der Kroef 1952, 229) headhunting raids appear to be the exclusive preserve of males, although the subsequent cleaning, preparation, and curation of trophy heads is very often assigned to the women of the community (e.g., D. Arnold and Hastorf 2008, 221–2). In certain cases, the relationship between heads and crops plays an even more overt role in the structuring of gender roles. Among the Shuar, for example, the rites associated with the arrival of a tsantsa into the village were believed to transfer its spiritual powers to the women of the community and thence, through their agricultural labour, into the crops (Harner 1973). This could be read as an appropriation of female productive power by male warriors – rendering female-dominated agriculture subordinate to male-dominated warfare.

The link with fertility and productivity may also help explain the lack of discrimination frequently shown in ethnographically documented headhunting raids. In the majority of cases there was little specific attempt to obtain heads of enemy warriors or chiefs. Instead, communities often displayed an 'any head will do' approach, in which heads belonging to low-status, marginal, or vulnerable members of the enemy community were considered just as desirable as those of respected warriors. Among the Nagas, for example, it was observed that 'any head counts, be it that of man woman or child' and 'as common a method as any was for a man to lurk about the water ghat of a hostile village, and kill the first woman or child who came to draw water' (Davis 1981 [1891], 258). Shuar heads could also be those of men, women, or children, although warrior's heads were sometimes favoured (Jandial et al. 2004), and similar examples can be found in virtually all regions where headhunting was practiced. In all such cases, it was not the individual from whom the head derives that was important but rather the intrinsic spiritual worth of a human head as a source of fertility, bounty, and success.

Grief, rage, and rites of passage

Not all anthropological studies have focused on the links between heads and fertility. Indeed, one of the most detailed and influential, carried out by Renato Rosaldo (1984) among the Ilongot of Luzon in the Philippines came to quite different conclusions. Despite the supposed abandonment of headhunting several decades earlier, the great majority of adult men interviewed by Rosaldo had themselves participated in headhunting raids. Like the Iban of Borneo, Ilongot men had decapitated Japanese soldiers during World War II, and the last internecine beheadings in the region were reported as late as 1972 (M. Rosaldo 1977, 168). Ilongot headhunting was highly unusual, however, in that the severed head was not retained or curated in any way but rather was cast away in the moments following decapitation. It was the act, not the head that was important, and in that respect, Ilongot practices support Needham's view that the head was not generally regarded as the precious container of some ethereal fluid. Despite their apparent disregard for the severed head as an object, however, the act of head-taking remained centrally significant to Ilongot social life.

Ilongot explanations for headhunting centred not on any perceived benefits in terms of fertility or agricultural success but rather on the need to give vent to individual rage or grief, often resulting from bereavement. Ilongot men were perceived to have 'passions' or 'anger' that needed to be quelled, and this could best be achieved through beheading a member of another community. The removal of these dangerous energies enabled the individual headhunter to mature spiritually, acquiring wisdom and self-control.

The case of the Ilongot highlights a persistent problem in anthropological research: the relationship between the explanations people themselves give for their actions (*emic*), and the sorts of interpretations that anthropologists might apply, from an outside (*etic*) perspective. Rosaldo (1984) himself, following his own bereavement, came eventually to accept and defend this indigenous, emic explanation. Yet, while a concentrated outpouring of grief and rage might indeed be a plausible explanation for extreme acts of violence, it does not really explain why the act of decapitation in particular should assume such importance. To make sense of this specific focus of Ilongot violence, we have to look at its broader context as part of a wider range of headhunting behaviours throughout Southeast Asia societies; and in many of these other societies, there is no question that headhunting is intimately bound up with cosmological and religious beliefs.

Robert McKinley (1976), in one of the more complex cosmological analyses of headhunting, proposed that the purpose of taking heads was to incorporate the wild, untamed outside world (as represented by enemy groups) into the fabric of the community. Bringing home their victims' heads transformed these 'semihuman' creatures into tamed spirits that could become part of the community. Journeys into enemy lands to obtain heads at the same time elevated the status of the headhunter, whose ability to move between 'worlds' took on a mystical quality, similar to the trance journeys of shamans. Thus, for McKinley (1976, 102), the headhunter is effectively a 'shaman on the march'. Although this interpretation has been heavily criticized (e.g., D. Freeman 1979, 233), it serves at least to highlight one striking aspect of the symbolism underlying headhunting, not only in Southeast Asia but also more widely – the role of heads as liminal objects, linking the worlds of the living and the supernatural. This liminal quality appears time and again in the ethnographic accounts, as severed heads are repeatedly implicated in rites of passage, moments of creation, and periods of transition.

In many societies, headhunting formed an integral part of rites of passage, particularly those involving the assumption of adult male status. The successful capture of a head, for example, was essential to this transition among the Asmat of New Guinea and the Iban (Zegwaard 1959, 1040; D. Freeman 1979, 238). Even among the Ilongot, successful headhunters earned the exclusive right to wear hornbill earrings, which greatly enhanced their prospects of marriage and their regard among other members of society. Headhunting was also often associated with particular moments of creation or transformation. William Archer, who worked as a colonial officer in northern India shortly after World War II, reported in his unpublished notebooks that the Sangtam Nagas required a head to consecrate a new log gong:

> As soon as the drum is in position, a human head must somehow be got – as until a head is offered, the drum must not be beaten. If a stranger happens to pass by the village he is immediately beheaded – this is very rare – as all the neighbouring villagers are on the alert. The village therefore has sometimes to go far afield and sometimes it takes many months. (Cited by Jacobs 1990, 120)

Similarly, in the Solomon Islands captured heads were used in ceremonies associated with the inauguration of war canoes (Hocart 1931, 303), and among the Asmat analogous rituals accompanied the installation of ancestral *bis* poles (Zegwaard 1959, 1028). Elsewhere in Southeast Asia, the taking of a head was frequently required to bring to an end a formal period of mourning (e.g., Haddon 1932, 214).

Headhunting and social power

In the popular imagination, headhunting is traditionally the preserve of remote tribal peoples with limited levels of social organization. Yet the ethnographic and historical literature demonstrates that the taking of heads was practiced by societies with widely varying degrees of social complexity. In keeping with the focus of anthropological study throughout the twentieth century, the most detailed accounts of headhunting describe small-scale communities, where social power was relatively fluid and seldom passed down through the generations. Among the Shuar, for example, power was acquired largely through individual warrior prowess. Successful headhunters gained considerable prestige and, after dispatching a certain number of victims, were awarded honorific titles. Yet success entailed exposure to risk; the more headhunting raids an individual attempted, the more likely he was to be killed by his enemies (Rubenstein 2006). The very basis of social power thus prevented its accrual, and warfare acted as a regulatory mechanism maintaining a broadly egalitarian community.

In other regions, however, headhunting became intimately linked to the establishment and maintenance of chiefly power. Janet Hoskins has attempted to define two quite distinct ideologies of headhunting that developed on the island of Sumba, now part of Indonesia in Southeast Asia (Hoskins 1996c). Although emerging from the same broad cultural and economic background, social forms in East and West Sumba had diverged markedly by the nineteenth century, partly as a result of their differing levels of access to trade and colonial contacts. In East Sumba, noble families came to control large estates farmed by servile labour. Trade with Dutch colonialists enabled them to accumulate significant wealth and consolidate their economic and political control. This part of the island was also characterised by frequent and bloody territorial wars, in which headhunting played a central role. Trophy heads were displayed to symbolise the subjection of defeated enemies and the ascendancy of victorious *rajas*. For Hoskins (1996c, 225), East Sumban headhunting embodied an 'ideology of encompassment', reinforcing the expansionist claims of noble families. By contrast, West Sumba, had no developed political hierarchies and few trade links, and power had to be won anew in each generation through personal charisma and success in war. Headhunting here formed part of a more traditional 'ideology of vendetta', where revenge killings perpetuated an ongoing cycle of violence (ibid.). It seems highly probable that this situation pertained across the whole of Sumba before the development of political hierarchies in the eastern part of the island. As a cultural practice then, headhunting was sufficiently malleable to serve the different social agendas under varying historical circumstances.

The nature of headhunting also varied across the island. In the west, the heads of elderly men and women were taken along with those of warriors; younger women and children were generally taken as slaves. Trophy heads were delivered to the headhunter's village in a raw and bloodied state and were subject to much mockery and abuse before being defleshed for display on the village skull tree (Ill. 3.5). In East Sumba, by contrast, only warriors from noble families were considered as suitable headhunting victims. Here the heads were defleshed prior to their arrival in the village as clean, 'pure' skulls (Hoskins 1996c, 228). This apparent deference reflects the status of these skulls as symbols of the valuable new dominions acquired by the victors.

Similar transformations have been charted by Shankar Aswani in his study of headhunting in the Solomon Islands. From the sixteenth to the nineteenth century, headhunting in the Roviana region increased in intensity as rival elites competed for social and political power. Although heads belonging to both sexes and all ages were taken, there was an increasing interest in the taking of prestigious heads, belonging to chiefs or notable warriors. This is reflected both in the organised attacks on hillforts (or *toa*) intended to 'secure the head of the chief and prominent warriors' and in the differential disposal of high- and low-status heads (Aswani 2000c, 61). In Roviana the period witnessed a transition from small-scale intercommunal headhunting, where 'any head will do', to large-scale headhunting raids led by ambitious chiefs. For Aswani, Roviana trophy heads gradually became a kind of capital, to be accumulated and displayed as a measure of chiefly status and a symbol of spiritual efficacy. I return to this case study in Chapter 6.

Physical traces

The beliefs and practices discussed here relate to societies observed at first hand by anthropologists, ethnographers, and others. Archaeologists, of course, must rely on the material

ILLUSTRATION 3.5. Early twentieth-century textile from East Sumba showing the skull tree motif. (Photograph © National Gallery of Australia)

evidence of past societies. It is worth considering, therefore, what sorts of physical traces headhunting might leave in the archaeological record. Generally, the material expression of headhunting might take two main forms. The most obvious (though by no means unambiguous) are the bones of headhunting victims, either in the form of detached, curated, or decorated crania or as incomplete burials with osteological evidence for the removal of the head. Here the main problem lies in distinguishing violent headhunting from the religiously motivated detachment of the head as part of a series of mortuary rites. This will be a recurring problem throughout the following chapters. More problematic still is the second type of evidence – the references to headhunting that appear, now and again, in the art and iconography of headhunting communities. Both types of evidence are found in the archaeological record of Iron Age Europe, and their analysis forms the basis for the rest of this book. First, however, it is useful to reflect on the materiality of headhunting in some of the ethnographically documented societies discussed in this chapter.

As a starting point, it is worth pointing out that headhunting, even where it is widely practiced and absolutely central to social identity, may be all but invisible archaeologically. The Ilongot, as we have seen, discarded the heads of their enemies as soon as they were severed, and the hornbill decorations worn by successful headhunters made no obvious iconographic reference to the taking of heads. Perhaps some indication of decapitation might be detected in osteological remains of victimized groups. Otherwise, without oral and literary testimony, the practice would entirely escape detection. The shrunken heads of the Shuar would similarly not be expected to leave much archaeological trace. The processing of the tsanta was begun as soon as the headhunting party had retired to a safe distance: it involved a complex process of defleshing, boiling, and heat treatment, which required many hours to complete. The discarded skull, which would bear characteristic marks of defleshing, was often thrown into a river where, conceivably, it might survive as the sole archaeological signature of the process.

Most documented headhunting societies did, however, retain the skulls of their victims, and these were often elaborately decorated, curated, and displayed, sometimes over long periods. During her fieldwork in the 1980s in Sumba, for example, Hoskins (1996c) was shown head trophies collected some 60 years earlier. In many cases, as among the Nagas, the crania of headhunting victims were displayed in significant numbers on wooden racks or other outdoor constructions. Elsewhere, trophy heads might be displayed within houses, ritual structures, or exposed on funerary platforms.

Anthropological and historical accounts provide a fair amount of detailed information on the ways in which trophy heads were displayed in the immediate aftermath of their capture. Unsurprisingly, however, they have much less to say about what happened to them in the weeks, months, or years that followed. One exception is an unusually detailed account of headhunting in southern New Guinea in the first half of the twentieth century, by the anthropologist Justus van der Kroef, which tracks the fate of trophy heads from capture to eventual disintegration. Strung up in the hot Melanesian climate, these skulls rapidly deteriorated; eventually, when the trophy was too far gone for display, the headhunter would retrieve what was left of the mandible and cervical vertebrae, which were retained and further curated, the remainder of the skull being discarded (van der Kroef 1952, 234). In such circumstances, despite the large scale of headhunting in the region, there would be little left for the archaeologist to find.

Oral histories from Roviana suggest that the skulls of commoners were eventually buried along trackways, where the trampling of feet above them would provide a final insult. More prestigious skulls might, however, be curated over considerable periods before being given up as offerings during funerary or other rituals (Aswani 2000c, 68). The Zemi Nagas buried captured heads around the log gongs that formed the ritual centres of their villages (Jacobs 1990, 98). Similarly, in East Sumba, the crania of prestigious headhunting victims were buried following the cessation of hostilities (Hoskins 1996c, 230). Where bone survives in the soil, such practices should be detectable archaeologically. The many thousands of heads set up on precolonial Mesoamerican skull racks, or *tzompantli*, were displayed for considerable periods. When they had begun to decay, the weathered skulls were taken down by ritual specialists and destroyed by burning (Mendoza 2007, 415) in a process unlikely to leave much in the way of diagnostic evidence for their earlier ritual use. In much of Southeast Asia there seems to have been less active curation, with bones becoming subject to gradual decay through weathering, scavengers, insects, and general decay. Where fragments became dispersed there is potentially a range of mechanisms by which they might find their way into archaeologically recoverable contexts, such as middens and pit fills. A whole range of taphonomic factors would determine the extent of any surviving remains, but assuming that soil conditions were conducive, we might imagine that some evidence would survive to demonstrate the former presence of human heads on these sites.

The only case where a significant programme of archaeological work has been conducted on an ethnographically documented headhunting society is in the Solomon Islands, where Peter Sheppard and his collaborators have attempted to use archaeological methods to trace the Roviana 'headhunting complex' from its nineteenth-century manifestations, back to its precolonial, prehistoric beginnings. As Sheppard and his colleagues (2000, 34) put it, 'the ancestral cult shrines with their skull deposits, and the canoe houses adorned with trophy skulls dominated the prehistoric settlements and bound the worlds of the living and the dead into a single whole'. Even in this region, however, where predatory headhunting and the veneration of ancestral heads were manifestly central to the nineteenth-century political economy, archaeological reflections of the phenomenon are hard to find. Surveys and limited excavation of certain high-status settlements, notably the hillforts of Nusa Roviana (thirteenth– nineteenth century AD) and Saikile (late eighteenth century), have produced evidence for shrines that still house fragmentary human crania belonging to venerated ancestors (Sheppard et al. 2000, 16). Yet the trophy heads of enemies killed in battle are rarely obvious in the archaeological record. The Roviana case thus suggests that even predatory headhunting on a grand scale may not leave more than a subtle trace in the archaeological record.

Aside from the bones themselves, whether elaborately decorated or simply curated, there are several instances where recent headhunting societies have a developed material culture associated with the practice. Usually, this springs from the role of headhunting in rites of passage and need not bear any overt visual relationship to the practice itself. For example, the wearing of hornbill earrings by an Ilongot warrior may have denoted his status as a successful headhunter, but the objects themselves have no obvious iconographic linkage to the severed head itself, except within Ilongot cosmology. Occasionally, however, the links are more obvious and would be suggestive even without access to the documentary sources. In Sumba, for example, elaborate textiles display the motif of the skull tree, hung with trophy heads, and framed by panels showing male warriors engaged in ritual performances (Ill. 3.5). Perhaps the most developed material culture of headhunting, however, existed among the Nagas where headhunters wore brass pendants depicting severed

ILLUSTRATION 3.6. Naga head trophies obtained from Yimpang village, North India, 1936. (Photograph © Christoph von Fürer-Haimendorf)

heads, and hairpins decorated with rows of head-shaped knobs denoting the number of their victims. Naga trophy heads were also at times hung on cords alongside wooden and woven facsimiles (Ill. 3.6). Here the human head formed a central symbol that recurred consistently in a range of social contexts.

Recurrent themes: A headhunting grammar?

More important perhaps than any parallels in the material expression of headhunting between recent and ancient societies is the potential for ethnographic writings to inform our understandings of past beliefs. Despite their diverse origins and agendas, there are sufficient consistencies in these accounts to suggest at least some potential ways of looking at headhunting in the past. What emerges most forcefully is the centrality of headhunting to cosmology, religion, and ideology. Indeed, the practice seems almost never to have been simply about trophy taking or the denigration of the dead. Far from being a by-product of violent conflict, headhunting was often the stated reason for war. This was not just the case in small-scale raids, but also, as in Roviana, for predatory headhunting expeditions involving hundreds of warriors and sea journeys over tens of kilometres (Aswani 2000c). Headhunting was implicated variously in origin myths, gender relations, rites of passage, funerary rituals, and in the establishment and maintenance of social authority. The head, whether of an enemy or an ancestor, has been a potent symbol with associations relating to power, fertility, gender, coming of age, and the acquisition of status. Understanding the control and deployment of this symbol can potentially tell us a great deal about wider social relationships, beliefs, and practices. Given this background, it would be rather surprising if Iron Age headhunting was simply a by-product of war.

Beyond its basic centrality to social life, however, the range of culturally specific practices associated with the human head is considerable. Even at the local scale, significant variation can be seen in the processing, treatment, and display of heads. Nonetheless, certain recurrent themes are useful in structuring our thoughts about the role of headhunting in the more remote past. It is useful to summarise a few of these here. In many regions, societies that captured the heads of their enemies also venerated the heads of their own kin. This was true, for instance, among the Maori and the Nagas and in the Solomon Islands. This is perhaps unsurprising since, in cosmologies where the head is accorded particular significance, it will potentially become important in a range of social contexts. Frequently, ancestor heads and trophy heads were treated quite differently (although in ways that might be hard for the archaeologist to discern), yet the essential premise of the retention and curation of the head remains key.

The recurrent association with fertility is also striking, often embroiling the human head in the origin myths of headhunting societies. The wide geographical and chronological distribution of this link, from Southeast Asia to Central America, and from the pre-Columbian Moche to the twentieth-century Nagas, suggests that the cosmological connection between heads and crops can arise time and again in otherwise unconnected societies. The body and its component parts act consistently as metaphor for understanding the world; terms like 'heads of corn', 'heads of rice', and 'heads of sago palms' make reference to the anthropomorphising of the crop, and the pervasive notion that human bodies and cultivated or naturally abundant food sources are linked and mutually dependent. As we shall see in the next chapter, there are grounds for believing that this correlation was also made in Iron Age Europe.

Also widespread is the association of heads with moments of creation or transition. The most obvious are rites of passage, especially funerals and the assumption of adult male status, and also the inauguration of important buildings or objects with religious significance. Heads, as physical embodiments of the dead trapped in the world of the living, form an obvious link between the natural and supernatural realms. Again, the archaeological record of the European Iron Age will show us similar associations.

Finally, heads can form an important symbolic currency in the assumption and maintenance of social power. Here our ethnographic evidence is limited by the marginal status of many of the societies described, often after years of colonial exploitation and disruption of indigenous power structures. Yet, as in East Sumba and Roviana, it is clear that whatever the cosmological and religious roots of headhunting, the head formed a powerful symbol open to appropriation by aspiring elites. Possession of the heads of enemies symbolised the completeness of their subjection, while control over the heads of powerful ancestors harnessed and redirected their prestige to subsequent generations. With the growth of increasingly stratified societies, the identities of individual heads became important. Any head might have acted metaphorically to express bounty or fecundity, but as social hierarchies became increasingly developed, the heads of prestigious individuals became increasingly prized. The head was thus a malleable symbol, capable of expressing subtle and potent ideas concerning the actuality and legitimacy of power.

Beneath the enormous variability of culturally specific practices, we can discern the persistence of certain ideas; an underlying grammar of head-taking, retention, display, and manipulation. Consistently we see human body parts implicated in strategies of power, dehumanisation, and victimisation, yet at the same time associated strongly with ideas of fertility, sanctity, veneration, and reification. From these building blocks a range of specific practices emerges at quite different times, places, and cultural contexts.

Ethnographic analogies from Southeast Asia and elsewhere need not of course be more relevant in understanding Iron Age headhunting than contemporary classical descriptions or the medieval tales derived from descendants a few generations removed from the people with whom we are concerned. But importantly, they open up the range of interpretive possibilities, and suggest some key themes that might usefully structure our examination of the Iron Age evidence. Critically, the ethnographic accounts can provide us with sets of ideas and concepts with which we can begin to approach the interpretation of Iron Age headhunting afresh, without resorting to blanket attributions of shared Celtic ethnicity.

CHAPTER 4

Pillars, heads, and corn

If we do not get a head every year, the crops will be bad, the pigs and cattle will not increase, our children will get ill. . . . We cannot say why this should be. It has always been like that.

Attributed to Konyak Naga informant, from private notes of W. G. Archer, 1940s, cited by Jacobs 1990, 120

Phallic stones and monkey bones

The enclosure known as Raffin Fort occupies the summit of a small hill, overlooking a large area of County Meath in eastern Ireland (Ill. 4.1). The enclosure itself is hengiform, meaning that its ditch lies inside its bank, rather than outside as we might expect if this were a defensive site. It was built around AD 380–570, but activity on the hilltop began much earlier, during the Late Bronze Age (Newman et al. 2007, 351).

The peculiar form of the enclosure links Raffin to the better-known 'royal sites' of Iron Age Ireland, such as Tara and Navan Fort (Raftery 1994, 65–82; Waterman 1997). These latter sites, of which only a few are known, seem to have been the focus for large-scale communal rituals, including the inauguration of regional chiefs or kings. Although prehistoric in origin, they preserved this role into the early Christian period. According to the early literary accounts, inauguration rituals in Ireland involved a symbolic marriage between the new king and the land itself, often personified by an earth goddess, such as Medb Lethderg at Tara, or Medb of Connacht (MacKillop 1998, 288–90; Waddell 1998, 352–3). Places like Raffin may have served a similar purpose, though at a smaller scale, for 'local, autonomous tribal units' (Newman 1993, 23).

Within the Late Iron Age enclosure at Raffin was a low stone pillar, around 0.6 m high (Ill. 4.2). According to the excavator, this had been 'intentionally selected because of its natural phallic shape'. Underneath it was a rather unusual deposit. In a shallow pit, around 0.4 m deep, lay a 'bed of charcoal' made up of several layers. On top of this someone had placed, face upwards, the frontal part of a human skull, probably belonging to an adult male (Newman 1993, 22), along with an animal rib and part of a pelvis. Curiously, although lying directly on the charcoal, the skull fragment was unburnt. The pillar stone had been laid directly over these fragments.

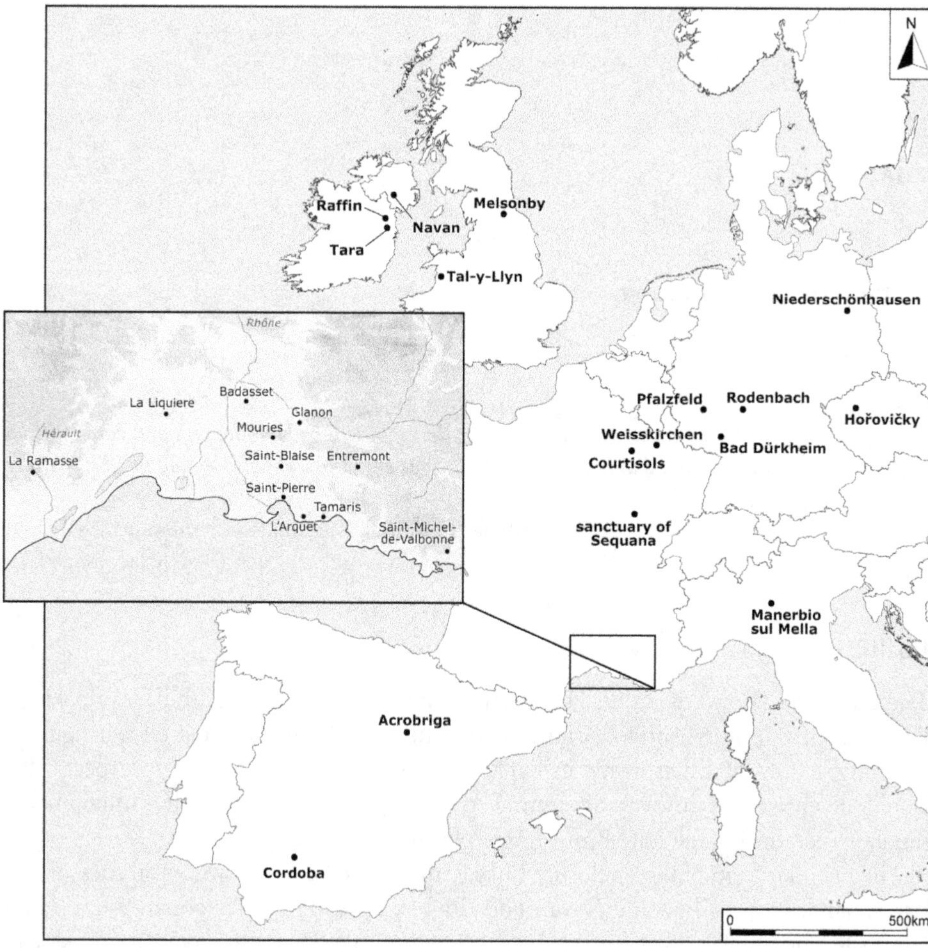

ILLUSTRATION 4.1. Map showing the main European sites discussed in Chapter 4. (Drawn by Rachael Kershaw)

Detailed analysis of the pit's contents revealed further complexities. The charcoal was found to consist of at least five layers, each burnt *in situ*, and probably laid down in rapid succession, but each of a quite different composition. Indeed, the excavators have suggested that particular wood species may have been 'specifically chosen for symbolic reasons'. Most striking was the prominence of alder, particularly young wood, which seems to have come from 'flower- or fruit-bearing branches' (Newman et al. 2007, 363). Since alder is a difficult wood to burn, it seems unlikely that it was selected for purely practical reasons. In early Irish and Welsh mythology, however, alder is recurrently associated with the idea of rebirth, perhaps because of the characteristic transmutation of the wood from white to red when cut (e.g., MacKillop 1998, 11).

Equally striking was the dating evidence from the pit. Three radiocarbon dates from the charcoal layers suggested that they had been deposited in the late third or fourth century AD – rather earlier than the construction of the surrounding hengiform enclosure. More

ILLUSTRATION 4.2. Stone pillar at Raffin. (Photograph by Conor Newman)

importantly, however, a radiocarbon date obtained from the human bone placed it earlier still, in the period 110 BC–AD 140 (Newman et al. 2007, 351). In other words, this individual had been dead for 100 years or more before his remains were deposited in the pit. During the intervening period, spanning several generations, his skull had apparently been carefully curated, as there is no evidence of rodent or carnivore gnawing. It had also been subject to repeated handling, since its edges had been worn smooth (Newman pers. comm.).

It is hard not to see this fragment of an ancient human head, planted like a seed below this phallic pillar, as a symbol of fertility and renewal. Along with the animal bone, and the charred remains of the carefully chosen wood, the skull fragment forms the residue of a meticulously choreographed ritual that culminated in the emplacement of the pillar. The precise meanings of this ritual will continue to elude us, but the association of the phallic stone, the young, burnt alder branches, and the skull fragment itself suggest that they involved concepts of fertility and rebirth. The setting of this small monument, in a ritual complex associated with the ongoing fertility of both tribe and land, further reinforces this association.

The association of skull and pillar at Raffin recalls the practice, among the North Indian Konyak Nagas in the early twentieth century, of placing the skulls of the dead in small, specially carved, phallic pillar stones (Jacobs 1990, 124). Among the Nagas, this treatment was restricted to particular members of the in-group and was quite distinct from the forms of display accorded to trophy heads obtained from outsiders. Unlike the Raffin example, these Naga skulls were placed on view within specially carved niches in the lower part of the stone, but the association of ideas is suggestive. Among the Nagas, heads were intimately associated with concepts of fertility (119), and these head pillars were one of several ways in which this idea was materialised. I will argue that the Raffin skull represents one instance of a similar association of ideas in Iron Age Europe.

The Raffin skull fragment is by no means an isolated find. Human calvaria have also been found associated with cattle skulls in ditch deposits at the Late Bronze Age enclosure of Stramullin, also in County Meath. Other apparently votive deposits from the same ditch included intact saddle querns (Ní Lionáin 2007, 20), again perhaps evoking an association between human heads and agricultural productivity. Work on the larger 'royal sites' has revealed analogous discoveries. Fragments of human skull have been found, for example, in the lower fills of the ditch around the Rath na Ríogh at Tara, County Meath, dating to around the turn of the first millennium BC (Roche 1999, 27). Although these may derive from disturbed burials (Roche 2002, 59), the disproportionate occurrence of cranial and mandible fragments may be suggestive of selective redeposition.

Several further instances of head deposition, in its broadest sense, come from the Navan complex in County Armagh, in the north of Ireland. At the King's Stables, an artificial pool lying below the Late Bronze Age hillfort known as Haughey's Fort, trial excavations revealed the frontal part of a human skull (Lynn 1977). The remains have been identified as the facial bones (excluding the mandible) of a young adult male. Close examination suggested that the facial portion may have been deliberately cut from the rest of the skull (Delaney 1977, 61). Among the accompanying animal remains were disproportionately high quantities of red deer antlers and dog bones, as well as articulated pig bones, suggesting that these were votive offerings rather than domestic rubbish. Interestingly, the excavator, Chris Lynn (1977, 55), has pointed out that the human remains were in far poorer condition than animal bones from the same deposits. This suggests that, as at Raffin, the human skull may have been curated for some time before its eventual deposition in the pool.

An analogous context of deposition comes from the natural lake of Loughnashade, which lies immediately below Navan Fort itself and seems to have formed a focus for votive deposition during the Iron Age. As such, it may have taken over the role of the earlier King's Stables pool. During the late eighteenth century, workmen digging drainage ditches, recovered a series of fine bronze trumpets, decorated in La Tène style. According to reports from the time, these elaborate instruments were accompanied by animal bones and human skulls (Mallory and McNeill 1991, 149).

Finally, from Navan Fort itself, it is tempting also to consider the complete skull of a barbary ape (*Macaca sylvanus*) found within the wall slot of one of the central buildings (Napier and Jenkins 1997). A combination of stratigraphic analysis, radiocarbon dating, and dendrochronological dating suggests that the Navan ape skull was deposited between 250 and 100 BC (Lynn 1997). Although most of the bone from the slot was apparently domestic rubbish, the condition and fragility of the ape skull suggest that it had been carefully deposited, perhaps as a votive offering. By any standards, this was an extraordinary discovery. Barbary apes, as the reader will be aware, are not native to County Armagh. In fact, the species, during the Iron Age, would have been confined to North Africa. We do not, of course, know if the ape ever lived at Navan, although it is pleasing to speculate on the impact that a diminutive, hairy, tail-less, humanlike creature might have had, swinging around the timber structures of this windswept Irish drumlin. Possibly, however, it was only ever the head that arrived at Navan. Either way, the animal, or its remains, must have been transported a considerable distance, by land and river across the continent, or else by boat around the Atlantic coasts. These apes, which can be trained to perform tricks, were prized as exotic pets in the medieval period, and this ability probably also explains their occasional occurrence during the Iron Age (Raftery 1997). They would have been appropriate gifts between individuals of high status, a practice that Jim Mallory (pers. comm.) has compared to the 'panda diplomacy' practiced by China during the 1970s. It is striking that nothing apart from the skull of this exotic, anthropomorphic animal was found, suggesting that this may, once again, have been curated.

In several of these instances, we see an association between the deposition of human (or anthropomorphic) skulls and monuments associated with kingly inauguration rituals, performances that were closely related to ideas of fertility and renewal. In the case of Raffin, the links between concepts of fertility and the human head seem especially pronounced. As I have already suggested, in highlighting the analogy with the Naga head pillars, these snapshots evoke some of the associations discussed in Chapter 3 between human heads and fertility in a range of ethnographically documented societies. For the remainder of this chapter, I want to examine some of the ways in which these ideas might play out in one specific regional context – the Iron Age of southern France and, in particular, the lower Rhône valley.

Iron Age encounters in southern France

The Iron Age archaeology of the French Midi forms the backbone of this and the next two chapters. In each case, I want to use the unusually rich contextual detail from this region to structure a wider discussion about certain aspects of the curation, display, and representation of the human head. Although there is a degree of chronological progression from chapter to chapter, there is also a good deal of overlap. I begin here with a brief summary of the archaeological background to the Iron Age of southern France, before moving on to discuss material from the later part of the Bronze Age and the early centuries of Iron Age, the period from roughly 850 to 500 BC.

The Iron Age in the lower Rhône valley displays an extraordinary constellation of material relating to headhunting and veneration, including literary, skeletal, and iconographic evidence. Some of the iconographic material is well known, at least visually, in the general literature on Celtic Europe. The seated figures, Janus head, and head-niche pillars from Roquepertuse (Ill. 4.3) make regular appearances in popular and academic books on the Celts. The cross-legged warriors from Entremont, holding severed heads in their laps, are equally familiar (Ill. 6.4). Usually, however, these images are abstracted from their specific regional context and used simply to illustrate the supposed pan-Celtic interest in heads. So habitual is their appearance that it is easy to forget that there has never been a detailed study of this material in English. Indeed, much of what is available (e.g., Benoît 1975) has been rendered obsolete by more recent research.

Greeks, Celts, and Ligurians in the French Midi

> Quite apart from the impressive seated figures, the iconography illustrates *truly Celtic* themes in the horse friezes, the sculptured decapitated heads, and in the provision for real ones for display in monumental settings. The groves of Mona destroyed by Suetonius may have witnessed no more barbarous rites than those enacted in these great sanctuaries of the Saluvii. (Powell 1961, viii, emphasis added)

ILLUSTRATION 4.3. Front and rear views of one of the seated warrior statues from Roquepertuse. (Photograph © L. Damelet and P. Groscaux, Centre Camille-Jullian/CNRS)

ILLUSTRATION 4.3. *(continued)*

Sometime around 600 BC, Greeks from the city of Phocaea, on the Anatolian coast, established a new colony on the southern French coast near the mouth of the river Rhône. The name given to the new settlement was Massalia (the modern port city of Marseille). These colonists did not, however, enter wholly unknown territory. Local communities had been involved for at least a generation in trade with Etruscan and subsequently Greek merchants. At the early fortified settlement of Saint-Blaise, levels dating from as early as 650 BC contain sherds of wine amphorae and Etruscan black bucchero pottery (Rolland 1964a; Bouloumié 1992). Massalia, however, seems to have been the first permanent colonial presence in the region.

For some five centuries after its foundation, during which it became established as a major centre of economic and political power, Massalia coexisted with the indigenous populations of the French Midi (Hodge 1999; Hermary et al. 1999; Bats 2003). Exploiting the major communications route formed by the Rhône-Saône corridor, it formed, at least during the sixth and early fifth centuries BC, an important trading link between the Mediterranean and the communities of temperate Europe (e.g., Collis 1984; Cunliffe 1997). Yet, throughout this time, Massalia remained essentially a sea power, with satellite colonies established along the Mediterranean coast, and the precise extent of its inland territory, or *chora*, remains unresolved (e.g., Goudineau 1983). Even at the height of its power, however, independent native settlements were still being established within a few miles of the city (e.g., Bernard 2000, 158; Rothé and Tréziny 2005, 124–30).

Beyond the Massaliote chora was a mosaic of indigenous polities. Traditionally, the native populations of the hinterland were thought to have been a mixture of Celts and Ligurians, the latter being supposedly part of a once-widespread cultural grouping that had extended along the Mediterranean coast into northern Italy. The Celts, or Gauls, by contrast, were thought to have arrived as invaders at some point during the Iron Age. Thus, the Iron Age populations of the region have come to be regarded as hybrid 'Celto-Ligurians'. The evidence for this culture-history, however, is limited to a series of often ambiguous textual accounts (Bats 2003), and there is clearly a significant danger that part at least of the conventional picture derives from the inconsistencies of the classical authors (see Collis 2003 for an excellent historiographical account of the problem, and Williams 2001 for the background to Roman writings on the Gauls). Whatever the truth or otherwise of the various ethnic ascriptions, there is nothing in the archaeology of the region to reflect a cultural divide between Ligurians and Celts. Neither is there evidence for a marked cultural break that might signal a Celtic invasion. People undoubtedly did move in and out of the region, but not apparently on a scale sufficient to disrupt the clear cultural continuities that span the whole of the local Iron Age. As Terence Powell (1961) suggests, the most evident cultural links of these southern French communities were with populations to the north and west, rather than with the heavily urbanized societies of the Mediterranean. For present purposes I am going to avoid making any assumptions about the ethnicity of these southern French populations. Terms such as 'Celto-Ligurian' will be used only in relation to the theories of other writers.

The 'Hellenisation' question

The terminology of colonialism implies a certain relationship between the colonizers and the colonized. Until quite recently it was generally imagined that the Greeks were the dominant partners on the basis of their assumed cultural superiority. A process of 'Hellenisation' was invoked by which the supposedly self-evident benefits of Greek culture were absorbed and imitated by their barbarian neighbours. Thus, over time, the native populations were gradually civilized.

In recent years, however, such views have come to be recognised as simplistic (Hodos 2006). There was clearly no reason for the dynamic and successful indigenous cultures of southern France to give themselves over to alien lifeways, simply because of the proximity of a culturally Greek outpost on their coastline. Michael Dietler's work (e.g., 1995, 1997, 2005) has shown that the indigenous groups, and in particular the more dominant or aspirant individuals within them, took from the Greek colonists those items of material culture that were most useful for their own social and political goals. Because competitive feasting and (more particularly) drinking were key social institutions, local communities were quick to trade with the Greeks for wine and the material accoutrements that came with it, primarily ceramic serving and drinking vessels (e.g., Dietler 2005). Some aspects of urban planning and defensive architecture were also adopted at certain times and places, but there seems never to have been any concerted attempt to imitate Greek models (Garcia 2004, 179). Literacy and coinage, having apparently little perceived value in the Iron Age cultural context of the interior, were adopted only after several centuries of availability (Garcia 2004). Clearly, this prolonged colonial encounter did not represent a simple process of Hellenisation.

Despite the proximity of Massalia, the indigenous communities of southern France maintained both their cultural and political independence for almost half a millennium. Autonomy was finally lost only with the annexation of southern France by Rome at the end of the second century BC (Ebel 1976). Within a generation of this annexation, it was this region that provided the source material for Poseidonius's well-known account of Celtic headhunting.

Art, influence, and the problem of chronology

Elements of Greek culture were indeed adopted by local communities, but these were transformed within their new cultural setting. For present purposes, the most important of the early borrowings from Massalia were the techniques of three-dimensional stone sculpture. Yet, as we shall see, indigenous sculptors had no intention of emulating the aesthetic prescriptions and thematic programmes of their Greek neighbours. Instead, the motifs that dominate from the earliest times are the twin images of the warrior and the severed head.

The best-known English-language account of the southern French sculpture is contained within Fernand Benoît's (1975) posthumously published summary of his excavations at Entremont. Benoît was the dominant figure in the Iron Age archaeology of southern France from the end of World War II until his death in 1969, and his was certainly an authoritative statement. From a present-day perspective, however, it has two major shortcomings. The first is the tendency, running through all of Benoît's work (e.g., 1955a, 1957, 1965), to derive virtually all aspects of the sculpture from Mediterranean prototypes. Thus the carved stone bird from the sanctuary at Roquepertuse became, for Benoît (1975, 254), 'the bird of the swamps of the Styx, according to Graeco-Italic symbolism', while the braided hair of some of the severed heads from Entremont 'is suggestive of the "golden hair" which attaches the mortal to life and which the Fates severed in Virgil's account of Dido's death' (256). Benoît's interpretations did not go unchallenged, even at the time, and many archaeologists felt that he had gone much too far in his invocation of Greek influences (e.g., Hatt 1970, 131). For Hodson and Rowlett (1970, 190), for example, there was 'nothing Greek in these sculptures either in style, content or spirit'. Nonetheless, Benoît's legacy took many years to fully shake off.

Subsequent generations have tended to reject Benoît's vision and have instead stressed the links between the symbolism of the Provençal statues and aspects of Celtic iconography from regions to the north. For François Salviat (1993, 212), for example, 'the severed head is a *leitmotiv* of the art of Entremont, and more generally ... of Celtic art, manifestly situated in Celtic culture and its spirit'. After years of being essentially Greek, the Iron Age sculptures of the Midi were now basically Celtic. For many years, this cultural tug-of-war has continued with little prospect of conclusion. As André Rapin (2004, 14) has recently put it, 'These works seem to lend themselves to only one form of interpretation: influence'. Yet these sculptures patently need to be understood on their own terms and within their unique cultural context rather than as colonial extensions of either Greek or Celtic culture.

The second major problem with Benoît's work (e.g., 1975, 250) was his presentation of the material within a compressed chronology centred on the period from the mid-third to the late second century BC, and largely devoid of any internal sequence. Influential earlier writers, including Salomon Reinach and Joseph Déchelette, had been content to place at least some of the Provençal sculpture as early as the fifth century BC, contemporary with the archaic phase of Greek art. Indeed, when Déchelette (1914, 1533) incorporated the seated figures of 'La Roche-Pertuse' (Roquepertuse) into his celebrated *Manuel d'Archéologie*, under the heading 'Graeco-Ligurian and Graeco-Celtic Sculptures', he proposed a date in the sixth or fifth century BC on stylistic grounds. But, by the mid-twentieth century, the simplicity and formality that Déchelette and others had seen as reflecting the contemporary influence of archaic Greek statuary had come to be regarded as no more than emblems of ineptitude – the work of poor barbarian copyists who had simply failed to master developing Greek styles. These were, in other words, the telltale signs 'of provincial backwardness which cannot serve as a basis for an early dating of the statuary' (Benoît 1955a, 47). Instead, Benoît and his contemporaries looked, not unreasonably, to the accumulating archaeological evidence as a basis for their chronology.

Some of the most important sculptural assemblages, including those from Roquepertuse and Entremont, were sealed by destruction layers that could be fairly securely dated by pottery typology, and which were often associated by their excavators with specific, historically attested events. Thus the Entremont statuary seemed to have been destroyed during the late second century BC, most likely by the Roman army, while the Roquepertuse material seemed to have met its end in an earlier, unrecorded conflict. Similarly, the seated warriors from Glanon seemed to have been associated with an elaborate shrine destroyed at the end of the second century BC, an event linked by the excavator to the Teuton incursions of 102 BC (e.g., Rolland 1968, 22). In other words, the chronology of these statues was determined by the dates of their destruction, which seem mostly to have occurred during the third and second centuries BC.

Dissatisfaction with this compressed chronology has come to the surface only in the past few years, most notably in a series of papers by André Rapin and Patrice Arcelin, which have promoted a radically extended chronology (e.g., Arcelin and Rapin 2003; Rapin 2003, 2004). Arcelin and Rapin recognised that the indigenous statuary had little in common with the styles that would have been current at nearby Massalia during the third and second centuries. If they were to be seen as attempts at copying Hellenistic styles, then they were very clumsy attempts indeed. Yet, for the most part, they are not clumsy statues. As Rapin (2004, 15) has pointed out, the 'hieratism and "clean" style' of sculptures, like the seated figures from Roquepertuse, has much more in common with Greek sculpture of the archaic period than the anatomical realism of their supposed Hellenistic contemporaries.

Yet few sculptures have been found stratified in early contexts. The earliest instance is probably the fragmentary bicephalic (or Janus-headed) bust from the oppidum of Marduel à Saint-Bonnet-du-Gard, which has been stratigraphically dated to the mid-sixth century BC (Py and Lebeaupin 1994, 257, fig. 55). The bust was associated with numerous stelae, and the whole assemblage has been tentatively reconstructed by the excavators as a contemporaneous collection, originally displayed on a terrace within the oppidum (259, fig. 57). The context is not dissimilar to that of the rather later assemblage from Roquepertuse. This mid-sixth-century date can probably be extended, on stylistic grounds, to the similarly bicephalic bust from Beaucaire and the helmeted busts from Sainte-Anastasie, Gard (262; Ill. 5.5).

Although the Marduel dates are early in relation to the traditional chronology, they are by no means the earliest sculptures in the region. Recent typological and stylistic reanalysis has pushed back the dates of many of the southern French statues, in some cases by several centuries. For the earlier statues in particular, the chronological shift can be considerable. One of the earliest is the 74-cm-high limestone figure known as the Warrior of Grézan, found in 1901 on the southeast outskirts of Nîmes. Only the upper part of this statue survives, but this is enough to show that it depicted a male warrior in a rigid pose, with his arms clamped against his sides. Although the face has been largely removed, the strongly marked ears and a thick neck survive. The figure wears a heavy hooded helmet with a distinctive dorsal plume, and a torc around his neck. His breastplate is elaborately ornamented both back and front, despite heavy damage. Everything about the piece conveys solidity and weight and 'seems intended to give an impression of power' (Py 1990, 813). The Warrior of Grézan had long been dated to the fourth century BC, but this attribution has been based on what now appears to be a mistaken typological interpretation of his belt buckle (Py 1990, 813–15; Rapin 2004, 19), which Arcelin and Rapin (2003) have reclassified as an Italic type dating to the eighth or seventh century BC. This early date also provides an acceptable context for his torc, cuirass, and distinctive helmet (*casque-capuche*). Conceivably such details could be consciously archaic, and a later dating is still favoured by some authorities (e.g., Dietler and Py 2003, 789). Nonetheless, it is hard to believe that such outmoded equipment would have been faithfully transcribed in stone several generations after its currency in the world of the living. Instead, its seems more probable that the Grézan figure dates no later than the seventh century BC, and thus predates the influence of Massalia.

Some time later, probably during the sixth century BC, the earliest of the distinctive seated figures appear at Nîmes and Glanon. Their anatomical styling and, more specifically, their acroteria (or plinths) show direct Greek influence for the first time (Rapin 2004, 21). According to the revised chronology, the seated figures from Roquepertuse were carved slightly later, perhaps around 450 BC, based on the detailing of their plinths (21 and fig. 7). Despite the obvious external influences that became apparent at this time, the iconographic content remained wholly indigenous.

A similar stylistic approach has been applied to the whole body of sculpture from the region. The process has been helped by new discoveries, notably the small bronze figure seated on the rim of a beaked flagon from the Glauberg tomb in Hessen, southern Germany (Weber 2002, 242). This cross-legged warrior, seated in the same pose and clothed in a strikingly similar fashion to the Roquepertuse warrior statues, can be securely dated to the mid-fifth century BC. The richly detailed armour, weaponry, and jewellery of the seated warrior statues from Entremont, similarly form a coherent typological group centred in the first half of the third century BC (Rapin 2004, 19). The unavoidable implication is that some of these statues were curated and displayed for many centuries before their final deposition. In the case of Glanon, for example, figures that Rapin believes to have been carved around 500 BC seem still to have been on display during the late second century (Roth-Congès 2004).

This new chronology allows for a combination of influences in the origins of the sculptural traditions of the Midi. On the one hand, the influence of Greek traditions of figurative three-dimensional sculpture is obvious and can now be better understood without the imposition of the artificial time lag that kept the barbarians an arbitrary couple of centuries behind their Massaliote neighbours. But the new chronology also allows for a more direct continuity with the indigenous stelae and statue menhirs of the southern French Bronze Age. The revised chronology remains imperfect and will, no doubt, be subject to further revisions. Nonetheless, it marks a major step forward and will underpin many of the arguments in the following chapters.

Human debris

The special status of the human head can be traced deep into the prehistory of the Midi. At the Chalcolithic site of Cambous à Viols-en-Laval, in Hérault, for example, excavation of a single large house produced cranial fragments from three adults, mixed up with butchered and cooked animal bone, pot sherds, and general household debris (Canet and Roudil 1978). In one instance, a fragment of the mandible was also discovered, perhaps suggesting the former presence within the house of a fleshed head. Otherwise, the only other human bones recovered from the site were a small femoral fragment and the skeleton of a baby, just a few months old, buried in a pot (Poulain 1978, 188). This village was abandoned many centuries before the period that most concerns us here. Nonetheless, this striking combination of human heads and infant burial in domestic contexts is one that we encounter time and again during the Iron Age.

During the Bronze Age, human skulls appear in a variety of overtly ritual contexts. Perforated oval objects known as *pendeloques*, probably intended as amulets, manufactured from fragments of human parietal have been found in cave deposits dating to the Early or Middle Bronze Age at Suzon I and the Late Bronze Age (Bronze Final II–III) at Landric à Saint-Baulize (Dedet 1992, 16–17). But from the Later Bronze Age onwards, it is the occurrence of human bones on settlements that is most striking.

Some of the most important Late Bronze Age assemblages come from a series of excavations on small lagoon-side villages around the étang de Mauguio, southeast of Montpellier (Dedet et al. 1985; Dedet and Py 1985; Prades 1985). In each case, excavations yielded considerable quantities of ceramics and other material, but little in the way of house plans or architectural detail. One such site was Camp Redon, where remains from at least nine individuals were recovered, the human bone fragments forming more than 10 percent of the total bone assemblage from the site. Of these nine individuals, eight were represented solely by fragments of the cranial vault, the remaining bone being (as at Cambous à Viols-en-Laval) a femur fragment (Duday 1985, 117–19; Py 1990, 799). The bones, belonging to seven adults, an adolescent, and a child of five or younger, came from sediment layers containing domestic debris such as ashes, animal bone fragments, and sherds of pottery. Most were from the settlement layer dated to Bronze Final IIIB (c. 850–725 BC), although some could be a little earlier. Other settlements around the étang also produced human remains, including three cranial fragments from the Bronze Final IIIB layer at La Rallongue (Duday 1985, 121), and a single adult cranial fragment from domestic occupation material in layer Bronze Final IIIB at Tonnerre I.

The Late Bronze Age sites around the étang de Mauguio present us with a remarkable concentration of evidence for what must have been a strong local tradition. Clearly, human heads were a not uncommon sight on these settlements. So how did they come to be there? Very occasionally, human remains on settlements may represent disturbance to earlier burials, as may be the case in the rather later Iron Age oppidum of Les Castels at Nages (Py 1990, 799–801), but this situation appears to be rather rare. It would certainly not account for the overwhelming dominance of skull fragments. So, were these the heads of venerated ancestors, or perhaps of decapitated enemies? This is a question we will return to at various points, and it is a very hard one to answer. For now, we can simply make a few preliminary observations. Firstly, at none of these sites do we find evidence for what we might recognise archaeologically as a votive or 'special' deposit. In part, this may be due to a lack of clear contextual evidence, which is unsurprising given the small scale of many of the excavations. However, it does appear that most of the human remains we encounter were simply strewn around with a mix of undifferentiated domestic debris. This does not seem particularly suggestive of veneration. Secondly, the age profile of the skeletal remains from Camp Redon, meagre as our evidence is, at least suffices to show that the heads of young children could be treated similarly to those of adults. This again does not immediately suggest that these were the heads of venerated ancestors, although we should not exclude the possibility that the bones of children could 'stand for' the ancestral community in some symbolic way. Equally, if these were the heads of outsiders, then they were not exclusively obtained from enemy warriors.

If any pattern is to be teased from such a small dataset, then we might suggest that the mix of ages, and the lack of apparent care in deposition, might reflect the treatment afforded to the heads of outsiders, obtained through raiding or ambush. Perhaps displayed or curated for a while, these were eventually discarded or left to disintegrate and eventually, as fragments, become absorbed in the general domestic waste of the settlement. In support of this idea, we might note that many of the fragments from around the étang de Mauguio displayed clear striations, made up of many curved and parallel lines of variable width. Henri Duday (1985, 121) believes that, while these do not suggest deliberate human action such as defleshing, they probably result from exposure and scouring, which would be consistent with the idea that they were left in the open for lengthy periods.

The sorts of practices represented in these Late Bronze Age sites continue into the Early Iron Age. Fragments of human skull, again suggesting the presence of fleshed heads, were found in two locations associated with a series of partially rock-cut houses at La Liquière, a village set on a plateau dominating the plain of la Vaunage, Gard. The village was occupied from around 625 BC until around 550/500 BC and seems to have been unenclosed, although it was afforded natural protection by steep slopes to its north and east (Py 1984). Michel Py sees this Early Iron Age settlement as the seasonal base for a semisedentary group practising a traditional and largely self-sufficient way of life; one that was soon to come to an end with the appearance of Massalia and the beginnings of large-scale economic interactions between the indigenous communities of the Midi and the Mediterranean world.

Within House LIB at La Liquière, the excavators found a fragment of the right parietal of a young adult, which appears to have been deliberately broken rather than subject to natural post-depositional breakage. From the same building came two joining fragments forming a near-complete human mandible, probably male. All of the bones may have belonged to the same individual (Duday 1984, 351). On the basis of the stratigraphic evidence, these bones would appear to date to around 575–550 BC (Py 1984, 213). From Zone 4, on the same site, came part of a second adult mandible, this time from a different individual. This particular fragment was found among ashes on a built clay hearth, along with abundant pot sherds, again dating to between 575 and 550 BC (Duday 1984, 351–4; Py 1984, 47 and 68). Although the osteoarchaeological report on the bones makes no mention of any obvious alteration of the bones caused by heating or cooking, it is possible that the association of a human mandible with a hearth indicates some form of treatment of the head aimed at its preservation, perhaps through smoking or boiling. However, the apparently deliberate breakage noted to the cranium of the individual in House LIB is more suggestive of the extraction of the brain, perhaps for consumption. If so, the apparently casual incorporation of leftover fragments into floor deposits, as in the earlier sites, does not suggest any particular reverence in the process.

Funerary sites do not figure much in my discussions of the southern French Iron Age, because for much of the region they effectively disappear after the sixth century BC. Where they occur, however, early in our period, they do provide some interesting comparisons. In a few tombs dating to the Early Iron Age in the Garrigues, in eastern Languedoc, for example, crania appear to have been removed from graves sometime after burial (Dedet 1992, 273–5). Numerous adults represented by teeth and elements of the postcranial skeleton, but entirely lacking the cranium, have been found from several tombs at Cazevielle, Saint-Martin, Peyrescanes, and elsewhere. In some cases, the only teeth present are those which are the first to fall out of the mandible or maxilla. This suggests that crania were removed after the corpse had undergone a period of decomposition but before it had become entirely defleshed. In other words, some considerable time after the burial and associated funerary rites, the grave was reopened and the partly fleshed cranium removed for some other purpose. Exactly the same phenomenon can be seen at the La Tène cemetery of Mont Trotté in the Champagne region of northeastern France (Rozoy 1965). Indeed, remarkably similar practices can be seen in quite unrelated contexts, at other times and other places. In the Neolithic Near East, crania were retrieved from burials under house floors to be plastered, painted, and used in various domestic rituals before finally being buried in small communal caches (Kuijt 2008). To facilitate the process, the position of the head was sometimes marked with red paint on the white plaster floors above the burials. Kuijt (2008, 186) has interpreted the plastered skulls as one stage in a long journey by which a deceased individual was embraced within the collective memory of the group. Do similar rites explain the human skulls from domestic sites like La Liquière and Camp Redon? This remains a possibility, but it does not fully accord with the condition or treatment of the heads from these settlements. More likely perhaps, the tombs of the Garrigues simply provide an insight into another, linked set of practices involving the curation and manipulation of the human head.

Clearly the human head was of considerable importance in the Later Bronze and Early Iron Age periods in the south of France. The recurrent presence of cranial and mandible fragments in domestic contexts and the near-complete lack of postcranial material testifies to that. The clear signs of tampering with the crania of the dead in various tombs, suggest that certain beliefs and practices were centred on the heads of ancestors. Yet the skull deposits on settlement sites do not immediately resonate with this interpretation. There is little sign of reverence in the handling and deposition of the material, most of which seems to have found its way into deposits of otherwise casually accumulated material. At La Liquière there is some suggestion of smoking, heat treating, even perhaps consumption of the brain. As yet, we see no clear signs of the mechanisms of display or, indeed, any monumentality of construction associated with the skulls fragments.

The examples I have considered so far come from west of the Rhône. But this apparent concentration of cranial fragments on Languedocien as opposed to Provençal settlements may be as much a product of the research priorities of Michel Py, Bernard Dedet, and their collaborators as it is of a genuine distinction in Iron Age practice. Where analogous sites have been excavated east of the Rhône, similar material has been found. Human remains from the Late Bronze Age lakeside settlement of L'Abion, just west from Martigues, for example, included vertebrae, pelvic, and lower limb bones as well as cranial fragments, from both young and old individuals (Chausserie-Laprée 2005, 41). But, as we shall see, the bulk of the relevant material from Provence consists not of heads themselves, but of representations of heads.

Pattern and purpose

Within their revised chronology, Arcelin and Rapin identify an early group of stone carvings that they date to between around 800 and 550 BC (Arcelin 2004a; Arcelin and Rapin 2003, 186; Table 4.1), the same broad period as our Languedocien skull deposits. The main examples come from three sites in Provence – Saint-Michel-de-Valbonne, Entremont, and Badasset – although I also discuss some other examples that appear to be similar in form. They depict carefully arranged groupings of carved human heads, sometimes in association with other motifs. It is worth looking at the key examples in some detail.

Saint-Michel-de-Valbonne

Perhaps the earliest iconographic expression of headhunting in Provence was also among the first to be found. Between 1864 and 1867, the Duc de Luynes excavated around the twelfth-century chapel of Saint-Michel-de-Valbonne, which crowns the summit of one of the westernmost peaks of the Maures, north of the modern coastal town of Hyères (Bats 1990; Brun 1999, 462–3). Here, among many later architectural fragments, he unearthed a crudely shaped sandstone pillar, carved with a series of enigmatic motifs, including what appear to be human heads. Modern excavations, by Jean-Pierre Brun in the early 1990s, have helped put these early finds in context, as well as bringing to light a second pillar, also displaying carved heads (Brun 1999). Both pillars are very roughly shaped, and both appear to have been freestanding, because it is hard to see how they could have been incorporated into any larger structure. They are clearly distinct, in terms of their technology as well as their iconographic content, from the many later stone fragments on the hilltop.

Aside from the pillars themselves, however, there is no obvious sign of human activity on the site until much later in the Iron Age. During the second century BC, the hill was enclosed by a rampart, equipped with square towers, creating a small oppidum with the earlier sanctuary at its core (Brun 1999). By the succeeding Gallo-Roman period, the sanctuary seems to have been dedicated to an indigenous deity known as Rudianus. Dedications to this god are also found at a number of other Gallo-Roman sites in the region, and his name is linked in certain inscriptions with the Roman god Mars. This has led to the suggestion that Rudianus may have been an indigenous war god, later twinned with his Roman counterpart. Despite its martial origins, the sanctuary appears to have survived as a location for worship throughout the early Christian period, culminating in the construction of the medieval chapel.

The iconography of the pillars suggests that the sanctuary may have had an association with violence and war long before the Gallo-Roman period. At 2.25 m tall, the larger pillar would have been an imposing sight, though its irregular shape suggests that it must have been partially buried in the ground in order to stand upright. It bears the outlines of five human heads and the roughly carved figure of a horse. There is also just the hint of a horseman, in the form of a single vertical line where the rider's back might be expected (Ill. 4.4; see also Benoît 1955a, plate XXVI, right). Four of the heads are arranged in two vertical pairs below the horse, while the fifth is set, roughly centrally, above it. Each is

TABLE 4.1. *Chronology and location of some of the principal objects discussed in Chapter 4*

Location	Object
Seventh century BC	
Provence	Saint-Michel-de-Valbonne, pillars
Sixth century BC	
Languedoc	La Liquière, human skulls
Provence	Entremont, pillar
Fifth century BC	
Provence	Entremont, *bloc aux épis*
Languedoc	La Ramasse, lintel
Bohemia	Hořovičky, phalera
Western Germany	Pfalzfeld pillar
German Rhineland	Rodenbach, ring
Fourth century BC	
German Rhineland	Weisskirchen, mounts
Southern Spain	Cordoba, pillar
Northern Germany	Niederschönhausen, fibula
German Rhineland	Bad Dürkheim, mounts
Third century BC	
Eastern France	Courtisols, torc
Eastern France	La Charme, bracelet
Second century BC	
Catalonia	Sant Martí Sarroca, pillar
Northern Ireland	Navan, ape skull
First century BC	
Northern Italy	Manerbio sul Mella, phalerae
North Wales	Tal-y-Lyn, mounts
First century AD	
Central France	Sanctuary of Sequana, votives
Northern England	Melsonby, mounts
Iberia	Arcobriga, pottery vessel
Second century AD	
Northern England	Corbridge, pillar
Third century AD	
Eastern Ireland	Raffin, pillar stone

Note: Most of the dates are typological or art historical and are thus approximations at best. For some sites (e.g., Badasset), even this level of approximation has not been attempted. Where appropriate, the detail of chronology is discussed in the text. Selective and approximate as it is, the table nonetheless shows how similar ideas and associations reemerge wherever Iron Age iconography survives, for example, in Provence and Languedoc throughout the early part of the Iron Age, in the heartlands of Early La Tène art during the fifth to third century BC, and in the British Isles during the late pre-Roman Iron Age and early Romano-British period.

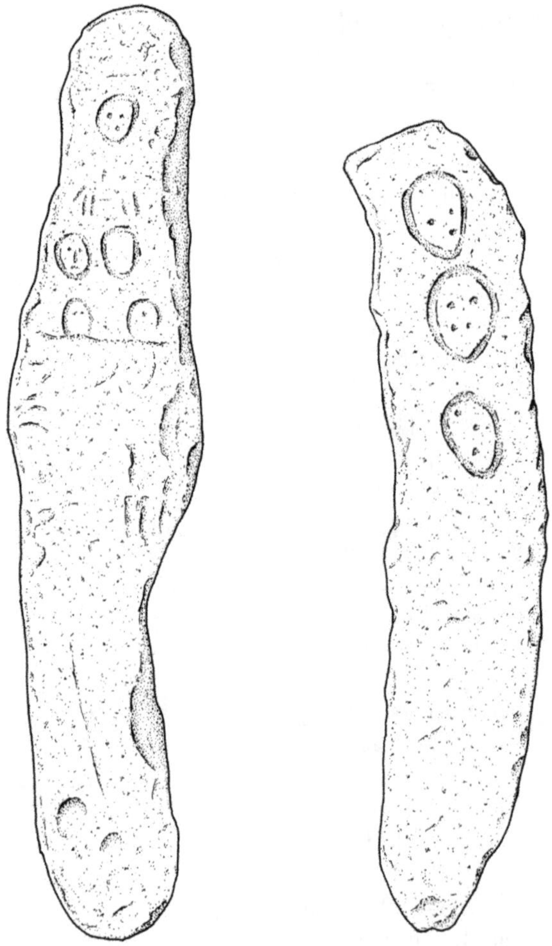

ILLUSTRATION 4.4. Views of the two complete pillars from Saint-Michel-de-Valbonne, Hyères, Var, showing the heads and horse motif on the larger pillar. (Drawn by Dan Bashford after Brun 1999, figs. 481 and 483)

outlined by a deep, broad line, formed by pecking. Although the pillar is very badly weathered, where internal features survive, the eyes are represented by simple dots, the noses by rough vertical lines, and ears are absent. One of the heads has a short slit mouth. All appear to be upright. It is quite possible that further heads were originally present lower down the stone, but the relevant areas are too badly damaged to judge.

The smaller pillar depicts a further three battered and roughly executed heads of the same basic form (Brun 1999, 462; Ill. 4.4). These are set one above the other, with the upper two nearly touching. In each case only the deeply pecked outline survives, and any internal facial detail seems to have weathered away, except for a very slight trace of the eyes of the middle head. As on the larger pillar, each head appears to be upright. A further detached block, probably from a third pillar, was also recovered during the Duc de Luynes's excavations. This bears the deep outline of a single head, with no surviving facial features (Brun 1999, fig. 482).

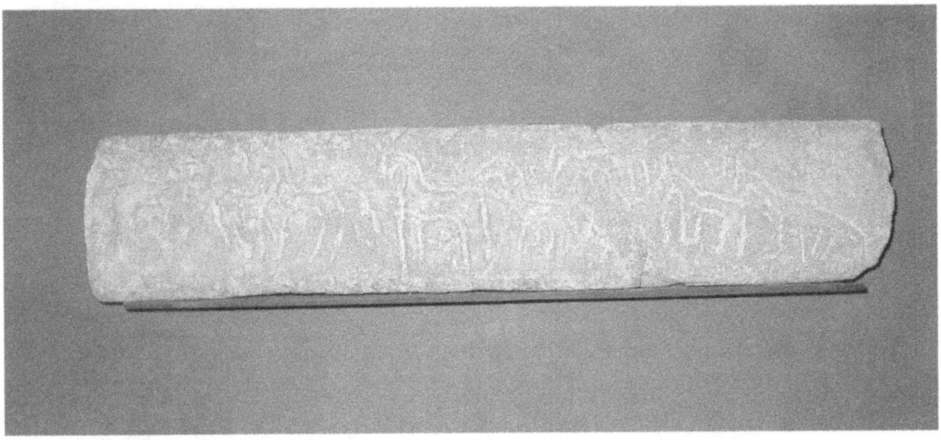

ILLUSTRATION 4.5. Early Iron Age stele from Mouriès, showing the characteristic depiction of horses. (Photograph by author)

Brun does not venture a date for the pillars, simply noting that the sanctuary may have preceded the second-century BC oppidum (Brun 1999, 462). There is little question, however, that the rendering of the horse on the larger pillar, although hard to discern in its present battered condition, links the sculpture to the profuse assemblage of Late Bronze Age or Early Iron Age horse-and-rider imagery from the oppidum at Mouriès (O. Coignard et al. 1998; Ill. 4.5) and to similar examples from Glanon (e.g., Paillet and Tréziny 2000, 190). In particular, we can single out two specific stelae, one each from Mouriès and Glanon, which may relate even more closely to the Saint-Michel-de-Valbonne pillars than has previously been realized.

The first is a broken stele, recovered from the foundations of the second-century BC monumental gateway (the *porte charretière*) at Glanon (Paillet and Tréziny 2000, 190; Ill. 4.6). Two faces of the stele bear roughly pecked images of mounted warriors (one holding a spear or javelin, the other unclear). All that remains of the single head (if head it was) is the bottom half of a pecked oval outline, floating above the horseman on the better-preserved face. In technology and style, it closely resembles the head depictions on the pillars from Saint-Michel-de-Valbonne, though it has not previously been interpreted in this way at Glanon. The second image can be found on a slightly better-preserved stele found in the rubble of the second-century BC rampart at Mouriès (Marcadal 2000, 193). Above the clear figure of a javelin-wielding horseman, pecked into the stone, is a faint and much-abraded circle (Ill. 4.6). No detail remains within it, but once again, it is tempting to link the putative composition of horseman and head, with that on the large pillar at Saint-Michel-de-Valbonne. In no case do we have a clear rendition of the overall composition, but the evidence is tantalising, suggesting a recurrent image of a large, disembodied head, looming over a mounted warrior.

Dominique Garcia (2004, 109–10) dates the Mouriès and Glanon stelae broadly to between the seventh and fifth centuries BC, while Patrice Arcelin (2004a, 71) opts for an even higher dating for both Saint-Michel-de-Valbonne and Mouriès, suggesting an origin for the carvings in the period from c. 800 to 650 BC. Their milieu seems to relate closely to the indigenous Bronze Age tradition of erecting freestanding stelae, which predates any

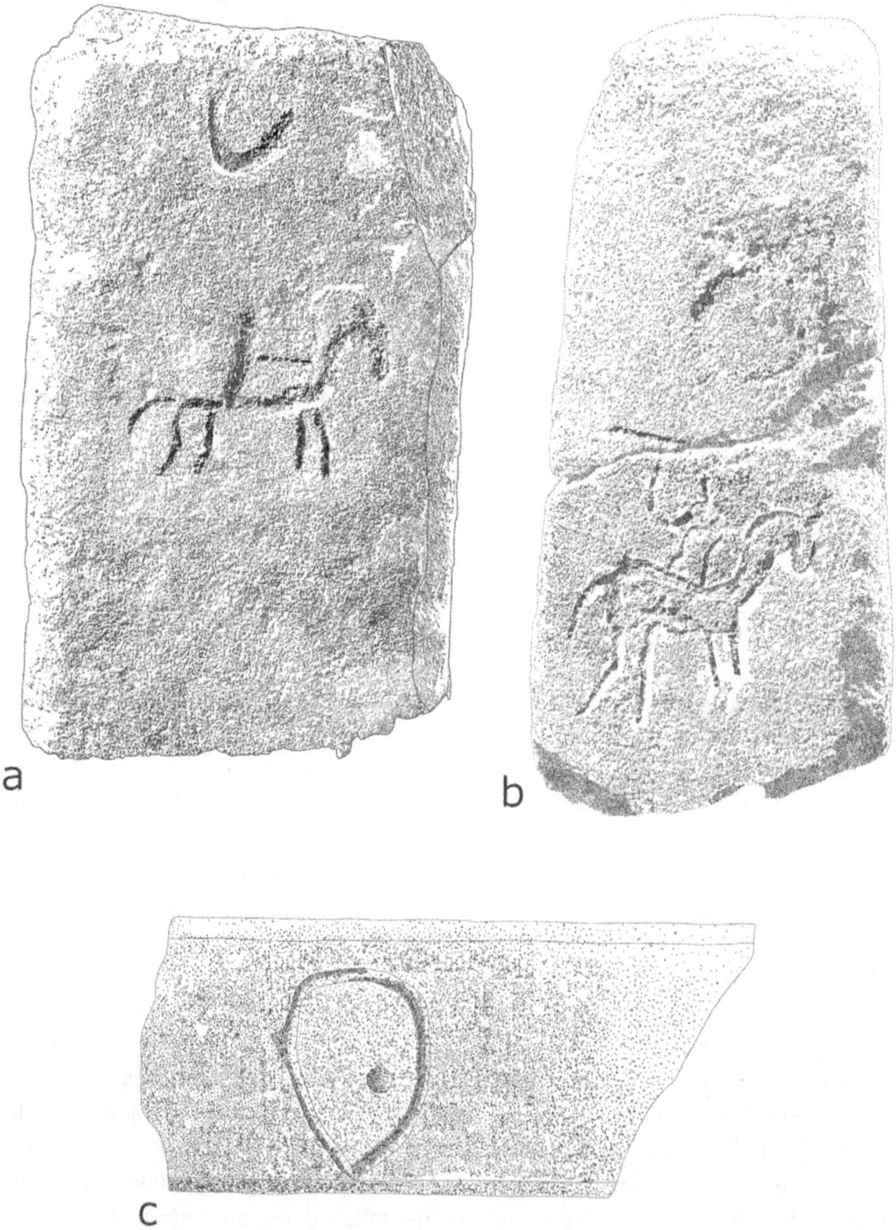

ILLUSTRATION 4.6. Stelae from (a) Glanon and (b) Mouriès, showing images of horsemen below what might be carved representations of heads; and (c) the lintel fragment from La Ramasse. (Drawn by Rachael Kershaw after (a) Paillet and Tréziny 2000, 190, (b) Marcadal 2000, 193, and (c) Garcia 1993, 304; various scales)

technological influence from the Greek world. At any rate, there is little doubt that they date to the end of the Bronze Age or the earlier part of the Iron Age.

To these can be added a geographical outlier from the oppidum of La Ramasse, Hérault. Excavations here yielded a group of earlier stelae that had been reused in the rampart built around 375 BC (Garcia 1993, 308–9). Among these was an unusual piece of worked sandstone, with neatly chamfered edges (303, *Ramasse* 15), which seems to have formed part of a lintel (Ill. 4.6). One face bears a very simple engraving in the form of a head-shaped oval, some 30 cm high with a slightly pointed base, rather similar to the simple head carvings from Saint-Michel-de-Valbonne. Off-centre within the carved oval was a small hole some 3–5 cm in diameter. As there was no trace of occupation on the site before the late 6th century BC (Garcia 1992, 164), it seems likely that the monument of which this lintel formed part was in use sometime between the late sixth and the early fourth centuries. I discuss this particular piece again in Chapter 5.

Entremont

The oppidum of Entremont lies on the northern fringes of the modern city of Aix-en-Provence, some 30 km north of central Marseille, where it overlooks the junction of two ancient communications routes: the east-west inland route from Iberia to northern Italy and a major north-south route leading from the Mediterranean to the Alps (Benoît 1975, 227). The site was first fortified during the early second century BC, when it appears to have become a capital of the Saluvii, a powerful confederacy of indigenous tribes. But the substantial assemblage of statuary from the site contains much older elements (Arcelin 2006). The archaeological context of this early material is poorly understood, because it was found entirely in secondary contexts within the second-century levels of the oppidum, but the stylistic evidence of the sculptures themselves provides a broad chronological sequence. I will look in more detail at the archaeological sequence from Entremont in Chapter 6, but for now I want to focus on three of the earliest sculptured stones.

The first is a simply shaped, rectangular pillar, found in 1953, which had been reused as a threshold stone in the doorway of a monumental building. The building dates to the period of major expansion of the oppidum in the mid-second century BC, but the pillar, which bears a series of carved human heads, is much older. When originally upright, the Entremont head pillar would have stood well above head height, at just over 2.5 m (Salviat 1993: 211, no. 29; Ills. 4.7 and 4.8), just a little taller than the larger pillar from Saint-Michel-de-Valbonne. It may have formed part of a larger structure, most likely a portico of some kind, and it was probably painted in bright colours. The pillar does not seem to have been bedded in the ground and so probably stood on a pad, plinth, or hard floor. There is no evidence of any fixing on its upper end, but it was most probably held in an upright position by the weight of a timber or stone architrave, or by being bedded into a timber lintel. Only one of the four faces is decorated. This depicts an elaborate composition of 12 highly stylized human heads, or *têtes coupées* as they are usually known in the French archaeological literature.

The heads are arranged vertically, in a series of rows ascending the pillar. The bottom two rows have three heads apiece, the next has two, while the top four each comprise a single head. The head in the middle of the bottom row is, uniquely, upside down. All 12 are more or less life-size, the uppermost being the largest. All are virtually identical in execution, rendered using only two deeply incised lines. The first outlines the elongated head and squared chin, while the second continuous line forms the slit eyes and broad triangular nose.

ILLUSTRATION 4.7. Head pillar from Entremont, found in a secondary position, forming the threshold of a monumental building (the 'hypostyle') and dating to the mid-second century BC. The pillar itself is much older and would originally have stood upright, some 2.58 m high, probably as an element of a larger structure. This photograph shows the replica that has replaced the original on site at Entremont. Clearly, the builders of the hypostyle made a deliberate choice, not just to use this already ancient stone as a threshold rather than simply as one of the many similarly shaped blocks that front of the building, but also to ensure that the single decorated face remained visible. (Photograph by author)

The absence of mouths has been taken to signify silence and death (Benoît 1964; Enright 2006, 95), an impression reinforced by the eyes, which are clearly closed, and the missing ears. This suggestion finds support in other French Iron Age contexts. A rare fragment of figurative decoration incised on pottery fragments from Clermont Ferrand, for example, shows a mounted warrior, with a severed head hanging from the neck of his horse (Périchon 1987; Ill. 2.2). The living horseman is depicted with a mouth, whereas the severed head is not, although the artist has otherwise depicted the two faces in the same style and in the same detail. Casting further afield, we might consider the early medieval Welsh tale of Branwen, daughter of Llŷr, where we find a 'magic cauldron' capable of reviving dead warriors (Gantz 1976, 71). But although these risen warriors regained their former physical strength, they were unable to speak. Several millennia earlier, mouths are also absent on the Late Neolithic anthropomorphic stelae of the same region (d'Anna 1977, 233), and this again has been interpreted as signifying death.

If we take a closer look at the carved outline of the heads on the Entremont pillar, a further detail emerges. Although the heads appear initially to have been defined by

ILLUSTRATION 4.8. Detail showing the decorated portion of the head pillar from Entremont. (Drawn by Libby Mulqueeny)

deep incision, the sculptural technique was, in fact, a little more complex. The shape and breadth of the incision is such that each head seems to emerge from within a slight recess or niche, in a technique 'half-way between engraving and incised bas-relief' (Salviat 1993, 211). In other words the depictions on the pillar clearly make reference to the practice, which

ILLUSTRATION 4.9. The *bloc aux épis* had been reused as part of the outer face of the east tower in the first rampart at Entremont, which was probably built during the early second century BC. This photograph shows the replica that has replaced the original on site at Entremont. In this case, the later builders made the decision not to display the face with the engraved heads but elected instead to display the side depicting two ears of wheat. (Photograph by author)

we shall discuss a little later, of displaying human heads, skulls, or faces, in niches cut within stone pillars and lintels.

The second piece from Entremont is a carved stone known as the *bloc aux épis* (the block with the ears of corn). It was found built into the early second-century BC rampart that represents the first fortification of the site. Like the head pillar, however, it is much older than the archaeological context in which it was found.

The name *bloc aux épis* refers to the depiction on this stone of two lightly incised ears of wheat (Ill. 4.9). This image had been known since Benoît's excavations of 1956 when the outer face of the stone, built into the east tower of the earlier rampart, was revealed. Benoît (1975, 245) noted that the single visible face of the block displayed 'a double ear of corn' and surmised that it had once formed part of a pillar. Following the completion of excavations, the block was left *in situ*, and that, for a time, was that. It was not until forty years later, in 1996, when routine inspection revealed that rainwater erosion had exposed an engraved *tête coupée* on the lower face of the stone. Rather than leave it vulnerable to further erosion, it was decided to take the stone under cover and replace it with a cast. The removal of the block, however, revealed an extraordinary set of images. Two adjoining

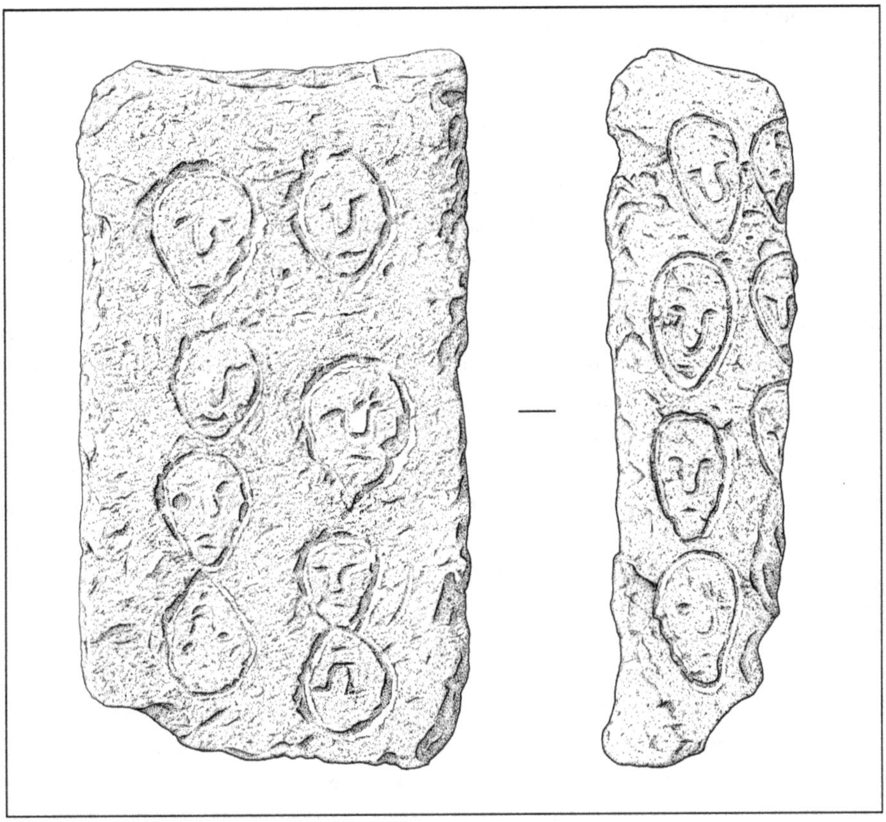

ILLUSTRATION 4.10. Two of the decorated faces of the *bloc aux épis* from Entremont. (Drawn by Dan Bashford after Arcelin and Rapin 2003, figs. 4b and 4c)

faces, previously obscured by the stone's position in the wall, were each found to display a composition of eight *têtes coupées* (Congès 2000, 104; Ill. 4.10).

The *bloc aux épis* had been subject to rather rougher treatment than the head pillar. One face had been completely reworked, which removed any decoration that might formerly have been present and truncated some of the heads on the side panel. In addition to this, the whole block had been truncated at the base, presumably to facilitate its reuse in the later tower. As it survives, the block is more or less square in section, around 0.65 m high, narrowing slightly towards the slightly concave top, which may have been shaped in this way to enable its integration into some form of timber structure. Conceivably, it may have formed the upper part of a much larger pillar, although it need not necessarily ever have been much taller than it survives today. Other than the missing face, it does not appear that the truncation of the stone has necessarily deprived us of much of the decorative scheme. The front panel seems complete and coherent in itself, although, as Congès (2000, 104) notes, the decoration may originally have been restricted to the upper part of a much taller block.

The engraving of the *bloc aux épis* is less skilfully executed than that of the head pillar, although the stone had been smoothed and shaped to a degree before the carvings were added. On the front face are two adjacent 'columns', each formed of four human heads. Like that at the base of the head pillar, the bottom head in each column is upside down, its chin joined to that of the head above.

Although stylistically very similar to the depictions on the head pillar, the heads on the *bloc aux épis* display some significant differences. The lines forming the eyes and nose on the *bloc aux épis* are less regular and more rounded than those of the head pillar, in some cases forming an inverted Ω-shape. They also lack the squared outline of the head pillar examples, tending instead to terminate in a point under the chin. Another important difference is that, in most cases, mouths are shown by an incised line. Where mouths are absent, it is not clear if this is intentional or simply a product of weathering and rough handling. This variation in the detailing creates, intentionally or otherwise, a greater range of expression in these heads than is the case for the head pillar (one even appears to be smiling). Nonetheless, the intended overall effect is clearly one of repetition of a standardised motif. As with the head pillar, the heads on the *bloc aux épis* are all composed in essentially the same way, with two basic lines, one forming the outline of the head and the other forming the closed eyes and the nose. And again, although it is less clear than on the head pillar, the lines defining each head are carved in such a way as to suggest that they rest within niches cut into the stone.

The layout of the decorative scheme on the *bloc aux épis* contains some evident peculiarities. I have already mentioned that the heads are relatively crudely carved, but this impression is accentuated by the seemingly haphazard positioning of some examples, especially those on the front face. On the left column (for the viewer), the heads are in more or less vertical alignment, although the gaps between them are somewhat irregular. On the right column, however, there are marked discrepancies in alignment and spacing. Whereas the bottom two heads are joined at the chin, the next one up is clearly separate, although joined to the head below it by a carved diagonal line. The uppermost head is separated still further from the one below it, and this time there is no line to link it. The overall effect, at least for the modern viewer, is to give an impression of movement, as if the top two heads were 'floating' away, balloon-like, from those at the base. What impression it might have given to the Iron Age onlooker, of course, is entirely unknown.

The composition of the side panel, showing eight further carved heads, is harder to judge since it is heavily truncated down one side. Again there are two adjacent columns, each of four heads, although this time all the heads are upright. As far as can be judged, the spacings here are more regular, although the heads in the left and right columns do not quite align. The most noticeable peculiarity, however, is that the uppermost head on the left-hand column leans markedly off-centre, again suggestive of movement. It is unclear whether its heavily truncated partner in the right-hand column did the same.

A certain basic level of skill would have been required to carve the *bloc aux épis*. In particular, the upper, concave surface would have needed some subtlety in execution. It would seem a fairly simple step, once the desired shape of the stone itself is achieved, to mark out the intended pattern on the stone with some degree of precision before carving the heads. Yet, either the precise positioning of the heads was considered unimportant, or perhaps what seem to us as irregularities were in fact significant in themselves. In the first case, the heads may have been added to the stone sometime after its initial emplacement, perhaps by an individual (or individuals) whose primary qualification was not his or her ability as a stone-carver. One might envisage a situation, for example, where individual heads were added to commemorate some special occasion or significant event. In such a case, they may have been carved by ritual specialists, dedicants, chiefs, or mourners. Yet there is, quite clearly, an overall scheme and a symmetry that belies the idea of piecemeal accumulation. And there is some level of skill in the carving of the heads, perhaps most obviously in the shaping of the carved outlines, to mimic the appearance of niches. So it seems inescapable that what may appear to us as maladroit quirks of spacing and alignment, were in fact intentional, and thus part of the way these motifs were intended to be viewed. If this is the case, then it marks a distinction between the *bloc aux épis* and the much more regularly structured head pillar. The overall original layout of the pillars at Saint-Michel-de-Valbonne is harder to assess, given the problems of preservation, but there are hints at least in the smaller pillar that it may have been more like the *bloc aux épis*.

Before leaving Entremont, I want to look at one additional sculpture that relates closely to the two I have already discussed. Yet it also introduces another theme that I will develop further in the next chapter; the display of actual human remains, set in stone. The so-called lintel from Entremont is actually a neatly squared, though badly damaged, limestone block, with carvings surviving on three sides. It is usually thought to derive from the same monument as the head pillar (e.g., Salviat 1993, 211; Arcelin and Rapin 2003, 213), though there is no evidence for this beyond the stylistic similarity of the head carvings and the discovery of both pieces reused in the same later building. The best known (front) face displays a carved human head, virtually identical in design and execution to those on the head pillar. The sole differences are its somewhat less squared jaw and the vague hints of a mouth (although the relevant portion of the stone is so badly damaged that it is impossible to be sure). The style of depiction, like the heads on the pillar, suggests that it is set in a niche. To either side of this head are the two roughly worked oval niches that set this piece apart from the other carvings discussed so far. On the basis of evidence from other sites, notably Roquepertuse (see Chapter 5), we know that such niches could be used to display human heads. This suggests that the heads carved on the head pillar and elsewhere were not simply 'aspirational' but could be intimately associated with the display of real heads.

Badasset

A recent chance find has added one further example to this small corpus of early carved heads. During 1999, deep ploughing on a recently deforested plateau in the hills above the Vallon de Badasset, around 20 km northwest of Entremont, revealed a roughly shaped block around 0.3 m square (Bringer and Dumont-Castells 2000, 105). The findspot lies in an area of Gallo-Roman settlement, but it is also close to an indigenous promontory fort where the stone may have originated.

Whatever its original location, the fragment had clearly once formed part of a larger architectural block. Unlike the examples discussed so far, the motifs on the Badasset fragment are carved in low relief, rather than by incision. On one face are two adjacent pairs of opposed heads, joined at the chin (Ill. 4.11). The similarity with the arrangement of the lowest heads on the *bloc aux épis* is striking. The detailing of the facial features also has

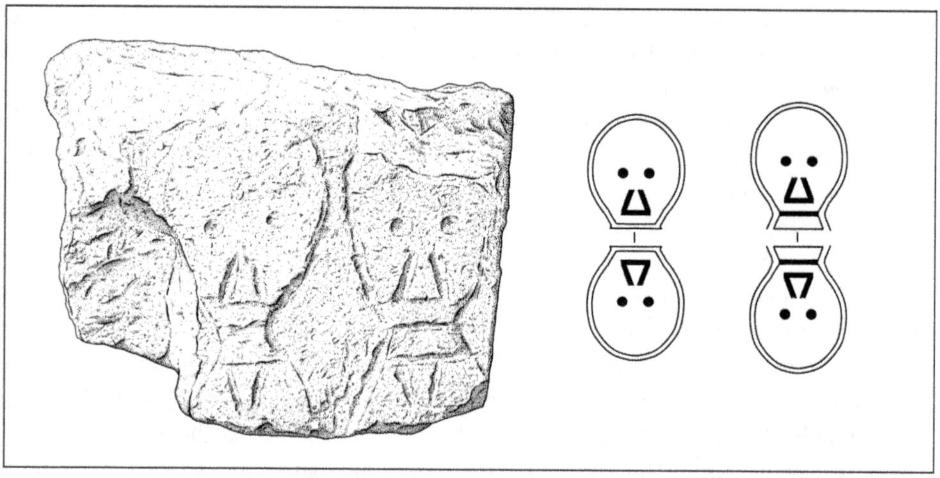

ILLUSTRATION 4.11. Carved block from Badasset showing two sets of opposed heads, conjoined at the chin (*left*). The heads can be interpreted in two ways. A literal reading (*middle*), pulling each pair of heads apart, would suggest that all are mouthless. However, as an alternative (*right*) we can interpret the lower two opposed heads as merging, sharing a single mouth and the lower part of the face. Comparison with Ill. 4.19 shows that artists in Iron Age Europe did not lack the capacity or sophistication for such visual trickery. (Drawn by Dan Bashford based on photograph in Bringer and Dumont Castells 2000, 105)

elements in common with the pieces already discussed. There are small dots for the eyes, as at Saint-Michel-de-Valbonne, and large triangular noses, as at Entremont. The square, narrow jaws are similar to those on the Entremont head pillar. Mouths appear to be absent, although the images can be read in at least two ways (see Ill. 4.11). A suspicious hollow, rather like a head niche (see Chapter 5), clips the edge of one of the carved heads (Garcia 2004, 104). If intentional, it is clearly later than the heads themselves and may relate to the modification of the stone for some secondary use (Bringer and Dumont-Castells 2000, 105). Like the lintel from Entremont, it suggests a link between the representation of the head and the reality of head curation and display.

Chronology and context

It is impossible to date any of these stones precisely, and there is sufficient variation in their style and execution to suggest that they may span several centuries. Nonetheless, the simplicity of the carvings, their architectural context, and the absence of direct Greek influence combine to suggest that they are early in the southern French Iron Age sequence. On a crude typological basis, one might suggest that the roughly shaped, pecked pillars from Saint-Michel-de-Valbonne are the earliest. Indeed, the parallels between the rendering of the horse on the larger pillar from Saint-Michel-de-Valbonne and those from Mouriès and Glanon suggest that this stone at least might date to the end of the Bronze Age or the earliest part of the Iron Age. The head pillar and *bloc aux épis* from Entremont and the stone from Badasset, however, display more sophisticated execution. Arcelin and Rapin (2003, 188) had dated all of these examples to the period before 550 BC, but more recently Patrice

Arcelin has suggested that the *bloc aux épis*, with its irregular incision, may predate the more rigidly structured head pillar. The former, he suggests, probably dates to the period between 700 and 450 BC, while the latter is more likely to have been carved between c. 500 and 400 BC, although he holds open the possibility that it may be even later (Arcelin 2006, 134). It is worth noting that even this more cautious dating would place the carving of the Entremont stones more than 200 years before the creation of the oppidum on that site (see Chapter 6). The dating for the lintel from La Ramasse also centres on the fifth century (Garcia 1992, 1993). Despite the chronological imprecision, it is nonetheless justifiable to associate this suite of carvings with the broad chronological context of the early Iron Age in the region. As such it is useful to examine the wider context of the period.

The period of the sixth and fifth centuries BC is characterised by Dominique Garcia (2004, 53) as one of increasing sedentism and settlement organisation. For the first time, on any significant scale, southern French settlements were enclosed with substantial defences, houses were grouped into organized 'islets' separated by formal roadways, and evidence for large groupings of population became more common in the landscapes of Provence and Languedoc. If we look specifically at Provence, where our early head carvings are mostly found, the key area during the period seems to be around Martigues, on the western edge of the étang de Berre, the largest of many coastal lagoons around the marshy mouth of the Rhône. In this region, a few miles west of Marseille, a number of large enclosed settlements, or oppida, appear to have been founded from the latter part of the seventh century onwards (Chausserie-Laprée 2005, 42). The earliest of these was at Saint-Blaise, situated on high ground some 4 km from the coast, but dominating a series of saltwater étangs (Bouloumié 1992). These small lagoons are known to have been intensively exploited for salt production from the late Iron Age to recent times, and they may have constituted one of the prime resources for Saint-Blaise even at this early period (e.g., Marty 2000). Certainly, the inhabitants seem to have traded with Etruscan merchants during the third quarter of the seventh century BC, as the discoveries of Etruscan bucchero pottery on the site suggest. Salt may have been one resource that enabled the inhabitants of Saint-Blaise to engage in long-distance trade networks; others may have included metal ores imported from the north and west, more local agricultural produce, and, almost certainly, slaves.

Only a little later, perhaps around 625 BC, the small oppidum of L'Arquet appeared, set on a promontory on the Mediterranean coast, around 16 km south of Saint-Blaise. Next came the strongly fortified oppidum of Tamaris, set on a neighbouring promontory just over a kilometre away to the east (Chausserie-Laprée 2005). These were exposed, wind-blown sites whose sole positive features appear to have been defensibility and command of good landing places. In other words, they suggest a proactive attempt by local communities to establish trading positions where they could engage directly with coastal merchants, whether from Etruria or elsewhere. The foundation of Tamaris, however, seems to have coincided more or less with the foundation of the Greek colony of Massalia just 25 km along the coast. It seems highly probable that the indigenous and intrusive centres were in competition from the start, and the sudden abandonment of both Tamaris and L'Arquet sometime between 575 and 550 BC would suggest that the Greek colony was an early victor. Although the abandonment of Tamaris and L'Arquet seems to have signalled the end of any indigenous attempt to develop specialized trading centres along the Mediterranean coast of Provence, it by no means brought an end to the incipient urbanisation of the region. As these early centres disappeared, they were quickly replaced by inland foundations, like that at Saint-Pierre, only around 3 km to the north, which was to survive seemingly unchecked into the Gallo-Roman period. Inland Saint-Blaise too, despite the uncertainties of its excavated sequence, seems to have continued into the late second century BC. Indeed, the number and density of oppida was to rise markedly from this period on.

Interestingly, despite fairly intensive excavations, these early foundations appear, so far, to lack any overtly religious buildings or indications of cult activity on any significant scale. Indeed, they have not produced any evidence immediately analogous to the carved heads discussed here. Both Tamaris and L'Arquet seem to have been places primarily of secular power, established with a clear intent to engage with nonlocal traders. The same probably applies to Saint-Blaise in its earliest period. Instead, the sites where these head carvings do occur seem to be divorced from contemporary centres of population. Although it was later to become a major oppidum of the Saluvii (see Chapter 6), Entremont during this period appears to have been devoid of any settlement. Its prominent, commanding location and spectacular views would always have marked it out as a special place within the landscape, and it appears most likely that it was the site of a small upland sanctuary during the period. An almost exactly analogous situation is found with Saint-Michel-de-Valbonne, where the oppidum long postdates the carved stones. The Badasset stone too, seems to have been unrelated to any contemporary settlement. In other words, we seem to be seeing, during this early part of the Iron Age, a separation of secular and religious power. Important secular centres can be identified, but those are quite distinct from the isolated sanctuaries suggested by the surviving iconographic evidence. There is nothing yet, however, to suggest that these two potential sources of power were deployed by the same groups or individuals, or with the same aim in mind. This situation was to change markedly in later periods.

Heads and fertility

We have lost a good deal of information through the batterings and truncations suffered by these images in antiquity. We should also remember that more detail may originally have been added to these stones in paint that has long since vanished (e.g., Barbet 1987, 1991). Nonetheless, the surviving iconographic content does allow us to make certain assertions about these sculptures. Despite differences in detail, these various stone carvings share a great deal in common. In each case, the dominant motif is patently the disembodied human head. The repetition, patterning, and organization of these highly stylized heads appear also to be of central significance. The recurrent upturning of the lowermost head suggests that these careful arrangements reflect a shared symbolism.

So what else can we say about these images? Clearly, they are intended to depict fully fleshed heads: the treatment of the eyes and nose, as well as the overall shape, in no way suggests that they are intended as skulls. Further, they are entirely disembodied, literally severed heads, with no indication even of a neck, far less a body. The depiction, in the Entremont examples, of fleshed heads resting in niches, further suggests that these images are, to a degree, skeuomorphs. That is to say, they are stone representations of real human heads, curated and displayed, with which the viewer was expected to be familiar. The recurrent presence of cranial fragments in domestic contexts may derive from exactly these sorts of displays. The lintel from Entremont shows that, in certain cases, carved images could be found interspersed with real human heads. In any case, the rendering of facial features, the disembodiment of the heads, and their arrangement relative to one another leaves little room for doubt that the heads belong to the dead and not the living.

What then is signified by these images of curated human heads? Clearly there has been no attempt to signify individual identity or ethnic affiliation or to differentiate social status. The specific identity of individual heads does not seem to be significant (in stark contrast to what was to come later). Instead, what is important is the way in which the groups have been composed. Despite the possible association with images of horsemen in what may be the earliest examples, from Saint-Michel-de-Valbonne and (perhaps) from Mouriès and Glanon, the emphasis throughout is on the heads and not their takers. If, as is often thought, these are images of trophy heads, taken in battle, then the martial aspects of their capture are patently rather downplayed. At Entremont, where the compositional schemes are most complete, there is no trace of the triumphant headhunter, whose victories produced these spoils. These images do not, therefore, seem to be celebrations of individual prowess.

Perhaps then they have nothing to do with headhunting. Various authors, including Benoît (e.g., 1975) and, more recently, Arcelin and Rapin (2003, 190–1), have suggested that certain of the sculptures, notably the Entremont head pillar, represent sacred monuments associated with an ancestor cult. Struck by the seemingly respectful depiction of these peaceful heads, their association with monumental architecture, and their sanctified settings, they suggest that these are the heads of venerated ancestors. One could go further and suggest the vertical arrangement of heads represents an explicit genealogy. Is the upturned basal head perhaps the founder of some lineage or dynasty? The ambiguous evidence from the cranial deposits on settlement sites is of little help. It seldom suggests much in the way of reverential handling, yet we know that the tombs of the recently dead in eastern Languedoc were sometimes revisited to remove the cranium (Dedet 1992). The perennial problem of how we distinguish enemy from ancestor is thus a factor in the interpretation of these images, and a question that I return to in Chapter 5.

Skull trees and trophies

Ethnographically, it is rare to find cases where the heads of ancestors were preserved in fully fleshed form, although there are exceptions, as among the New Zealand Maori of the eighteenth and nineteenth centuries (e.g., Walsh 1894, 611). More usually, the skull, or simply the cranium, is recovered at some stage during the funerary rites, perhaps after a period of initial burial. It is these 'clean' bones that, having completed the necessary funerary processing, may be retained as ancestral relics. It would also be extremely unusual to find ancestral heads on public display. The distinction between the 'public' role of trophy heads and the 'private' role of ancestral relics can be discerned to varying degrees in many ethnographic instances, for example, among the Maori and the Nagas. There are, of course, exceptions, the most notable being in Central America, where the curated heads of deceased Cocom Maya leaders were displayed in overtly religious settings. These were, however, first defleshed and then treated with decorated stucco to re-create the living features (Serafin and Peraza Lope 2007, 234); the shrivelled flesh of the ancestors was not part of the package. Indeed, I know of no ethnographic cases where groups of fleshed heads belonging to the members of the in-group were put on public view. There are, however, several ethnographic instances where the severed heads of outsiders were treated in just such a way – for example, among the North Indian Nagas and the South American Shuar (Harner 1973). The case cannot be proved, but it seems a strong possibility that the heads depicted on monuments such as the Entremont head pillar represent trophies obtained from outsiders.

There are also numerous cases in the ethnographic record where severed heads were placed on public display as part of carefully choreographed compositions, and a brief examination of some of these might help to expand our range of interpretative possibilities. The three most striking examples, from Sumba and Nagaland in the nineteenth and early twentieth centuries and from sixteenth-century Mexico, all involve the display of trophies taken from enemies.

The skull trees of Sumba in Indonesia are familiar from descriptions and sketches by early explorers and colonial administrators (Hoskins 1996c). These monuments, surrounded by circles of stone and hung with enemy heads, were associated with cult houses and were used to display the crania obtained from boiling freshly taken enemy heads (Hoskins 1987, 607–10). These crania, displayed as 'fruit' hanging from sacred trees, underscore the conceptual links between heads and fertility in Sumba and, indeed, throughout Southeast Asia. As well as their physical dominance within the traditional Sumbanese village layout, skull trees also form a common motif on textiles, where their association with other images allows us to draw out a little more of their meaning. In the early twentieth-century textile reproduced here, for example, the repeated skull tree motif is associated with male figures displaying exaggerated genitalia, further emphasizing the link between heads and fertility (Ill. 3.5).

The display of trophy heads was also common in North Indian Naga villages of broadly the same period. Among the Naga, preserved heads or skulls were displayed in a range of different ways: enemy heads might be strung up on 'head trees' not dissimilar to those of Sumba, or else laid out on bamboo 'skull racks', while the skulls of ancestors were reverentially placed in niches within phallic stone pillars (Jacobs 1990). Head trophies were often embellished with the horns of a mithan (a domesticated Southeast Asian ox), hair, and textiles. One particular photograph, of trophy skulls from Yimpang village, taken in 1936 by the anthropologist Christoph von Fürer-Haimendorf, shows a row of five such 'trophies', each suspended from a cord (Ill. 3.6). In each case, the single human head or cranium at the base is surmounted by a vertical row of 'imitation heads' carved from wood or woven from basketwork. As with the carved skeuomorphs of real heads at Entremont and the woven images displayed in the Sumbanese textiles, these show the translation of curated human remains into another medium. In the 'trophy' second from the right, for example, a human cranium is surmounted by a series of carved heads, with stylised facial features, decreasing in size towards the top. Above these are at least three simpler, featureless woven 'heads'. The cranium suspended second from the left in the same photograph, adorned this time with mithan horns, underlies a column of ten of these basketwork 'heads'.

Within Naga cosmologies (as cited by Jacobs 1990, 120), there is a strong connection between the human head and concepts of fertility. The taking and display of heads were regarded as essential to the survival and prosperity of individual Naga communities, like the one at Yimpang. The display of heads captured in von Fürer-Haimendorf's photograph gives material expression to these ideas. The trophy, brought into the village, sometimes adorned with symbols of potency like the mithan horns (Mills 1926, 206), is seen to give rise to a profusion of 'new' heads, symbolising its generative qualities. Despite the obvious lack of even the remotest cultural connection, the vertical display of these 'heads', with the basal head differentiated, is strongly reminiscent of the imagery present on the head carvings from Entremont, Badasset, and elsewhere.

Similar monuments can be found in other regions where headhunting was practised, including some societies with a much more complex social organisation than was present among the Nagas or Sumbanese. The *tzompantli*, known in various forms from several pre-Hispanic Mesoamerican civilisations, was characteristically a tall wooden rack for the display of heads, usually belonging to enemy war dead or sacrificed prisoners (Moser 1973; Ill. 3.3). They are known both from archaeological evidence and through their depiction in contemporary stone carvings, as at the Mayan centre of Chichén Itzá, as well as from illustrations in both indigenous and early Spanish manuscripts (V. Miller 2007, 168–73). The largest of several *tzompantli* at the Aztec capital of Tenochtitlan seems to have held somewhere between around 60,000 and 136,000 skulls at the time of its detailed description by sixteenth-century Spanish observers (Harner 1977, 122), although other examples would have been much smaller. Excavations at the Aztec city of Tlatelolco, for example, yielded a group of around 170 skulls, pierced for mounting on a *tzompantli* (González Rul 1963, 5). Further examples, dating as early as the third century BC, have been found at Loma de la Coyotera, Oaxaca (Spencer 1982, 234–42).

Both textual and iconographic references make it clear that the *tzompantli* was conceived as a gourd tree, with the skulls as its fruit (Serafin and Peraza Lope 2007, 233), and they are sometimes depicted as bearing leaves (Houston et al. 2006, 72). Indeed, the sophisticated architectural elaborations of the immediately precolonial period seem to have evolved ultimately from the practice of displaying trophy heads in the branches of trees (Mendoza 2007, 403). This is in accordance with wider Mesoamerican cosmologies, in which the severed head was associated with fertility and food production (e.g., Moser 1974, 37). Once again, the *tzompantli* and its representation in other media give material form to these underlying cosmological concepts.

In each of these ethnographic cases, the display or depiction of grouped heads is related to deep-seated concepts that link human heads to fertility. The various displays of severed heads, and their renderings in stone or textiles, materialize this concept and symbolise the fecundity of the community. The elaborate Naga compositions, centred on the display of trophy heads, cannot be regarded as 'genealogical', but the basal heads can nonetheless be regarded as 'generative' in a wider sense. In no case do these images overtly glorify individuals; the head-takers are absent and their victims anonymous. Instead they display bounty and plenty through the repetition and composition of nameless heads.

The sickle and the sword

> The harvesting of crops, I think, very quickly brings to mind the idea of attacks on necks and heads. (Enright 2006, 347)

The southern French carvings from the early part of the Iron Age clearly have some resonances with the material culture of headhunting in other parts of the world. The feature common to all these societies is a belief in the links between human heads and the fertility of crops, animals, and people. Indeed, this idea is central to the cosmologies of many of these groups. As we have seen in Chapter 3, this recurrent linkage of ideas generated a good deal of debate among anthropologists during the twentieth century. Can we then suggest that a similar concept existed among the Iron Age populations of southern France?

The phallic nature of the pillars upon which several of the southern French head carvings occur immediately suggests that there may be something in this link. Perhaps more compelling, however, are certain detailed aspects of the motifs. On the *bloc aux épis* from Entremont, the depictions of severed heads, on two sides, are complemented by the depiction of two ears of wheat on the third. This association of heads and crops on the same carved block is unlikely to be simply fortuitous and suggests at least the possibility of a conceptual link between the harvest of the fields and the collection of human trophies.

The metaphoric relationship between the action of the sickle, harvesting crops, and that of the sword, swiping off heads, occurs quite widely in the ancient world, although it is inevitably hard to detect in preliterate contexts. A range of surviving textual sources, however, does make symbolic links between the heads of humans and the heads (or ears) of corn. In a Mediterranean context, these include the Greek myth of Lityerses, who beheaded those he defeated in harvesting contests, as well as several references in the *Iliad* (Onians 1954, 113–14). Similar motifs also occur in earlier, Eastern texts, where, for example, the Sumerian god Ninurta is reported 'reaping like grain the necks of the insubordinate' (West 1999, 229). The same linkage appears in the early Irish text known as 'Cormac's Glossary', dating from around the ninth century AD, which contains definitions, or pseudo-etymologies, for numerous Irish terms. Among these is the phrase '*mesradh Machae*', which is said to mean the 'mast-feeding' or 'nut-harvest' of the war goddess Macha (MacKillop 1998, 281). The term is defined as 'the heads of men that have been slaughtered' (Stokes 1862, 29). In other words a conceptual link clearly existed in Early Medieval Ireland between the severed heads of dead warriors and the harvest of nuts from the trees. Metaphoric links between heads and the harvest are by no means restricted to ancient Europe. In the mythology of the Southeast Asian Iban, the war god Lang plants seed drawn from a trophy head, which results in a rich human crop that is then 'harvested' in another round of beheadings, perpetuating a cycle of fertility and violence (D. Freeman 1979, 244). In Sumba, Indonesia, the hair was described as the 'leaves' of the head, and scalps were hung with ears of corn and other agricultural produce as religious offerings to assure a good harvest (Hoskins 1996c, 230).

In Quichean Maya cosmology, the decapitated head of the fertility deity *Hun* Hunahpu is placed in the branches of a dead gourd tree, which then spontaneously bears fruit

(Mendoza 2007, 420–2). The equivalent deity of the classic Maya period is often seen as a personification of maize, and the myth of his decapitation is linked to the harvest and subsequent regrowth of the crop. A particularly dramatic illustration of this conceptual association comes from the so-called Red Temple at the central Mexican, Maya-influenced site of Cacaxtla, occupied from around AD 600 to 900. Here, within a complex of elite structures, a well-preserved mural depicts maize plants with the ears of corn replaced by human heads (Robertson 1985; Stuart 1992, 134–6). Pakal, the Mayan ruler of Palenque who died in AD 683, is depicted with hair cut to resemble the leaves on a maize plant, falling around his artificially elongated skull (M. Miller 2009). Elsewhere, the intermixing of maize cobs and human heads forms a persistent motif throughout Maya art (Taube 1983).

The Cacaxtla murals and other maize/head substitutions are particularly interesting in the context of this discussion, since they raise the possibility that symbolic linkages between crops and heads may be signalled through iconography. The rendition of the corn on the *bloc aux épis* is especially interesting in this regard, since it is not dissimilar to the rendition of human heads on the adjacent faces of the same stone (Ills. 4.9 and 4.10). Indeed, the shape of each grain echoes the shape of each head, while the paired vertical columns of heads mirror the arrangement of the individual grains on each ear of wheat. Conceivably, the lowermost, upturned head might represent a planted grain, reinforcing the generative qualities of the severed head. Such precise symbolic messages cannot of course be recovered with any confidence, but the associations of heads, fertility, and regeneration remains suggestive. Such an interpretation might even be extended to encompass some of the other motifs that decorated the Entremont stones. Among these is a serpent, carved in high relief on a block reused in the same second-century BC 'hypostyle' building that yielded the head pillar and lintel fragment. In many disparate cultures, snakes are commonly associated with ideas of rebirth and regeneration because of their ability to shed their skin. Their occurrence in a ritual monument associated with fertility and renewal of the community, its crops and its animals, would seem to be quite apt.

These links between the severed human head and harvested grain are perhaps not wholly surprising in an Iron Age context. Richard Bradley has recently written about the wider associations between the human body and the agricultural cycle in later prehistoric Europe. Bradley (2005, 210) points to numerous ways in which such ideas were given material expression during the Iron Age, including, for example, the burial of cremated human remains in ceramic urns modelled on granaries; thus, a symbolic connection was made between 'the death of the human body and the storage and regeneration of the crop'. Bradley's case studies deal generally with the treatment of the dead from within the community, but there is no reason to suppose that the bodies of outsiders were not similarly viewed. We might then suggest that the accumulation and display of trophy heads, both real and depicted, were viewed as a means to ensure the health and prosperity of the group.

Analogous objects

Having set out this broad cosmological nexus linking heads, fertility, and renewal, I want now to turn to a series of other Iron Age objects that display some similarities in iconographic content to the southern French carvings. Collectively, these suggest that the cosmological links discussed in the preceding section may have figured in regions far beyond southern France during the Iron Age. These objects comprise a collection of wooden figures from the source of the Seine in central France, some iconographic images from Spain, and the extraordinary Pfalzfeld pillar from western Germany.

A particularly intriguing set of parallels for the southern French carvings comes from the sanctuary of Sequana at the source of the river Seine, south of Châtillon in central France. Excavation of the area around these springs at various times since the mid-nineteenth century has exposed a complex of stone structures, enclosures, and open spaces. The most spectacular discoveries, however, came in the 1960s when several hundred votive wooden objects were discovered, along with even greater numbers in bronze and stone (Deyts 1983, 63). These votives, dating to the early Gallo-Roman period, seem to have been deposited by pilgrims hoping to heal various maladies and afflictions. Offerings include carved representations of assorted body parts and innards, which were seemingly intended to deliver quite specific messages regarding the nature of the sufferer's condition.

Among the wooden votives, however, is a striking series of eighteen wooden carvings that take the form of short poles, up to around 0.75 m high (Deyts 1983, nos. 138–51, 152, 153–5). Each is made up of between two to four superimposed human heads, many with slit eyes and mouths, reminiscent of the carvings from Entremont and elsewhere. Most of the heads are extremely stylized (Ill. 4.12), and even allowing for variable preservation, there seems to have been little attempt to convey expression or to portray specific individuals.

Of the examples shown here, the first comprises three heads carved from a single cylindrical trunk. Their flattened faces have heavily incised eyes, broad triangular noses carved in low relief, and short narrow lines to mark the mouth. The second example is made up of four heads, varying in shape from near cylindrical to well rounded. The facial features closely parallel those on the first example, although these have the addition of a fringe of hair on each head. Perhaps the most striking difference, however, is the rather greater vertical spacing between each head. Although these intervening gaps have been interpreted as necks, they could equally be read as indicating a timber post onto which neckless heads have been attached. This would seem to accord better with the disproportionate length and the shape as depicted on several of the carvings. If this interpretation is correct, then these votives might, like the southern French sculptures, represent iconographic renderings of human trophies displayed on posts.

Numerous individual wooden heads have also been found within the sanctuary. Their size and style suggest that some of these may have been cut down from multiple head poles similar to those described. This has tended to deflect interest from the multiple, repetitive nature of the complete poles, since it has been easy to disregard them simply as a stage in the production of single wooden heads. Nonetheless, head poles clearly were deposited as votives in their own right on numerous occasions and cannot, therefore, be dismissed as simply the stock-in-trade of some Gallo-Roman wood-carver. Indeed, one might rather see the individual heads as fragments, or 'tokens' of the multiple-head poles than as end products in the conventional sense.

Associated pottery and coinage indicate that the sanctuary of Sequana was in existence during the early part of the Gallo-Roman period, with the most intensive period of votive

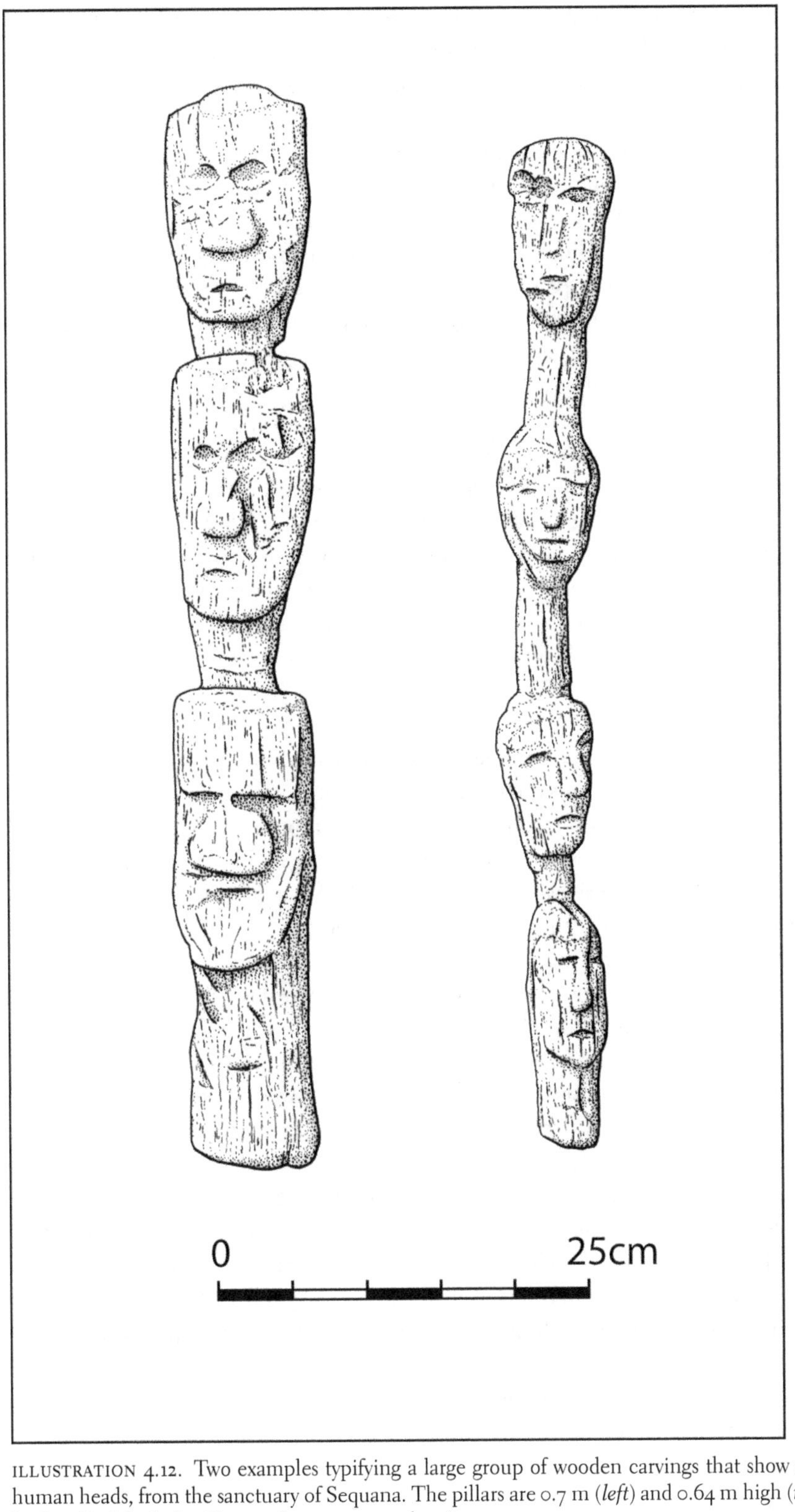

ILLUSTRATION 4.12. Two examples typifying a large group of wooden carvings that show superimposed human heads, from the sanctuary of Sequana. The pillars are 0.7 m (*left*) and 0.64 m high (*right*). (Drawn by Dan Bashford after Deyts 1983, nos. 139 and 140)

deposition ending before around AD 70. This gives a useful *terminus ante quem* for the wooden sculptures, but the date of their manufacture is much less certain (Deyts 1983, 63–4). Nonetheless, there is nothing to suggest that the images predate the first century AD, and, as Greg Woolf (1998, 218) has suggested, the whole complex can be interpreted in the light of Gallo-Roman religious practice. Despite the recovery of a large assemblage of animal bone, the sole human fragment was the distal end of a right femur (Poulain 1983, 69), so this was not apparently a site where trophy heads were displayed. Yet these superimposed wooden heads are not Roman in origin or influence and find no obvious parallels in the wider empire. They suggest not simply a continuing belief in the importance of the human head, but a conviction that the power it holds can be captured in images and enhanced by repetition, grouping, and order.

Further parallels for the southern French head carvings come from Iberia, where significant numbers of Iron Age stone heads have been found (e.g., Almagro-Gorbea and Lorrio 1993). Grouped heads can be found, for example, on a poorly dated stone from Puentedueme, La Coruna, which shows four adjoining heads carved in low relief, one on each face of a four-sided block (Lenerz-de-Wilde 1993, fig. 12). A single carved head on a pillar from Cordoba also shares stylistic similarities with the early carved heads from Entremont, its closed eyes depicted by a single straight line from which a stylized nose descends (fig. 21). Decoration on the pillar can be linked to Iberian pottery of the fourth to third century BC and may provide a broad chronological context for the stone (Taracena 1943, 169). The phallic shape of this pillar is typical of other pillars from Iron Age Iberia. There are similar objects of Early Iron Age date in Germany, notably the pillar from Hirzenhain (Schoppa 1958), though in these cases the body of the pillar appears to stand for the human body, and the pillars are probably intended as anthropomorphic.

Suggestive as these phallic stone pillars may be, however, perhaps the most striking link between the human head and concepts of fertility in Iron Age Iberia comes from a less obvious source. During early excavations at Monreal de Ariza, around 100 km southwest of Zaragoza in northeast Spain (thought to be the ancient site of Arcobriga), a painted pottery vessel was discovered dating to the very end of the Iron Age (Ill. 4.13). This small pot, no more than 15 cm high and 20 cm in diameter, bears two identical scenes (Sopena 2005, fig. 8). Each comprises a frontal view of a building or portico, framed by two elaborate columns with wide bases and capitals. The columns are decorated with crescentic patterns that have been taken to represent lunar symbols (Marco-Simón 1993), but which may in fact be vegetal motifs. A decorated arch links the two columns, central to which is an ill-defined dark blob. Two 'branches' bearing heart-shaped ivy leaves (Marco-Simón 2003, 123), or perhaps fruits of some kind, sprout from the column capitals. The whole structure is flanked on either side by standing birds (probably domestic fowl) and a peculiar motif similar to a bent spear, but which Francisco Marco-Simón (1993) has interpreted as a horned snake of the kind often found in Celtic iconography. Most importantly for present purposes, the whole edifice frames a human head, resting on some form of box or plinth, from the top of which grows a tree. The whole decorative scheme seems to embody a complex and highly formalised symbolism associated with fertility and growth, and with the severed human head at its centre.

ILLUSTRATION 4.13. This decoration, taken from a reconstructed pottery vessel from Arcobriga, appears to show a tree growing from a severed human head within a small building surrounded by symbols of various kinds. (Drawn by Rachael Kershaw after Sopena 2005, fig. 8)

The final object for consideration here is the Pfalzfeld pillar, from the Hunsrück region of western Germany, which dates to around 400 BC and is generally thought to have been a funerary monument, marking the grave of an important individual (Weber 2002, 318; Ill. 4.14). As it survives today, this sandstone block stands some 1.4 m tall, tapering towards the top, though early depictions suggest that it originally stood to around 2.8 m (Joachim 1993, 266). Its base takes the form of an upturned dome, which might have been buried in the ground to give stability, but the main body of the pillar is quadrangular, with decoration on each of the four sides. On the lower part of each face is a 'floating', disembodied head, bearing the *blattkrone* or leaf-crown, probably representing mistletoe leaves. The symbolism of the leaf-crown is far from clearly understood, but it does seem generally to have associations of high, or even divine, status. Each of the Pfalzfeld heads also bears a lotus motif on its forehead, set above a narrow band that might indicate that lotus and leaf-crown form part of a headdress of some sort. When first found, in the early seventeenth century, the pillar was thought to have been surmounted by a Janus head wearing a similar leaf-crown (Joachim 1993, fig. 4), though this has long since vanished.

In his detailed analysis of the stone, Hans-Eckhart Joachim (1993, 266) describes the heads as having moustaches but no mouths, which raises the possibility of a link with the imagery on the Entremont head pillar. The possible importance of Joachim's observation is enhanced by the very clear depiction of mouths on stylistically similar carvings, such as the life-sized warrior statue from Glauberg, which dates to the same broad period as the Pfalzfeld pillar, also wears a *blattkrone*, and is executed in a remarkably similar style

ILLUSTRATION 4.14. Drawing showing one of the four faces of the Pfalzfeld pillar as it may have looked when first illustrated, around 1608/9. It may have been surmounted by a Janus head. (Drawn by Rachael Kershaw after Joachim 1993, fig. 4, with amendments)

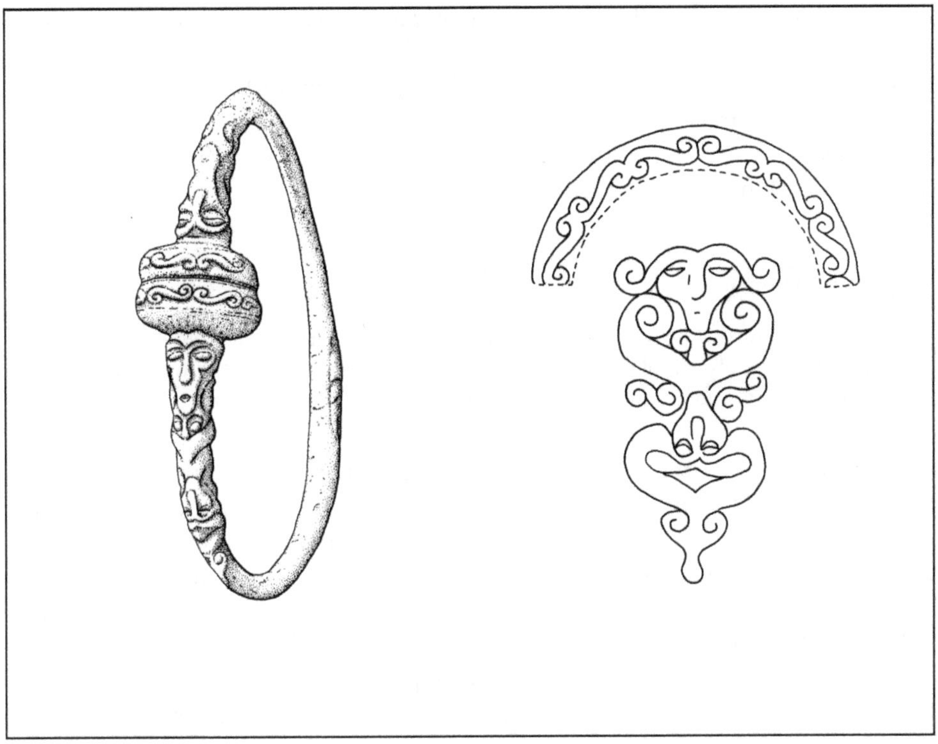

ILLUSTRATION 4.15. Bronze torc from a grave at Courtisols, Marne, in eastern France, displaying two sets of opposed heads with seemingly closed eyes. The upper head, in the 'rolled-out' detail shown on the right, appears to rest on some form of plinth. Another fragmentary torc from the same site shows a similar motif (Jacobsthal 1944, plate 126, no. 208). (Drawn by Dan Bashford after V. Megaw 1970, no. 119, and Stead and Rigby 1999, no. 1711)

(Weber 2002). Is it possible then that the Pfalzfeld pillar is intended to depict severed heads? There is certainly no indication of a neck, far less a body, attached to any of the heads. Indeed, each head appears to be perched on top of a peculiar tripartite arrangement of leaf shapes. In the past, these enigmatic objects have been subject to conflicting interpretations. Joachim (1993, 267), for example, described them as 'beards', whereas Michael Enright (2006, 34–5) has suggested that they represent tasselled torcs similar to excavated examples from the Glauberg. In fact, they closely resemble motifs on metalwork, such as the torcs from the late fourth- or early third-century BC grave from the cemetery at Courtisols in eastern France (Stead and Rigby 1999, no. 1711), where apparently severed heads rest upon some form of plinth (Ill. 4.15).

Ross (1958, 23) has suggested that Pfalzfeld and other sculptures of 'decapitated' heads set on pillars, such as the Romano-British examples from Corbridge and elsewhere, may have been phallic symbols. This interpretation could be extended to other objects under consideration here, including the Sequana head poles and many of the Iberian pillars. This association of disembodied heads with a phallic stone pillar echoes several ethnographic accounts, notably the Nagas practice, mentioned earlier, of constructing phallic stone cists to receive the skulls of the dead (Jacobs 1990, 124). As with the stone carvings of southern France, it appears that in Iron Age southern Germany, Iberia, and central France severed human heads were regarded as potent objects, bound up in wider concepts of fertility and renewal. Indeed, wherever significant traditions of Iron Age stone or wood carving survive in Iron Age Europe, the human head seems to play a prominent role. This leads us to a final discussion of the most widespread and pervasive corpus of Iron Age art in Europe: La Tène metalwork.

Above and below

> There is something fleeting and evanescent about these masks, which often are not even complete faces, only bits of faces. It is the mechanism of dreams, where things have floating contours and pass into other things. (Jacobsthal 1944, 19)

There are no obvious parallels in the extensive corpus of La Tène metalwork for the vertical arrangements of human heads seen in the southern French sculptures or the Sequana head poles. Yet there are, nonetheless, notable compositions involving multiple heads. Perhaps the most striking is a series of seventeen silver disks found in the 1920s at Villa Vechia, Manerbio sul Mella, near Brescia in northern Italy (V. Megaw 1970, 130). These disks, or phalerae, which would most likely have been deposited as grave goods, appear to have been elements of high-status horse harness, probably dating to the early first century BC (Ill. 4.16). A circle of identical disembodied heads surrounds the central area of each disk, their braided hair swept back from their foreheads, in stereotypical Gaulish fashion. All have bulbous, closed eyes surrounded by rings of small 'beads'. Moustaches are visible above their downturned mouths. Like the images discussed earlier, they seem to depict severed heads.

Although the Manerbio disks are unique, the concentric arrangement of heads has parallels much earlier in the Iron Age. Another phalera, this time in bronze, from a La Tène A barrow grave at Hořovičky in Bohemia, depicts two concentric rings of leaf-crowned heads around a central bulbous boss (V. Megaw 1970, no. 47). This piece probably dates to the mid-fifth century BC, around 400 years earlier than the Manerbio disks, yet the similarities in design are quite marked. The same could be suggested for the gold inlaid plaque (probably originally part of a brooch) from a German La Tène A barrow grave at Weisskirchen. This depicts four clean-shaven heads, eyes apparently closed, ordered concentrically around the central circular design (V. Megaw 1970, no. 46). In these scattered examples of high-status metalwork, we see the same concern with the careful grouping, arrangement, and repetition of identical anonymised human heads.

While such elaborate compositions are relatively rare, individual or paired human heads are much more common. Although these can be quite naturalistic, especially in the earlier stages of the art style (i.e., in the fifth and early fourth centuries BC), they are often semiabstracted, or partially hidden in complex vegetal motifs. These are the heads described by Lambrechts (1954, 20) as the *motif national* of Celtic art. Although the frequent incorporation of individual disembodied heads within patterns drawing on plant forms may suggest a further link between heads and regrowth (e.g., Szabó 1993), I want to focus here on the frequent occurrence of paired, or rather opposed, heads, since I will

ILLUSTRATION 4.16. The Manerbio phalerae vary in size, the largest (shown here) being around 19 cm in diameter and the smallest being around 9 cm. The heads are formed in repoussé, the motifs being hammered up from the back of each piece. (Drawn by Rachael Kershaw)

go on to suggest that these may have a more direct relationship to the cosmological ideas embodied in the southern French carvings (Armit 2010a).

By 'opposed' I mean heads that appear as reflections or transformations of one another along the same axis. Usually one is upside-down relative to the other, and they are typically joined at the neck, or by some object(s) in lieu of a neck (Ill. 4.17). An archetypal example is the gold ring from Rodenbach, in the German Rhineland, which bears two opposed heads emerging from intricate foliate decoration (Ill. 2.1). The motif can also be found on numerous bronze fibulae (e.g., Jacobsthal 1944, plates 156 and 157), and objects such as the third-century BC bronze arm ring from La Charme, Champagne (V. Megaw 1970, no. 140). One particularly interesting example of the opposed-head motif comes from Tal-y-Llyn, in Gwynedd, north Wales. Here, two small bronze plaques, perhaps originally shield mounts, were discovered as part of a metalwork hoard, dating to the late first century BC or early first century AD. Each bears two opposed human heads joined by an improbably long 'neck' (Green 1986, fig. 100). The heads are surrounded by a vegetal pattern, although they do not themselves form part if it. In each case, the eyes are closed and the mouths downturned, suggesting that they depict severed heads (Jope 2000, plate 96).

ILLUSTRATION 4.17. Bronze mount from the Tal-y-Llyn hoard, Gwynedd. The mount is around 10.4 cm wide. (Drawn by Rachael Kershaw)

The Tal-y-Llyn plaques are by no means the only examples that appear to depict the heads of the dead. I have already mentioned the motifs on the torc from Courtisols in eastern France (Ill. 4.15), where opposed human heads appear to be set on plinths. Other examples include an early La Tène gold bracelet from Dürkheim, in the German Rhineland, where three sets of opposed heads appear to sit atop columns or pillars (Jacobsthal 1944,

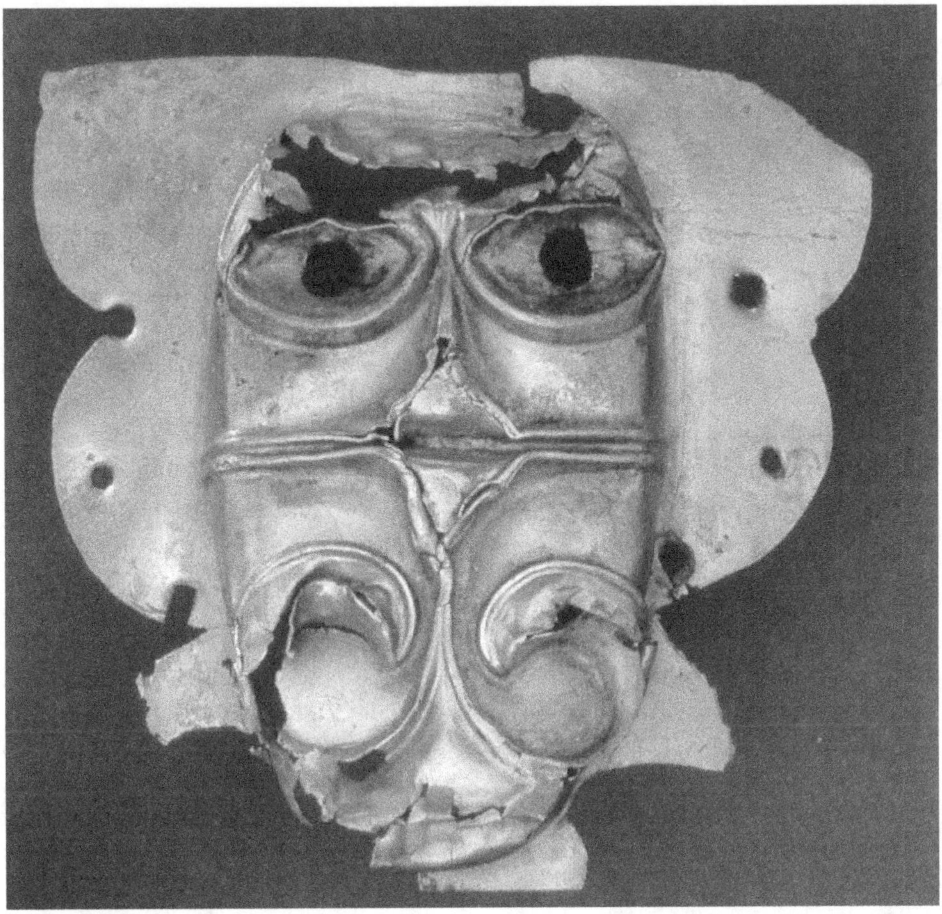

ILLUSTRATION 4.18. This bronze mount, one of three from the Melsonby hoard, North Yorkshire, has been interpreted as a single human face with a forked beard (e.g., V. Megaw 1970, 157; Jope 2000, 100). Seen in the context of the other opposed heads in Iron Age art, however, it seems more likely that its 'reversibility' was both meaningful and intentional. (R. Megaw and Megaw 1993, 211). The mounts are around 7 cm long. (Photograph © The Trustees of the British Museum).

plate 46, no. 57). These suggest that the opposed-head motif may have related quite specifically to severed human heads, rather than being some more generalised decorative device. The opposed nature of these paired heads also suggests comparisons with the southern French carvings, where the lowermost head, as we have seen, is frequently upturned. So what might these oppositions signify? To illuminate some of the possibilities, I want briefly to examine three objects that represent variants on this general theme – from Melsonby in North Yorkshire, Bad Dürkheim in the German Rhineland, and Niederschönhausen, near Berlin, in northern Germany.

The two bronze mounts from Melsonby come from a metalwork hoard found in the mid-nineteenth century, half a mile away from the Brigantian power centre of Stanwick (McGregor 1962; R. Megaw and Megaw 1989, 224–5). The mounts are small, less than 8 cm high, and depict two highly stylised, opposed human heads (Ill. 4.18). The repoussé

TABLE 4.2. *Some conceptual oppositions suggested by the opposed-head motifs in Iron Age art*

Youth	Age
Body	Soul
Wakefulness	Sleep
Life	Death
Living world	Underworld
Embodied	Disembodied

faces are dominated by large lentoid eyes, which would originally have been enhanced by the use of enamel inlay. By contrast the noses are tiny, terminating in what appears to be either a small downturned mouth or a stylised moustache. The upper head, as shown here, has eyes wide open, staring at the viewer. The lower head, by contrast, although damaged, has its eyes closed, perhaps representing sleep or death. This image seems to elaborate on the opposed-head motifs already discussed, in that it implies a conceptual as well as physical opposition between the heads above and below; an opposition perhaps between wakefulness and sleep, or between life and death (Table 4.2). The composition of the Melsonby hoard, which also included such luxury items as a suit of chain mail, suggests that it derives from a high-status cremation burial, probably deposited around the third quarter of the first century AD. Indeed, it has recently been suggested that the mounts may originally have been attachments for an iron-bound wooden bucket, which would have contained the ashes of the deceased (Fitts et al. 1999). The use of this particular motif thus seems intimately linked to the transition from life to death, and from this world to the next.

A related set of oppositions is represented by the second example – the two fragments of gold foil that probably formed part of a single mount from a late fifth- or early fourth-century BC grave at Bad Dürkheim (Ill. 4.19). Seen from one perspective, they appear to depict youthful, clean-shaven faces, adorned with a 'leaf-crown'. When reversed, however, they transform into the faces of aged, bearded men. Again the opposed-head motif, here in rather modified form, seems to suggest a conceptual relationship between distinct states of being.

The third set of opposed heads appears on an unusual cast bronze fibula from Niederschönhausen (Ill. 4.20). This is one of the rare cases in continental La Tène art where an entire human body is depicted. In this case the diminutive naked, possibly ithyphallic, male body, some 2.6 cm long, is attached to a disproportionately large bearded head, some 1.9 cm high. The end of the fibula is formed into a ram's head that curves back to loom menacingly over the human figure. Again the two heads seem to signify different states of being: one disembodied, old, perhaps dead, the other embodied, naked, young, and vulnerable.

Although there has previously been little attempt to interpret the meaning of these opposed heads, they have occasionally been linked to the Janus head motif, which also occurs sporadically in Iron Age art. Indeed, Jacobsthal referred to examples like the Rodenbach ring (Ill. 2.1) as 'a Janus head in flat projection' (cited in V. Megaw 1970, 68). Actual Janus heads, where two conjoined heads (or two faces on a single head) face out in opposite directions, occur very rarely in La Tène metalwork, as on a fifth-century bronze neck ring from Glauberg, and a first-century BC iron knife from Slovakia (V. Megaw 1970, 82, 127). They are, however, more common in Iron Age stone sculpture, with examples including the Holzgerlingen statue in Germany, from the sixth or fifth century BC (V. Megaw 1970, 48); the stone head from Badaczony-Lábdi, in Hungary (Szabó 1965); and a

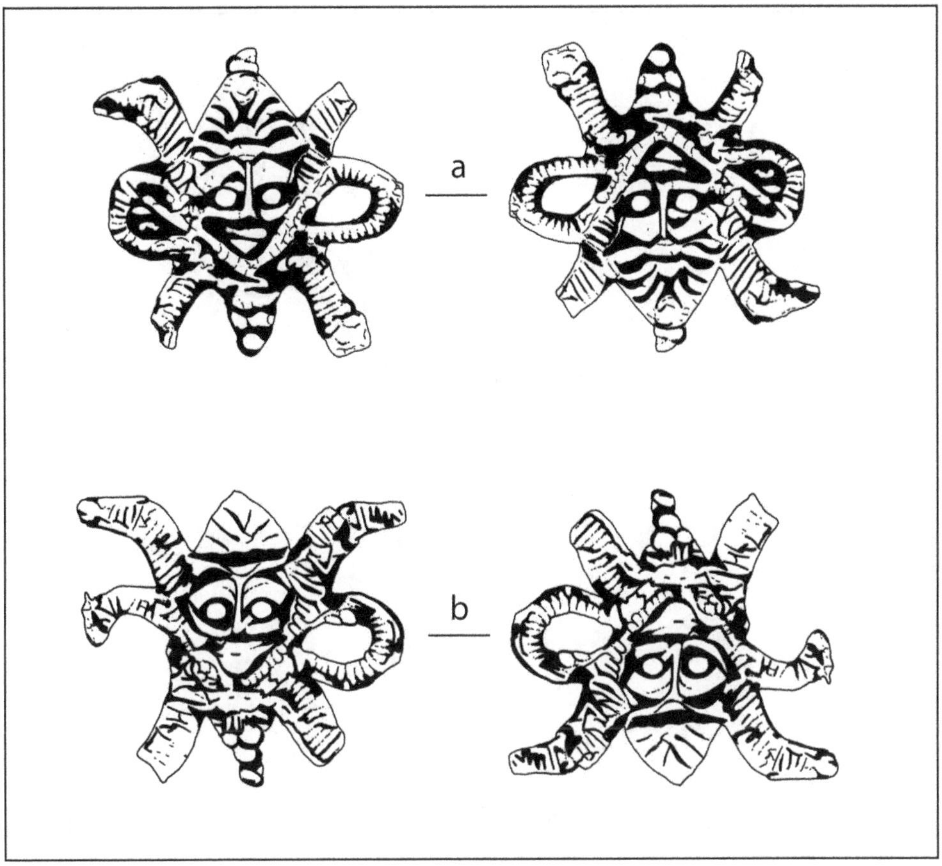

ILLUSTRATION 4.19. Drawings showing two views of the late fifth- or early fourth-century BC foil heads from Bad Dürkheim, in the German Rhineland. The drawings on the left show youthful faces wearing 'leaf-crowns'. When reversed, as seen in the right-hand drawings, these become the faces of old, bearded men. The mount would originally have been secured by nails, probably to a wooden object. The tiny heads are around 1.7 cm tall. (Drawn by Dan Bashford after Jacobsthal 1944, plate 28)

particularly striking example from Roquepertuse, which I will discuss further in Chapter 5. Janus appears to have originated as an Italic god of boundaries and doorways and is probably related to the Etruscan sky god Ani and the Greek Hermes. It is entirely likely that similar two-faced deities existed among the nonliterate societies of temperate Europe. In any case, it seems unnecessary to suggest any specific Etruscan or Italic influence for the appearance of Janus heads in La Tène art, and still less for the much more common opposed-head motif. Yet the association of the opposed-head motif, and severed heads in general, with boundaries and doorways is one that I explore further in Chapter 5.

I want to conclude this discussion of the opposed-head motif with a brief return to the sanctuary of Sequana to look at two further wooden votives omitted from the earlier discussion (Ill. 4.21). Each comprises two human heads, and both are clearly variants on the multiple-head poles from the same site. But rather than being stacked one upon the other, the heads in these images are opposed and joined at the neck. Effectively, these are

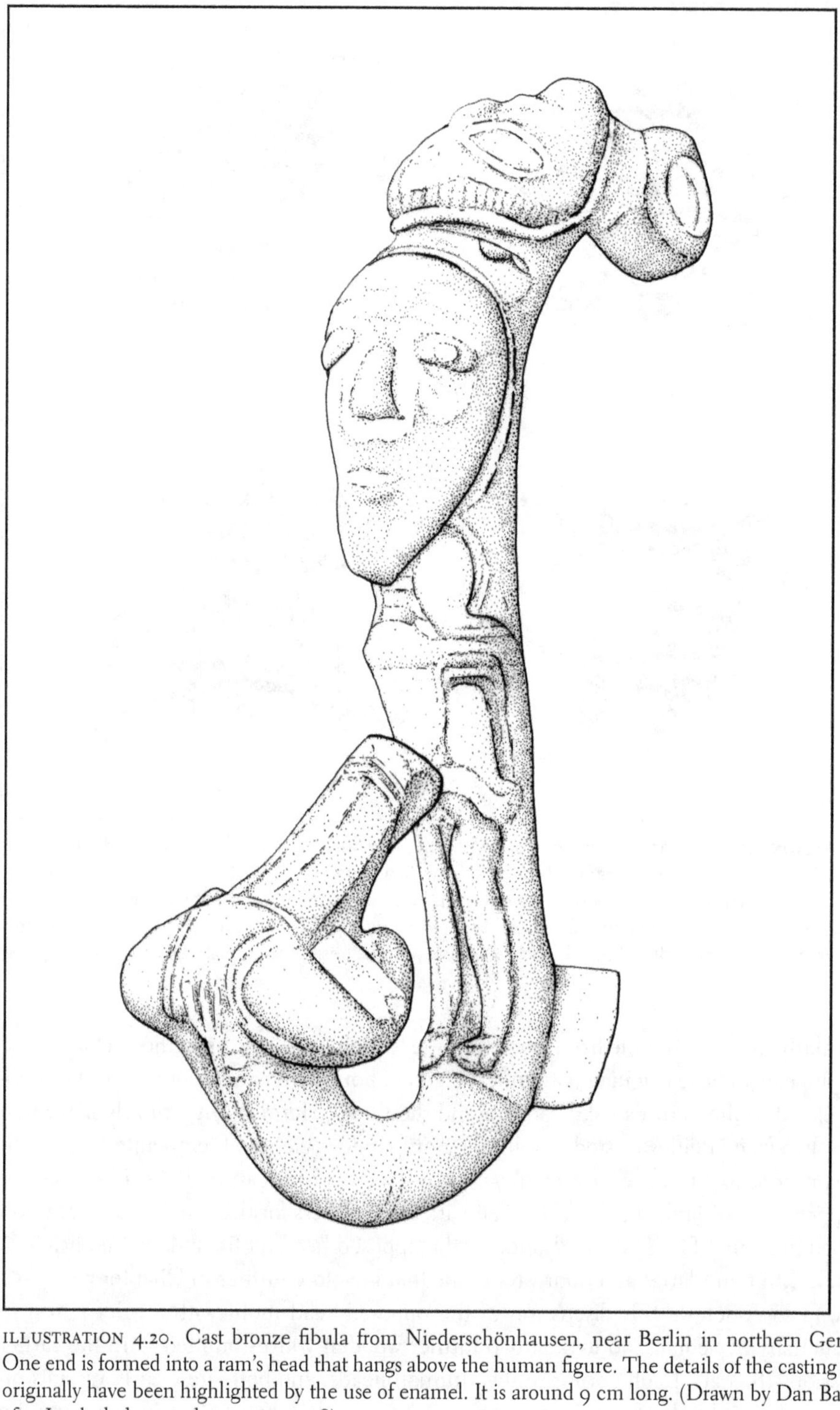

ILLUSTRATION 4.20. Cast bronze fibula from Niederschönhausen, near Berlin in northern Germany. One end is formed into a ram's head that hangs above the human figure. The details of the casting would originally have been highlighted by the use of enamel. It is around 9 cm long. (Drawn by Dan Bashford after Jacobsthal 1944, plate 157, no. 308)

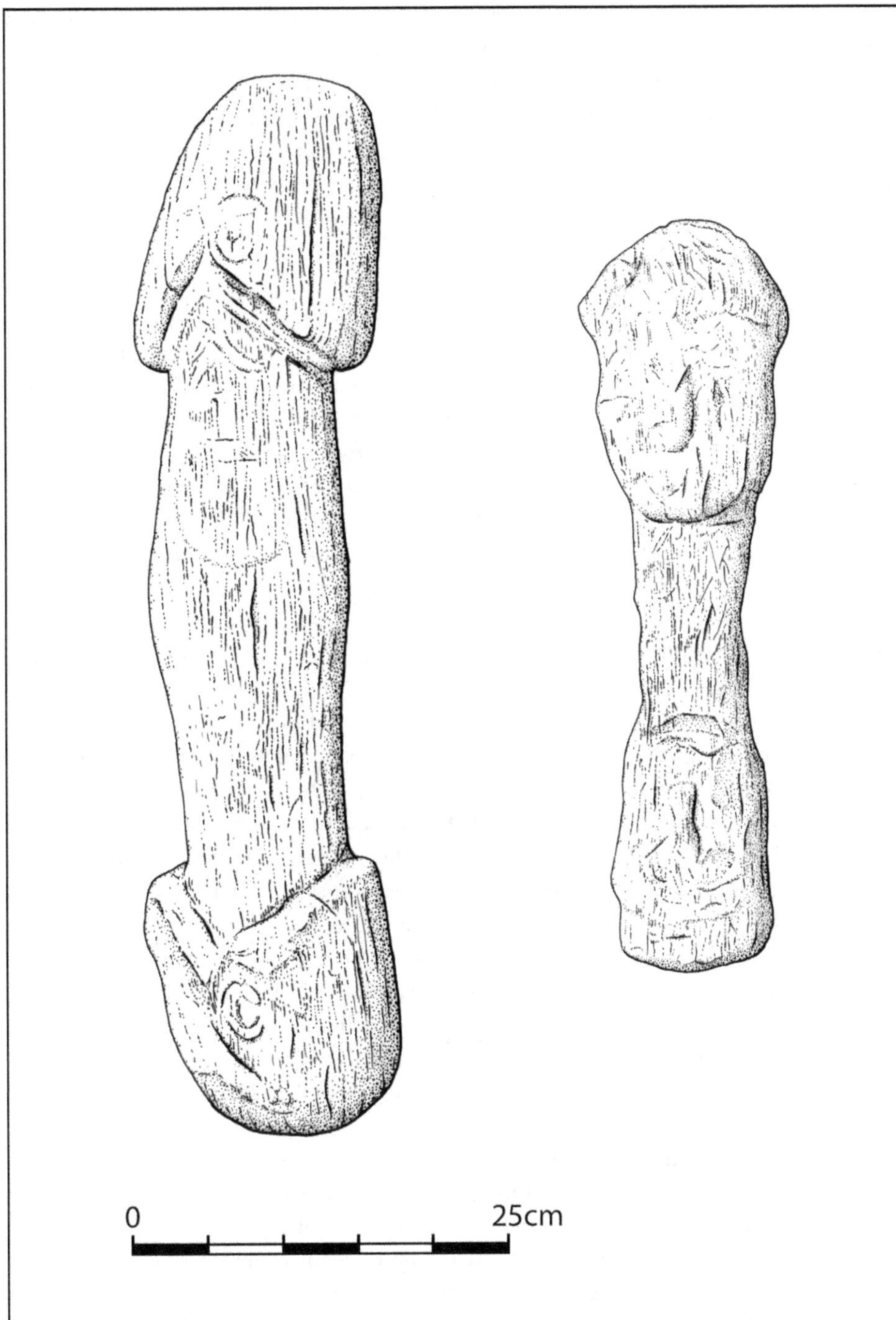

ILLUSTRATION 4.21. Two wooden votives from the sanctuary of Sequana displaying opposed heads, in one case at least clearly intended as phallic. The objects measure 0.68 m and 0.48 m respectively. (Drawn by Dan Bashford after Deyts 1983, nos. 156 and 157)

renditions in wood of the motif familiar from the Iron Age metalwork. They thus represent a further link between the indigenous iconography of the Gallo-Roman period and that of the earlier La Tène metalwork, suggesting that both drew on a wider body of cosmological understanding. As if to underline the symbolism of fertility and renewal associated with this complex of ideas, the visual punning between head and phallus could hardly be more pronounced.

Conclusion: Harvesting heads

Cosmological links between fertility and the human head, so widespread in the ethnographic literature, seem to have been equally deep-rooted in the European Iron Age. The recurrence of such ideas in societies so widely scattered in space and time should not perhaps occasion too much surprise. The universality of the physical body makes it a natural vehicle for the expression of cosmological ideas. Houses, for example, are often structured with reference to the human body, with various elements of the building being identified with different body parts (e.g., Parker Pearson and Richards 1994). All traditional societies similarly share a concern with the fertility of the animals and plants that support them and with the reproductive success of their own communities. Thus, the independent emergence of similar cosmological associations between heads and fertility in nineteenth-century Indonesia, ninth-century Mexico, Iron Age Europe, and elsewhere need reflect little more than similar preoccupations and similar tools for their expression.

From the rituals at Raffin, through the head carvings of Provence, to the sanctuary of Sequana and beyond, we find an intimate association between the human head and ideas of fertility and regeneration. Images of heads seem frequently to signal real, severed heads, taken in raids or warfare, and these heads themselves turn up frequently on Iron Age settlements. In Provencal stone monuments and the wooden carvings from the sanctuary of Sequana, their careful arrangement and organisation suggest that these heads gained in power through multiplication. If heads symbolised fertility, then these groups express bounty or an excess of health and fecundity. They seem to promote or celebrate the success of particular communities. Yet always the heads themselves are anonymous, stylized, and repetitive – objects rather than individuals. Their importance is simply that they are heads, not that they formerly belonged to fierce warriors or great leaders.

As in Southeast Asia or North India in more recent times, heads seem to have been sought for use in collective rituals, intended to promote or enhance the fertility of the community. Whether they were conceived as containing some intrinsic power or 'soul substance' that could be transferred through ritual from the dead to the living is unclear. As we saw in Chapter 3, understanding the conceptual link between heads and fertility is difficult enough for anthropologists studying more recent societies, and it is inevitably more or less impossible for the Iron Age. Yet it seems inescapable that such a link not only existed but was central to Iron Age understandings of the world. These ideas might be materialized in many different ways, through displays and performances involving the heads of enemies, ancestors, or symbolic heads carved into stone other media. They seem also to have impelled certain Iron Age communities actively to seek the heads of their enemies.

This recognition of the cosmological association between heads and fertility is an important first step in understanding the importance of the human head in Iron Age thought. Yet it does not in itself take us very far in terms of detailed interpretations of particular Iron Age societies. In the next chapter I consider in more depth how this cosmological background could give rise to some very specific and dramatic practices.

CHAPTER 5

Neither this world, nor the next

Holding us between repulsion and respect, terror and deference, we're still, it would seem, affected by these gutted husks. Much like the mask of the gorgon, the skulls belong neither to this world nor the next, but to that wavering interface, that intermediary realm between being and non-being, the living and the dead. Threshold figures, they command passage.

Sobin 1999, 122

All of these sculptures are extremely barbaric in character. One assumes that they should be attributed to the Saluvians, and indeed I cannot imagine who else could have done so poorly.

Mérimée 1989 [1835], 141

Betwixt and between

Standing at the mouth of the Sculptor's Cave, looking northwards across the wide, grey expanse of the Moray Firth, one can easily understand how this came to be regarded as a special place (Ills. 5.1, 5.2). The cave takes its name from a series of Pictish symbols that were carved around its entrance walls sometime during the mid-first millennium AD. It is a naturally gloomy spot, almost constantly in shadow, beneath a low cliff. At high tide, access from the land is cut off, adding further to its sense of isolation. The cave is entered along two parallel passages, each around 10 m long, separated by a narrow wall of rock. The passages are tall and wide enough to allow unimpeded access, and there is no need to crouch or crawl. Once inside, the roof rises to reveal an imposing internal space. When the original excavator, Sylvia Benton (1931, 177), first visited the cave, she found a floor still 'strewn with human bones', and it was this that prompted her to embark on her three seasons of excavation from 1928 to 1930.

Benton's excavations recovered evidence for two main periods of activity, both represented by distinctive metalwork assemblages. The earliest finds date from the Ewart Park phase of the Later Bronze Age, between c. 1000 and 800 BC, and include a number of fine bronzes such as hair rings, pins, and bracelets. The nature of the finds suggests that the cave was used primarily for ritual activity rather than simply for domestic occupation,

ILLUSTRATION 5.1. Map showing the principal sites discussed in Chapter 5. (Drawn by Rachael Kershaw)

as indeed we would expect from its inaccessible location. After an apparent gap in the sequence, there was a further concentration of finds from the Roman Iron Age, centred on the second to fourth century AD. These included pins, tweezers, beads, and a notable concentration of coins dating from around AD 330–60. Even at the height of the empire, this part of northeast Scotland lay well beyond the limits of direct Roman military control, but the elites within the region may have received periodic bribes to help keep the peace (Hunter 2003, 22; 2007, 24). It appears that at least some of this 'bribe-wealth' ended up in votive deposits in ritual places like the Sculptor's Cave. As in the Bronze Age, there is little to suggest any domestic occupation.

ILLUSTRATION 5.2. Entrance to the Sculptor's Cave, Moray. (Photograph by Rick Schulting)

Strewn with bones

Despite the fine quality of the bronzes, the most striking finds from the Sculptor's Cave were the substantial amounts of human remains (Ill. 5.3). Unpublished anatomical reports suggest that up to 1,800 human bones were originally recovered from inside the cave, although the vast majority was discarded after only a brief examination, on the instructions of the anatomist, Professor Alexander Low (Armit, Schulting, et al. 2011). Enough detail was recorded, however, to permit some limited observations. Most importantly, it appears that although some adult bones were present, the proportion of children was markedly high (Low 1931, 207). Intriguingly too, there was a clear under-representation of skull fragments, suggesting that the heads of these dead children had been removed or retained prior to deposition. Since there were no reports of articulated bones, it was probably defleshed bones, rather than complete skeletons that had been placed inside the cave, although the relative crudity of the excavation methods and the difficulties of working in such a dark and cramped environment may have contributed to a failure to recognise articulated skeletal material.

One particular group of bones was retained. This comprised a series of seven cervical vertebrae of which six show cut marks consistent with decapitation (Ill. 5.4). Although Low's terse remarks in the original report give the impression that these were children's vertebrae, recent re-examination has shown that most derive from adults (Armit, Schulting, et al. 2011). This recent work has shown that, in each case, decapitation was carried out

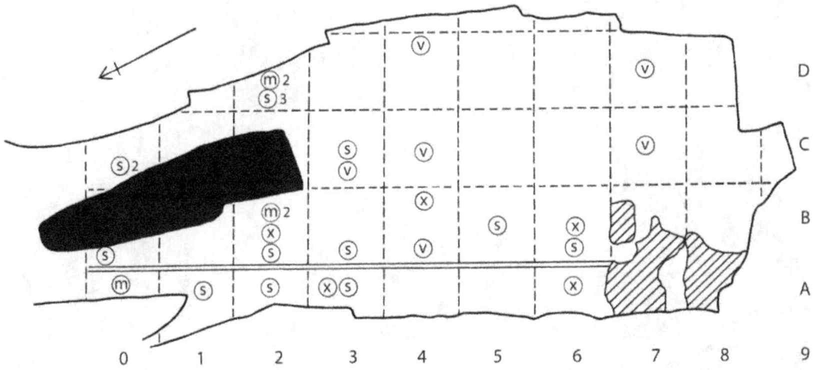

ILLUSTRATION 5.3. Plan of cave showing the distribution of human skull fragments according to Sylvia Benton's excavation grid from the 1929–30 excavations. Most are located only to grid square, although a few from the 1979 excavations, notably the mandibles, in the entrance area, are more precisely located. Much of the Benton material represented here is now lost. S = unclassified 'skull' fragment; V = cut-marked vertebra; M = mandible; X = maxilla. (Drawn by Dan Bashford).

using great force, with a sharp, heavy blade. In some cases there had been multiple blows to the back of the neck as the head was hacked off. One bore evidence of at least eleven hacking blows of this kind. Initially this pattern might seem to suggest that these were trophy heads, struck from the necks of dead or dying warriors on the battlefield. Closer inspection, however, shows that this cannot be so. Crucially, in one particular case, the blade injuries were on the upper, rather than lower, side of the vertebra. In other words, this vertebra would have remained attached to the victim's body following decapitation, rather than being carried off with the severed head. This is an important observation, because it means that this vertebra at least cannot have come from a trophy head, brought to the cave after decapitation elsewhere. Either the entire decapitated body was, rather improbably, dragged back to the cave along with the head, or the decapitation occurred *in situ*, within the cave itself or in its immediate vicinity.

Some fifty years after Benton's work, a further excavation in the Sculptor's Cave was undertaken by Ian and Alexandra Shepherd, focussing on the limited surviving deposits in the twin entrance passages (Shepherd 1995, 2007). As well as yielding further bronzes, the 1970s excavations recovered more human remains, including four mandibles, from individuals aged between six and nine years. These appear to suggest that the fleshed heads of young children had been hung up at the entrance to the cave and left to decay *in situ*, the lower jaws eventually becoming detached and incorporated into the cave floor deposits. The stratigraphic association of these mandibles with gilded bronze hair rings, raises the intriguing possibility that the heads may have been groomed and adorned with prestigious objects. Hair rings were 'highly prized and sought-after objects' and probably had an Irish origin (Waddell 1998, 248). Their discovery in association with curated heads

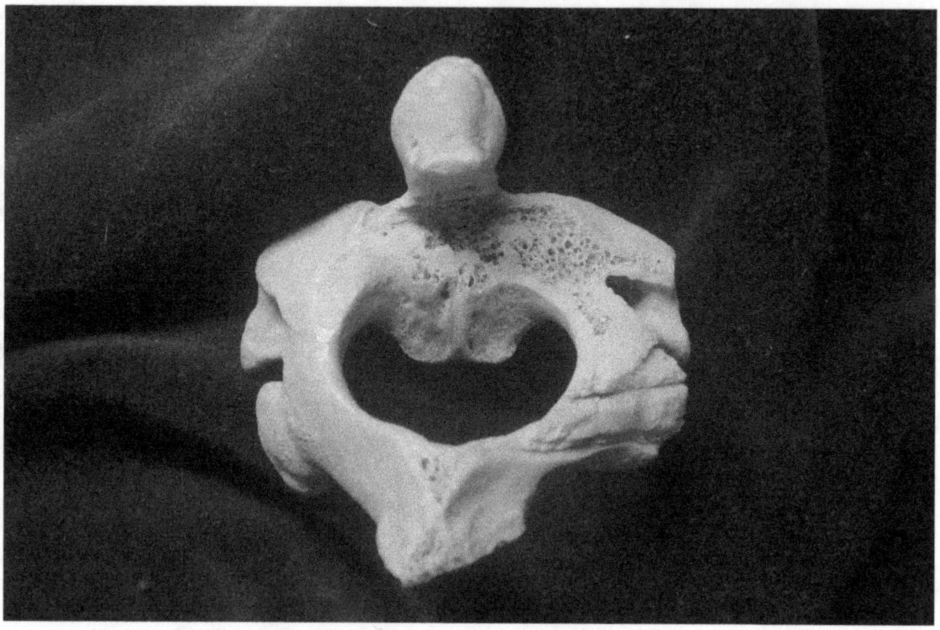

ILLUSTRATION 5.4. Human axis vertebra from the Sculptor's Cave, Covesea, showing cut marks consistent with decapitation. (Axis HM-159. Photograph by Rick Schulting)

raises two important points: firstly, it appears that the heads of these children were treated with considerable reverence and care; secondly, the fact that they were left to rot with their finery intact suggests a powerful social taboo against the removal of what must always have been valuable and exotic objects.

The Shepherds' excavations also revealed other aspects of the cave's history. Below the anthropogenic deposits, which lay up to 3 m deep in places, was a series of natural sand and clay layers, which seem to have formed when the interior of the cave was waterlogged. Indeed, there may even have been standing water inside the cave when human activity began. Excavations within the twin entrance passages also revealed the foundations of successive timber structures, which seem to have controlled access to the cave. These included a structure that Ian Shepherd has interpreted as a timber platform, from which offerings may have been cast into the sodden interior (Shepherd 2007, 195).

Unravelling the Sculptor's Cave

Until recently, it was believed that the human bones from the Sculptor's Cave were exclusively associated with the Late Bronze Age deposits (e.g., Brück 1995, 260; Shepherd 2007). A recent AMS-dating programme of the human and animal bone from the site has, however, revealed a much more complex and intriguing history (Armit and Schulting 2007; Armit, Schulting, et al. 2011).

Essentially the human remains appear to fall into two chronological groups. The bulk of the anthropogenic deposits seem to have been deposited between around 1100 and 800

BC, and two of the juvenile mandibles seem to be associated with this phase of activity (they each gave identical AMS determinations of 1120–910 BC). During the succeeding Iron Age, activity in the cave seems to have been much more sporadic, and we have no human remains dating to this period. Instead, the second group, which comprises the cervical vertebrae from the decapitated individuals, forms a very distinct chronological group centred on the third century AD. Given their clustered dates and their marked similarities in treatment, they may even represent a single violent event during the Roman Iron Age.

The human bones from the Sculptor's Cave thus represent two quite distinct periods of activity. In the Late Bronze Age, the heads of children were apparently displayed at the entrance to the cave during a period when it was much frequented. This may well be the same period that saw the bulk of the deposition of children's bones in the cave's interior and the construction of elaborate timber structures at the entrance. Many centuries later, long after regular activity in the cave had ceased, a second phase is represented by the remains of several decapitated individuals, all but one of whom were adults. Although there is no evidence for a significant accumulation of anthropogenic deposits, periodic visitations to the cave seem to have continued, resulting in a series of votive depositions of Roman and indigenous metalwork, including a small hoard of fourth-century Roman coins.

Neither set of bones from the Sculptor's Cave seems to represent the product of headhunting. Set-piece battles might be expected to produce a disproportionate number of adult males, while small-scale raids would produce a somewhat wider cross section of the community. The dominance of children in the Late Bronze Age deposits at the Sculptor's Cave, however, suggests that another mechanism was involved. Conceivably the cave was the focus of child sacrifice, with subsequent dismemberment, deposition, and display of the heads at the cave entrance. Alternatively, it may have been a place where secondary funerary rites, restricted to juveniles, were practiced.

Neither are the later decapitees suggestive of headhunting. As we have seen, the victims were decapitated either within the cave or close by. This may have taken the form of ritualised sacrifice, for example, of war prisoners, slaves, or social outcasts, or it may simply represent the more prosaic execution of transgressors. Although we have no evidence for the curation or display of these individuals, the broadly contemporary use of the cave for the votive deposition of coins and other valuable objects suggests that the former interpretation is more likely. The Sculptor's Cave was, after all, never an easy place to reach, and almost any activity carried out there could have been achieved more conveniently somewhere else. It is thus hard to escape the conclusion that this was a place of religious importance, stretching over more than a millennium, during which the nature and frequency of the rituals practiced in the cave changed markedly. Yet we have two extraordinary assemblages of material showing a special interest in the human head from widely separate points in this long sequence.

The mouth of Hell?

Whatever the precise nature of the rituals practiced at the Sculptor's Cave, it seems to have been a place where religious practice was formalised and controlled. The timber structures seen in the Late Bronze Age deposits suggest the restriction of access and the choreographing of movement into and around the cave. The apparent separation of the head for decoration, curation, and display, while other body parts were deposited in the watery interior, again suggests adherence to certain conventions of religious practice. The apparent maintenance (or perhaps reinvention) of the cave's religious importance and the recurrent importance of the head over many centuries hint at some form of conceptual continuity from the Late Bronze Age to the Roman Iron Age. So what was the purpose of this special cave? Patently this question can never be answered in any detail, but we can perhaps go some way towards understanding why this place, and not others, was considered appropriate for rites involving the appropriation, curation, and display of human heads, and particularly those of children.

The physical form of the Sculptor's Cave marks it out as a meeting point between the natural and supernatural worlds. During prehistory, the land on which it lies formed a promontory, separated from the mainland by the Loch of Spynie, which, although now no more than a shrunken vestige, was at that time a substantial sea loch. Cut off at high tide, even from this promontory, the Sculptor's Cave straddles the boundary between land and water, set apart from the day-to-day landscape of farms and fields. The cave itself forms a physical link between the world above ground and the world below. On many levels, then, the Sculptor's Cave is a liminal place, lying 'betwixt and between' (Leach 1977), belonging neither to this world nor to the next. The deposits within the cave echo this theme. The fleshed heads hung at its entrances embody a state of transition, caught between life and death, while the disproportionate number of children shows a concern with those yet to enter fully the social world of adults. The unusual form and distinctive location of the Sculptor's Cave made it a highly appropriate setting for religious practices involving mediation between the world of the living and the world of the dead.

There are several other British caves where human remains have been found in association with Late Bronze Age metalwork (e.g., Philips 1931; Britton and Longworth 1969), but the closest parallels for the activities at the Sculptor's Cave come from continental Europe. Perhaps the most important is the Trou de Han, in the Belgian Ardennes, through which flows the river Lesse, a tributary of the Meuse (Warmenbol 1996). Like the Sculptor's Cave, this extensive limestone cave system had a long history of human activity, with notable concentrations of artefact deposition in the Late Bronze and Iron Ages (Warmenbol 2007). Large quantities of prehistoric material have been recovered over several decades, many of them found by divers on the bed of the river Lesse itself.

During the Late Bronze Age, the underground river seems to have been a focus for the deposition of fine metalwork and other objects, including human remains. The bones of four children and an adolescent have been found intermingled with potsherds and food remains in the section known as La Galerie de la Grande Fontaine. An unstratified human cranium, discovered in La Galerie Belgo-Romaine, was also found to date to the Late Bronze Age or Early Iron Age (Warmenbol 2007). But human remains have not survived in any quantity from the riverbed deposits where most of the metalwork has been found.

In the mid-1960s, however, excavations by Marc Mariën, in a part of the cave, known as La Galerie des Petites Fontaines, recovered seven human mandibles clustered tightly together in the Late Iron Age levels (Mariën 1970, 1975). The bones lay on a bed of naturally deposited cave sediment, directly under a hearth. Indeed, one of the mandibles had been burnt by the fire, and it may well be that the deposition of these bones formed a foundation deposit for the overlying structure. Distinctive cut marks on the ascending rami of three of the mandibles (Mariën 1975, 256–9) show that these individuals had been decapitated from behind by an assailant armed with a sharp, heavy blade. Yet, radiocarbon dates from all seven mandibles range quite widely, from before 200 BC to the first century AD (Warmenbol 2007), and it is unlikely that they derive from a single event. Rather, it appears that they represent fragments from curated heads, accumulated and displayed in the cave over many generations, which were finally gathered together for deposition at some time during the first century AD.

In the preceding chapter, I discussed the cosmological links that appear to have existed in Iron Age Europe between human heads and fertility. These links are less obviously in evidence at the Sculptor's Cave and the Trou de Han. What these extraordinary cave sites do highlight, however, is the role of severed heads as liminal objects, caught between the mundane world of the living and the otherworld of the dead. Eugène Warmenbol (1996) has memorably described the Trou de Han cave system as 'La Bouche des Enfers' or 'the mouth of Hell'. By this, I take him to mean that it was, for the prehistoric populations of the region, conceived as a transitional zone, where the living could transact with the dead. As at the Sculptor's Cave, rites focussing on the human head, and incorporating both decapitation and display, seem to have been recurrent, if not constant, over prolonged periods. In both cases, the disembodied human head, still fleshed and 'frozen' at the point of death, gave material expression to the fragile boundary between this life and the next.

Porticos, pillars, and severed heads: Roquepertuse and beyond

In Chapter 4 I wrote about the Iron Age iconography of southern France with little reference to the locations in which it was displayed. As the Sculptor's Cave and the Trou de Han starkly remind us, however, the specific location of ritual practice can be crucial. In returning to southern France then, it is important to consider places as well as objects. While lacking the drama and immediacy of the north European caves, many places within the landscapes of southern France seem to have been accorded a special status during the Iron Age. As we shall see, activity at sites such as Roquepertuse and Glanon suggests the emergence of increasingly formalised religious beliefs and the choreography of ritual practice within what were essentially natural locations elaborated and lent emphasis by built structures of various kinds. Here the discussion moves from the realm of broad-scale cosmology to the more specific and localised religious beliefs and practices through which Iron Age communities engaged with the supernatural.

In the early part of the Iron Age, perhaps from the eighth century BC, sanctuaries seem to have been small, dispersed, and lacking formal definition (Garcia 2004, 109). Many were located on hilltops, on the margins of the cultivated lands where people would have lived their day-to-day lives. The earliest population centres, like the coastal oppida of Tamaris and L'Arquet, which flourished during the sixth century BC, have no clear evidence for sanctuaries or religious areas (Chausserie-Laprée, 2005). Yet areas of votive deposition are

ILLUSTRATION 5.5. Warrior bust from Saint-Anastasie. (Photograph by author)

known, in areas away from these centres of population, and numerous statues and sculptural fragments have been found in locations with no apparent trace of settlement.

This is the case, for example, with the two warrior busts from Saint-Anastasie, Gard, dated, stylistically to before 550 BC (Py 1990, 816–18; Arcelin and Rapin 2003, 186). These two, near-identical, busts were found in the 1920s, buried side by side in a field around 12 km north of Nîmes (Ill. 5.5). Follow-up excavations by the landowner apparently found the remains of a few walls, but, typically, there is nothing to suggest the presence of a formal sanctuary on the site. In the centuries that followed, however, this pattern was to change quite radically.

Undoubtedly the best archaeological evidence we currently have for the development of a Provençal sanctuary comes from Roquepertuse, near the village of Velaux, Bouches-du-Rhône (Boissinot 2004; Lescure 2004). For the remainder of this chapter, I will use the archaeological sequence from this site to structure a wider discussion of religious practices in the southern French Iron Age. As before, the human head has a prominent part to play.

Discovering Roquepertuse

The reconstructed portico from Roquepertuse in the Musée de Vieille Charité in Marseille is formed from a series of stone pillars, joined by lintels, each bearing niches carved to hold human skulls (Ill. 5.6). In front of the portico sit two cross-legged male figures in elaborately sculpted clothing (Ill. 4.3). Reassembled as a group, they form one of the most visually striking creations of the European Iron Age.

The American poet Gustaf Sobin's comments (1999, 122) on seeing the portico, quoted earlier, convey something of the power of these sculptures and underline the role that preserved human remains can play in giving material expression to the boundary between the worlds of the living and the dead. Such is the visual impact of the Roquepertuse statuary, however, that the extraordinary location of the site itself is often neglected. The sanctuary, which comprises a rising series of walled terraces, is set into the gentle southeast slope of an otherwise sheer-sided plateau overlooking the valley of the river Arc. Although the white limestone cliffs are nowadays concealed by thick tree cover, the plateau would have been a prominent landmark during the Iron Age, rising 20 m or so above the valley floor (Ill. 5.7). The sanctuary itself, however, would always have been invisible from the valley.

The first statues were found during agricultural operations in the mid-nineteenth century, but it was Henri de Gérin-Ricard's excavations from 1919 to 1927 that marked the first archaeological investigation of the site (Gérin-Ricard 1927, 1929). Although standards were fairly good for the time, these excavations lacked close stratigraphic control and left many uncertainties over the precise contexts from which the sculptural fragments derived. Thus, although successive museum displays and graphic depictions of the sanctuary at Roquepertuse have tended to combine the sculptural material into a single, monumental ensemble (e.g., Benoît 1955b, fig. 4; Lescure 1990, 167), there is no stratigraphic evidence to support this arrangement. In fact, recent stylistic analysis suggests that the creation of the various sculptural elements spans almost two centuries, even if they may eventually have been displayed together.

In recent years, our understanding of the archaeology of Roquepertuse has been transformed by a renewed programme of excavations led by Philippe Boissinot. As now understood, the site can be divided into four main areas (Ill. 5.8): firstly, the naturally fortified summit, which bears surface traces of close-spaced, rectangular house foundations; secondly, the 'sanctuary-area' (Terrace 1), from which much of the statuary has been recovered, and which is fronted by a sizable stone rampart with two round towers; thirdly, an extramural zone of densely arranged buildings (Terrace 2); and, finally, the 'Plateau des Amandiers', connected to the main plateau by a narrow ridge of rock. This whole area covers rather less than a hectare, although a series of small trial trenches has identified

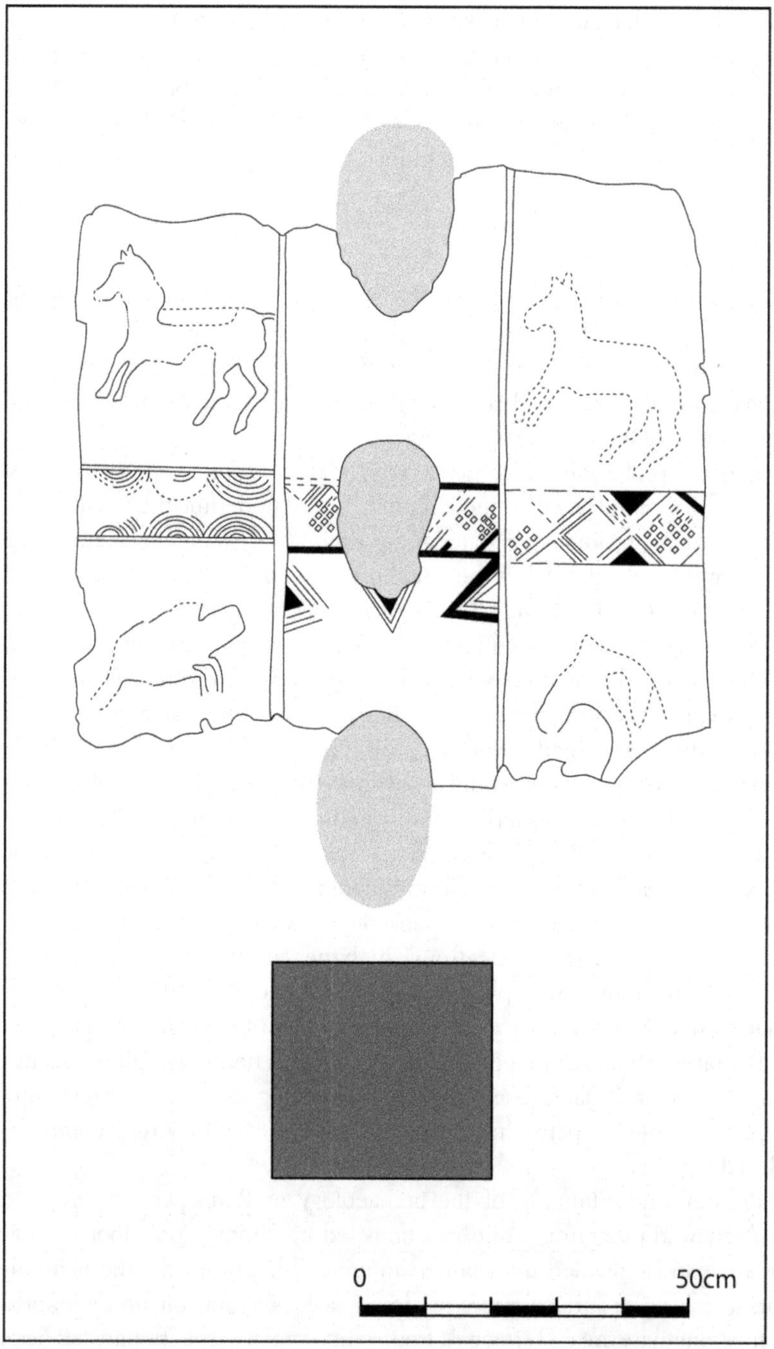

ILLUSTRATION 5.6. Simplified drawing indicating the broad decorative scheme of Pillar C from Roquepertuse. The central face contains three superimposed skull niches, while the two side faces are divided into registers containing painted images of horses and a carrion bird. Much of the original decoration is, of course, lost or obscured. (Drawn by Dan Bashford after Barbet 1991, 73, fig. 17)

ILLUSTRATION 5.7. Photograph of the limestone peak of Saint-Propice gives some impression (albeit on a slightly grander scale) of what the neighbouring peak of Roquepertuse may have looked like from the Vallée de l'Arc, before it became obscured by modern afforestation. (Photograph by author)

archaeological deposits extending further to the south and west of the extramural zone (Boissinot and Lescure 1998, fig. 2). Boissinot's excavations have focussed on the sanctuary and the extramural settlement, and little is yet known of the character or development of the structures on the two plateaus.

Recent excavators have been at pains to distance themselves from Gérin-Ricard's view of Roquepertuse as a primarily religious centre populated only, if at all, by priests and their families (e.g., Boissinot 2003, 240). Indeed, there is little doubt that from its initial fortification at the end of the fourth century BC, a settlement developed that was more than simply an appendage to a pre-existing sanctuary. But there is a danger, nonetheless, of underplaying the religious importance of Roquepertuse. As Boissinot (2004, 50) notes, the site is far from convenient for settlement, being highly exposed to both wind and sun, and with awkward slopes, vulnerable to erosion by rainwater. Neither do its fragmented surfaces and limited areas of level ground lend themselves to fortification or house construction. These factors, combined with the wealth of statuary, suggest that it was most likely the religious significance of Roquepertuse that led to its creation, its subsequent growth, and perhaps to its final destruction. Nonetheless, it is now clear that, rather than the isolated sanctuary envisaged by Gérin-Ricard (1927, 39), Roquepertuse was a fortified centre associated with a substantial range of domestic buildings (Boissinot 2004).

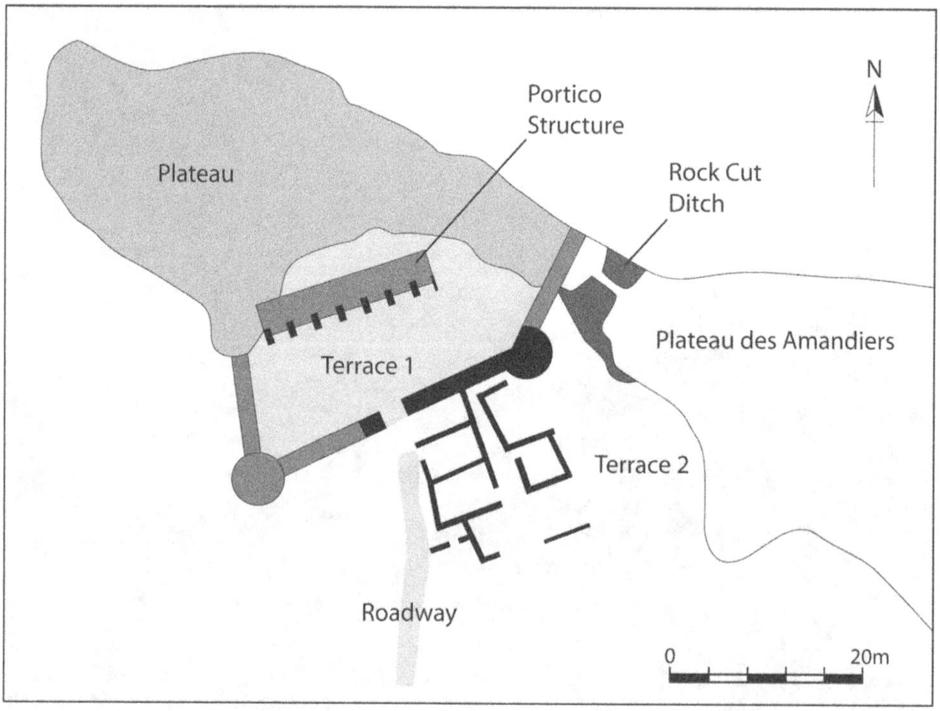

ILLUSTRATION 5.8. Roquepertuse, showing the main areas of the site. (Drawn by Dan Bashford after Boissinot 2004)

Boissinot (2004, 59) has defined 14 successive 'periods' of occupation, reflecting the immensely complicated stratigraphy of the extramural settlement in particular, but for present purposes it is convenient to group these into three main phases; the early sanctuary; the monumental sanctuary; and the destruction and subsequent reuse of the site.

The early sanctuary

As is often the case in the southern French Iron Age, the earliest activity at Roquepertuse is hard to trace archaeologically. The oldest objects are a series of around 30 small stone pillars, or stelae, which had mostly been reused in structures built during the 'monumental' phase of the sanctuary, during the third century BC (Lescure 2004, 46). These simple carved stones probably date to the Bronze Age (e.g., Gantès 1990, 164, nos. 10–11) and suggest a long history of ritual activity on or around the plateau, prior to the physical formalisation of the hilltop.

Stelae such as these appear to be a characteristic feature of early sanctuaries in the region, perhaps straddling the end of the Bronze Age and the earliest part of the Iron Age. They are almost always found in secondary contexts, usually incorporated in the walls of later ramparts or houses. Although they are often rather plain, some bear carvings of horses and riders, as at Mouriès (see Chapter 4), or occasionally other animals or objects. These stelae were extremely varied in shape, size, and quality of workmanship, suggesting

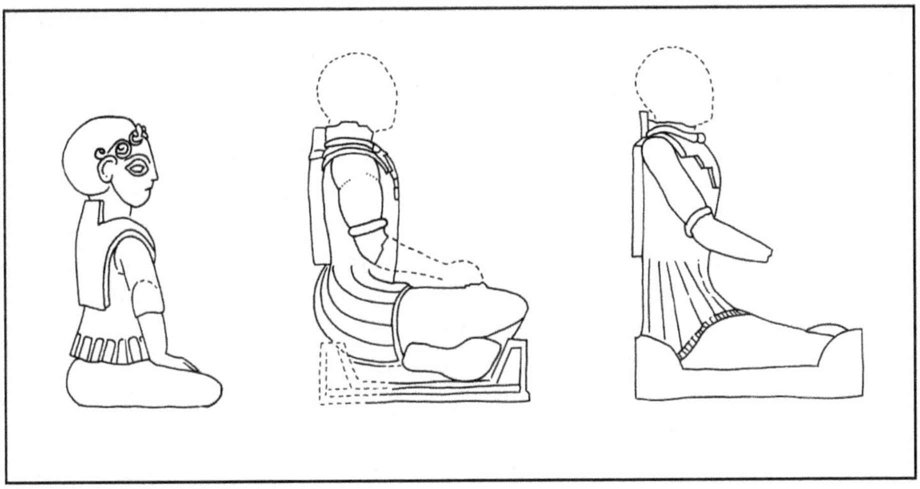

ILLUSTRATION 5.9. The tiny human figure, perched on the rim of a mid-fifth-century flagon from a wealthy grave at Glauberg, southern Germany (*left*), is remarkably similar in both pose and costume to the near life-size seated warrior statues from Glanon (*middle*) and Roquepertuse (*right*). Not to scale. (Drawn by Dan Bashford after Rapin 2004, fig. 7)

small-scale local manufacture. Soft sandstones or limestones were usually selected, which could be easily carved into shape, but they were often brought significant distances, of up to around 20 km, from their quarry site (Garcia 2004, 105).

There are some hints that such stelae may have represented or commemorated individuals, and the occasional shield motifs suggest that these individuals may, at least sometimes, have been conceived of as warriors. Most explicit in this context is the example from the oppidum of Sextantio, Hérault (Garcia 2004, 109), which bears images of a spear, sword, and shield. Elsewhere, the association of stelae with warriors is even more overt, as, for example, in the Iberian Bronze Age (R. Harrison 2004). Although there are no such militaristic motifs on the Roquepertuse stelae, it remains possible that the stelae prefigure the seated warrior statues of the Early Iron Age. As such, they may well have belonged to an early phase of the sanctuary, long predating any of the surviving structural evidence.

Recent stylistic analysis of the seated warrior statues themselves now suggests that they too were created some considerable time before the construction of the terraced sanctuary (Arcelin and Rapin 2003). Key to this reinterpretation is their unusual clothing. Each of the warriors wears a short tunic, over which is drawn a distinctive cuirass with a stiffened backplate, surmounted by a square neck guard. André Rapin (2004) has identified this cuirass as a type of body armour characteristic of both northern Italy and parts of temperate Europe during the fifth century BC. It is paralleled, for example, on another squatting figure – a little warrior perched on the rim of a mid-fifth-century bronze flagon from the Glauberg tomb in southern Germany (Ill. 5.9). Both cuirass and tunic at Roquepertuse are decorated with patterns formed of lightly incised squares, which would have been brightly painted (Barbet 1991, 1992), and these geometric designs would also seem to support the early dating. Taking all the stylistic evidence into account, Arcelin and Rapin (2003, 206) propose a date during the fifth century BC for the creation of these statues.

If we judge from the surviving fragments, there may originally have been as many as 10 of these seated figures at Roquepertuse (Arcelin and Rapin 2003, 185), but only 2 are sufficiently well preserved to enable a reasonably detailed description of how they must have appeared. Both are a little under life-size and represent male figures seated cross-legged on a square plinth. One of the plinths has upturned corners, as if representing a cushioned seat. The figures are slim and sinuously carved, with a high degree of technical skill (Ill. 4.3). The left arm of the better-preserved figure is bent, with the hand clutched against the left side of the chest. An armlet is visible just above the elbow. Although most of the right arm is missing, it seems to have rested on the right knee, where the remains of a hand hold the stub of an iron object of some kind (Arcelin and Rapin 2003, 204, fig. 28a). What survives of the second figure is almost identical, although in this case the left arm is lowered (Lescure 1990, no. 3). Again, he wears an armlet above the left elbow, and this time enough survives of the neck to show the remains of a torc. Close examination of the plinth shows the lightly incised heads of two horses, one in profile and one frontal view, which would again have been highlighted using paint (Barbet 1987, figs. 7 and 8). Analysis of the stone-working techniques employed on these statues suggests that they were carved by at least two different sculptors, both well schooled in the use of specialist tools (Bessac 1991, 50).

To the modern eye the seated figures appear to radiate calm and control, and the pose has sometimes been described as 'buddhic' (e.g., Benoît 1955a, 42). It is probably this peaceful air that has led to their frequent interpretation as 'indigenous divinities or priests, draped in sacred vestments' (Déchelette 1914, 1533). Although this idea has had a long currency (e.g., Gilles 1873; Benoît 1955a, 42–3), there is now little doubt that the sculptures were intended to portray warriors. As well as their protective clothing, three sculpted scabbard fragments recovered from the recent excavations suggest that at least some of the statues carried sheathed swords (Boissinot 2004, 58). The roles of warrior and priest, or warrior and divinity, are not of course mutually exclusive, but we must at least allow that the Roquepertuse imagery had an explicitly martial dimension.

The Roquepertuse warrior statues are by no means unique. A further series of at least three male figures, in the same pose and similar costume, comes from the sanctuary of Glanon, some 45 km to the northwest, in the range of low hills known as the Alpilles. The best preserved of the Glanon series is strikingly similar to the statues from Roquepertuse. Life-size, he sits in the same cross-legged position. Over a tunic he wears the same distinctive cuirass, once brightly painted with squares and chevrons, and like the Roquepertuse figures, he wears an armlet on his upper left arm. As with all the others, he has long since lost his head, though a torc is again visible around his neck. Interestingly, the short tunic of the Glanon figure, which stops midthigh, is laid open to reveal his genitals. This is an apparently unique instance of phallic depiction on a human figure from the southern French Iron Age and seems to provide a conceptual bridge between these serene warriors and the more overtly fertility-conscious head pillars discussed in Chapter 4.

Although there is little superficially to distinguish them from the Roquepertuse examples, Arcelin and Rapin regard the distinctive plinths of the Glanon figures, with pronounced upturned corners, as a direct borrowing from Mediterranean traditions (Arcelin and Rapin 2003, 205; Rapin 2004, fig. 7). Accordingly, they suggest a slightly earlier date of

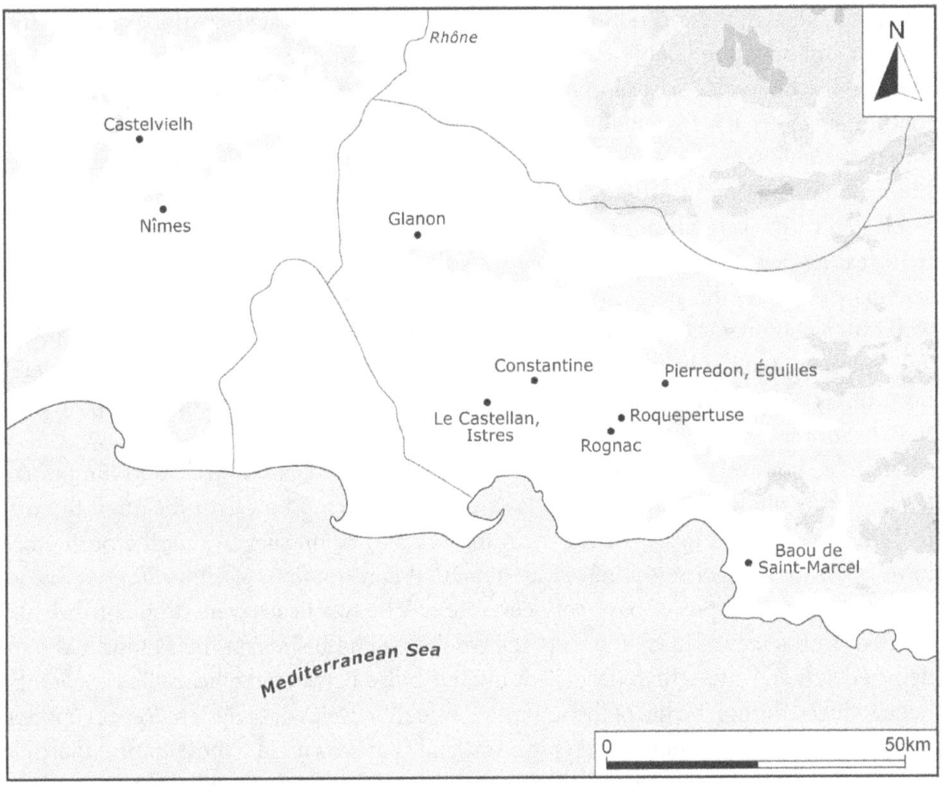

ILLUSTRATION 5.10. Map showing sites in southern France with warrior statues in the Roquepertuse/Glanon style. (Drawn by Rachael Kershaw)

around 500 BC, despite the dearth of archaeological evidence for activity of this period at Glanon. Because the site's later development was based around its healing waters, it seems probable that the warrior statues were displayed somewhere close to the spring, long before the settlement grew up around it. A series of rock terraces and artificially enhanced rock shelters overlooking the spring seems the most likely location for this early sanctuary. The statues themselves, however, were in use for many centuries thereafter and were probably moved several times. One was discovered in the foundations of a building constructed around 125 BC, while another seems to have been deposited in its final location as late as the mid-first century BC (Roth Congès 2004, 38). At both Glanon and Roquepertuse then, these statues had a lengthy use-life spanning several centuries, and the date of final deposition would appear to have little or no bearing on the date of their creation.

Fragments of similar statues, with the same distinctive costume have also been found elsewhere in southern France (Ill. 5.10). These sites are mostly in Provence and were generally to become fortified later in the Iron Age, as at Le Castellan, Istres (Marty 2000; Armit et al. 2008), and Pierredon (Arcelin and Rapin 2003, 205). The strong similarities within this group suggest an increasingly homogeneous iconographic repertoire not seen in the earlier sculptured stones. They demonstrate a degree of formalisation in religious imagery, which might be taken to imply a parallel formalisation in religious belief and practice.

Despite all this, there is nothing in the imagery of these seated warrior statues to suggest any link with the 'Celtic cult of the head'. Even so, the long-presumed association between the Roquepertuse examples and the portico, with its embedded skulls, has meant that these statues have generally been regarded as part of the same cultural phenomenon (e.g., Arcelin et al. 1992). This view has been reinforced by consideration of the later statues from Entremont, where seated warriors in similar pose are quite unambiguously associated with severed heads (see Chapter 6). This association need not, however, apply to the earlier period. As we shall see, the recently revised chronology for the Roquepertuse sanctuary separates the creation of warrior statues from the construction of the portico by well over a century. The same chronological disjunction applies at Glanon, where pillars and lintels containing head niches are again much later in date than the statues. Indeed, none of the early warrior figures has any reliable association with severed heads, whether sculpted or real.

There is, however, one early piece from Roquepertuse that does relate to the human head. This is the finely carved stone Janus head discovered by Gérin-Ricard at the rear of Terrace 1 (Lescure 1990, 171, no. 7). A mortise and tenon suggest that the piece once formed part of a larger sculptural or architectural construction, of either stone or wood, although no joining pieces have yet been found. The two heads were originally brightly painted, with traces of black hair on their foreheads and beards on their chins. Eyes are depicted as living rather than dead. A distinctive bulge between the heads had apparently been reduced during the life of the sculpture, by either accident or design. Recent opinion suggests that this protrusion represents a vestigial 'leaf-crown', like those on the Pfalzfeld pillar discussed in Chapter 4, which would suggest a date in the middle or late fifth century BC (e.g., Garcia 2004, 144). Janus heads associated with the leaf-crown symbol are also found, for example, on the Holzgerlingen statue dating to the fifth or fourth century BC (Weber 2002, 318).

The Roquepertuse example is not the only Janus head from southern France. From the early levels of the oppidum of Marduel à Saint-Bonnet-du-Gard, in Languedoc, for example, came a fragmentary 'bicephalic' bust associated with fragments of stelae and pillars. The excavators suggest a date of around 550 BC for the statue (Py and Lebeaupin 1994, 258). Also reminiscent of the Janus head from Roquepertuse is what is usually described as a tricephalic pillar found in 1912 at Beaucaire, near Nîmes (Py 1990, 819–21). The three heads form the top of a four-sided limestone pillar some 77 cm tall. The central face, if such it was, has been erased, seemingly deliberately, leaving a more conventional Janus-like arrangement of two heads with rather benevolent-looking faces. Traces of a mortise on the top suggest that it too may originally have had some form of crown affixed to the tops of the heads, as at Roquepertuse (Py 1990, 821).

Aside from the sculptural evidence, there are few archaeological indications of activity at Roquepertuse during this early period. The earliest securely dated structures from Boissinot's recent excavations are a series of oval, clay-walled buildings built in the middle of the fifth century, which makes them broadly contemporary with the carving of the warrior statues and the Janus head. These structures appear to have been irregularly spaced around the base of the plateau, with no obvious order or planning. Yet even this relatively minimal evidence for settlement seems to vanish altogether in the fourth century BC

ILLUSTRATION 5.11. This rock shelter, cut into the limestone cliffs at Roquepertuse, seems to have been artificially enlarged. It is around 3 m high with a fairly level floor, allowing easy access and standing room for an adult. Directly above it is a small circular niche, similar in size to those carved into the Roquepertuse pillars and lintels. (Photograph by author)

(Boissinot 2003, 238). This makes it hard to judge where the statues might have been displayed.

The Roquepertuse statuary is made of soft local Coudoux limestone that, while easy to carve, is prone to rapid erosion. This being so, the crisp state of statues tells us that they must have been kept sheltered from the elements during their lengthy period of use. This implies the existence both of roofed shrines and of an impressive continuity of reverence and care. It is possible that the statues may have been housed in one or more of the oval buildings found by Boissinot. Alternatively, they may have been set within rock shelters cut into the cliff walls. These probably began as natural erosion hollows that have, in some cases, been artificially enlarged. A particularly striking example is encountered on the south edge of the cliffs in a prominent position that would have overlooked the likely approach to the sanctuary from the valley below (Ill. 5.11). The shelter is fairly shallow, but nonetheless large enough for an adult to stand upright on its flat floor. Cut into the rock above it is a small circular niche, tantalisingly similar in size and shape to the head niches carved in the pillars and lintels of the portico. Is it possible then that the early sanctuary at Roquepertuse was not a formal, built sanctuary but rather a lightly modified 'natural' place, with some special meaning in the landscape? There are some suggestive parallels from Glanon.

The core of the site at Glanon is strung out along a narrow pass through the Alpilles. At one particularly constricted point, just before the valley opens out to the north, is a spring, which was later to become the monumental 'nymphaeum' of the Hellenistic-period town (Verdin 2003, 566). This area appears to have been the focal point for the early sanctuary. The most likely original location for the warrior statues is the hilltop 'rock sanctuary' that overlooks the spring at the narrowest point of the pass. At the summit is a cave, approached by a series of rock-cut steps and artificial terraces, which seems to have formed a focus for ritual activity. At a similar elevation on the other side of the pass is a second small cave, artificially enlarged and with a seat or bench at its furthest recess. Looking out from here one's view is focussed on the summit of the 'rock sanctuary' (Ill. 5.12). As at Roquepertuse, these enhanced natural features formalised and gave emphasis to particular views, pathways, and patterns of movement within the landscape.

Whatever the details of their original siting, it seems likely that the warrior statues at both Roquepertuse and Glanon were originally placed within primarily 'natural' sanctuaries, accompanied at most by a few small buildings. In each case, subsequent settlement grew up around places of primarily religious significance (Garcia 2003, 230). At both sites, warrior statues sculpted according to quite specific iconographic conventions were displayed at special places in the landscape – a healing spring in the case of Glanon, and a prominent geological feature in the case of Roquepertuse. Indeed, this strikingly homogeneous distribution of statues seems to signal a shared set of conventions applied to the veneration of particular locations or the spirits associated with them. As initially conceived, these statues may have been quite unrelated to the monuments discussed in the Chapter 4. Both, however, were subsequently to have dramatic associations with the curation and display of severed heads.

The monumental sanctuary

Some time around 300 BC, more than a century after the creation of the warrior statues, the sanctuary at Roquepertuse went through a radical transformation. This centred on a series of large-scale building projects that created the major structural features that dominate the site today. A substantial rampart, with two round towers, was drawn across the head of the small valley running down from the southeast side of the plateau. This effectively turned the main plateau and sanctuary area into a minuscule oppidum, with the sanctuary occupying nearly half of the usable space (Ill. 5.13). As well as defending the only viable access to the summit, the rampart also formed a retaining wall for Terrace 1, which now became the ritual focus of the site. Around the same time, or perhaps shortly afterwards, the stone portico, with its display of curated heads, was created, forming part of a structure at the rear of Terrace 1 (Boissinot 2004, 59). Outside the rampart, a series of rectangular stone buildings was constructed, arranged into regular *insulae* or 'islands' divided by narrow streets. There is nothing to suggest that this outer settlement was enclosed. What had been an irregular straggle of houses, set around the cliff-bound plateau, seems now to have become a far more formal, even monumental sanctuary, with an associated complex of domestic buildings.

ILLUSTRATION 5.12. The earliest activity at Glanon seems to have focussed on a cave on this prominent peak overlooking a spring within a narrow pass through the Alpilles. This image is taken from a second, artificially enlarged, cave on the opposite side of the pass. (Photograph by author).

ILLUSTRATION 5.13. View from the Plateau des Amandiers showing the monumental terracing at Roquepertuse, occupying a hollow in the hilltop. (Photograph by author)

The spatial distribution of sculptural remains, insofar as it can now be reconstructed, shows a clear concentration on Terrace 1 (Boissinot 2004, fig. 5). As well as being the focus of the early excavations, however, this terrace has suffered badly through illicit later digging and was thus largely devoid of meaningful stratigraphy by the time of the recent excavations. Nonetheless, Boissinot did succeed in identifying a series of postholes, at the rear of Terrace 1, which seem to have supported the pillars of the portico.

The pillars and lintel fragments from Roquepertuse are well known through their frequent reproduction in general works on the archaeology of the Celts. The three best-preserved pillars, usually referred to as A, B, and C, are very similar in size and design. Each is quadrangular, around 32–6 cm in thickness, and each bears between one and three carved niches approximating the shape and size of a human head. Detailed analysis of the stone-working techniques has shown that B and C were carved using the same tools, whereas A was probably manufactured either by a different sculptor or at a different time (Bessac 1991, 50). The same analysis suggested that the best-preserved lintel was not originally associated with any of the preserved pillars.

These elements must represent only a small proportion of the original structure, and despite various attempts to reconstruct them, we have really no idea how they were configured within the original monument. The surviving fragments appear to form parts of perhaps 3–5 pillars and 2 lintels, while the postholes revealed in the recent excavations suggest a structure founded on between 6 and 12 pillars and a corresponding number of lintels (Boissinot and Gantès 2000, fig. 15). The most likely scenario is that most of the structural elements were originally of timber, and this idea gains support from the presence of carbonised wooden structural fragments in both the early and recent excavations. Whether these presumed wooden elements contained similar skull niches is impossible to say.

Various reconstructions of the original portico structure have been suggested (e.g., R. Coignard and Coignard 1991, fig. 11), but the currently most-favoured option envisages a complex, multilevel structure, founded at the rear of Terrace 1, but with its main floor resting on the slightly higher Terrace 5 (Ill. 5.14; Boissinot 2003, 240). This reconstruction gives a ground-plan of 17.5 by 3.5 m (Boissinot 2004, 55). Unfortunately, the floor area of this putative building has been excavated to bare rock by earlier generations of excavators, so nothing is known of its interior arrangements. However, this reconstruction does suggest that the existence of a very impressive double-height façade greeting visitors as they approached Terrace 1 via the entrance through the new rampart.

Traces of paint on the surviving stonework show that the portico building would originally have been brightly coloured (Barbet 1991, 1992). Only a few colours were used (black, white, red, yellow, and, very rarely, green), and these were not apparently mixed but applied in their pure forms over a white undercoat, giving a vivid series of designs. This may suggest a limited familiarity with the use of paint, and a consequent conservatism of technique. Alternatively, there may have existed a series of specific symbolic associations for each colour that would have been compromised by mixing and shading. While most of the pigments could have been obtained fairly locally, the green pigment had a distinctive chemical signature indicating importation from northern Italy (specifically from the area around Verona), despite the availability of lower-quality local alternatives (Delamare and Guineau 1991, 84–5). The survival of paint on some of the pillars suggests either that they were kept under shelter, frequently retouched, or that they were exposed for only a short period before being destroyed.

Animal bodies, human heads

Although dominated by the head niches and abstract geometric designs, the portico decoration also contains naturalistic representations of various animals. To appreciate the formality of these paintings, we can consider the design of pillar C, in which the overall decorative scheme is relatively well preserved (Barbet 1991, 73, fig. 17). On its frontal face, the pillar displays a sequence of three vertically arranged head niches, while its sides show zoomorphic motifs arranged in two registers (Ill. 5.6). The face to the right of the heads (for the observer) depicts a horse moving from right to left in its upper register, and a horse's head seen in profile in the lower; the left face similarly has a horse moving from right to left in its upper register, but the lower register here appears to show a bird, perhaps a raven or a crow, facing the same direction. The registers are differentiated by broad bands of geometric painted designs, placed in such a way as to run through the central head niche; the positions of the heads, therefore, do not coincide with the division of registers. This is inevitably a minimalist reading of the decorative layout; there are numerous

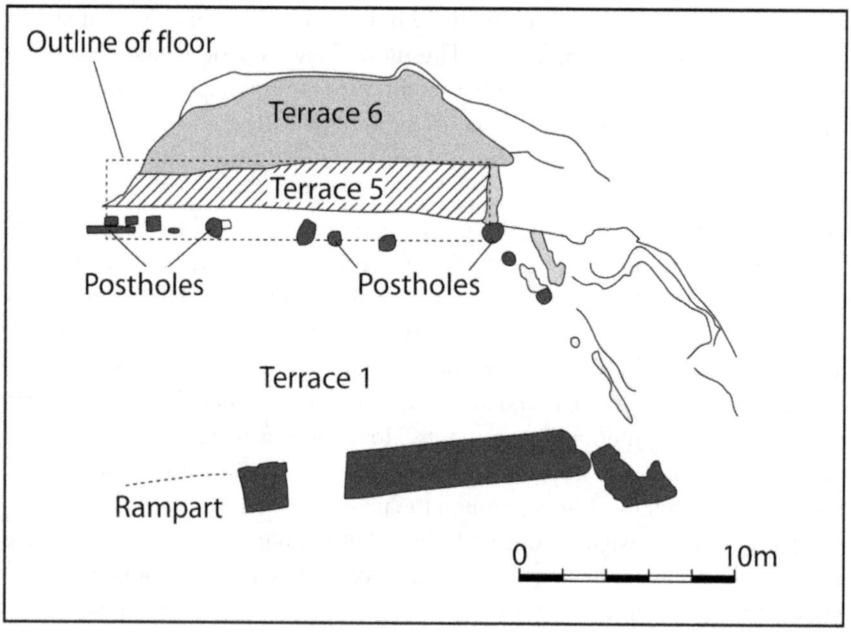

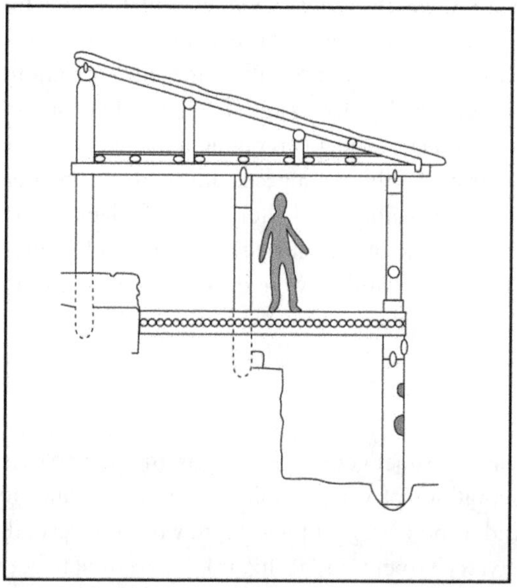

ILLUSTRATION 5.14. This schematic reconstruction of the Roquepertuse portico, drawn by Dan Bashford, is based on the work of Philippe Boissinot and Lucien-François Gantès (2000, 263, fig. 15), which attempts to reconcile the superstructure of the portico building with the ground plan revealed by excavation. According to this scheme the supporting pillars, standing around 2.1 m high (Lescure 2004, 46), would have rested in postholes at the rear of Terrace 1, supporting a series of lintels. These lintels in turn supported the projecting floor of a rectangular building at the level of Terrace 5, a couple of metres higher than Terrace 1. Although it is the current best guess, many questions remain unanswered in this reconstruction. Since Terrace 5 was entirely stripped bare by early excavators, we will never know anything about the floor plan of this presumed building. And it remains unclear what happened in the space between the pillars and the rear of Terrace 1. Was this simply 1 m or so of wasted space, or did it form an access, via a timber stair, to the upper part of the building?

other painted lines, blobs, and smudges that could represent vestigial motifs. However, there does appear to be some internal logic to the scheme as outlined.

The other two pillars, each of which bears a single surviving head niche, are less easy to decipher, but the broad division into registers and the relatively narrow repertoire of carrion bird and horse seem equally to apply. Among all the animal designs there is also a single preserved instance of a human figure, on the upper part of pillar A, lightly incised with its legs wide apart and arms apparently held together, sitting above a large disk (R. Coignard and Coignard 1991, 31). The only detailing is formed by faint horizontal lines that mark the eyes. The pose does not appear to replicate that of the cross-legged warriors and looks, if anything, more like the exhibitionist sheela-na-gigs from medieval Ireland and Scotland (Frietag 2004).

Some further information can be gleaned from the lintels. The largest of the lintel fragments bears three head niches interspersed with painted motifs that appear, in at least one case, to represent sea horses. The underside of the same lintel depicts a long sinuous snake. The second, smaller lintel depicts a frieze of four horse heads seen in profile, looking from right to left. At least one appears to be bridled (Ill. 5.15; Barbet 1991, fig. 22). Horses also appear on the base of one of the warrior figures.

Aside from these painted animals, a three-dimensional stone sculpture of a bird was also recovered from the early excavations (Lescure 1990, 171, no. 6). This creature, standing some 60 cm high, had to be reassembled from around 25 pieces scattered across Terrace 1, and had clearly been deliberately broken up. As well as showing traces of burning itself, the bird statue was apparently associated with the remnants of a burnt oak beam, which suggests that it originally sat atop a timber lintel or pillar. Although there was, for a long time, considerable confusion as to what sort of bird was represented (duck, swan, and vulture being among the preferred candidates), the discovery of a hooked beak fragment in the most recent excavations leaves little room for doubt that this was a bird of prey. It echoes, therefore, the carrion bird motif painted on the pillars. A further recently discovered fragment from Roquepertuse seems to represent the head and neck of a small three-dimensional bridled horse (Boissinot 2004, 57), complementing the painted horse imagery from the pillars and lintels.

The animals represented in the Roquepertuse decoration thus form a relatively restricted group comprising horses (and sea horses), carrion birds, and a snake. The importance of the horse is reinforced by discovery of the complete skeletons of two horses buried close to the portico (Gérin-Ricard 1927, 33–4). Many animals that might have been represented, and which were commonly encountered in the daily lives of the people of the region (e.g., cattle, goats, sheep, pigs, dogs), are ignored. In view of this deliberate selection, it is worth giving some consideration to the sorts of symbolic associations that these chosen animals might have carried.

Horses are the most commonly depicted animal not just at Roquepertuse but in the iconography of southern France as a whole. Painted horses can be seen, for example, on pillars displayed alongside with the Glanon warrior statues (Barbet 1987, 47), and the Sainte-Anastasie warrior bust has a frieze depicting two galloping horses and a bull (Arcelin and Rapin 2003, 196). Mounted warriors are also among the imagery present at Entremont. This association of horses and warriors is hardly surprising, but it does not

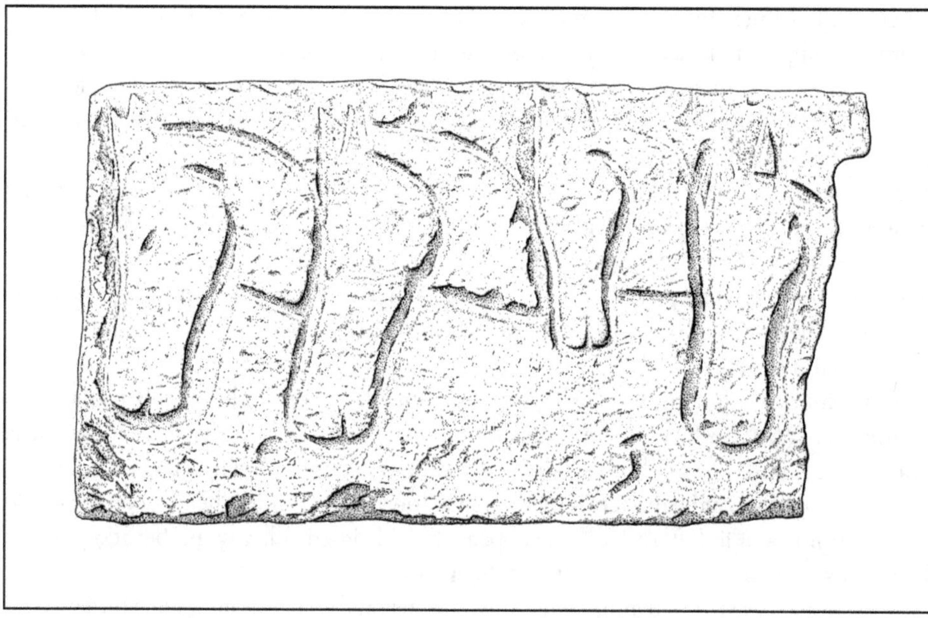

ILLUSTRATION 5.15. Lintel fragment from Roquepertuse depicting four horse's heads. (Drawn by Dan Bashford after R. Coignard and Coignard 1991, 30)

form the only locus for the display of horse imagery. Numerous clay objects known as 'chenets', or fire-dogs, decorated with horse heads have been found on various sites in the Languedoc (e.g., Py 1990, 793–9), while horses, ibex, and deer, are also found on terracotta firebricks in same region (Sintès 1996, 31). Two particular instances of Iron Age horse imagery in the region, from either end of the Iron Age, are worth considering in more detail. These comprise the group of stelae, pillars, and lintels from the oppidum of Mouriès and a decorated lintel from Nages.

Caisses à Mouriès is located on an inland promontory on the south side of the Alpilles (the opposite side from Glanon) and defended on three sides by steep cliffs. The occupation of the oppidum falls into two phases; an initial period in the sixth and fifth centuries BC, and a major reoccupation in the second to first century BC (O. Coignard et al. 1998, 67). The first stelae were found in 1939 by Fernand Benoît, but excavations during the 1980s have revealed many more, including pillar and lintel fragments, bringing the total to around 70, mostly from the later rampart. The stone for the stelae had been transported at least 8–10 km from the nearest source of appropriate fine-grained white limestone at Baux-de-Provence (O. Coignard et al. 1998, 81). Among the images are mounted warriors holding or waving spears above their heads (Ills. 4.5, 4.6), but the dominant motif comprises groups of unaccompanied horses. As discussed in Chapter 4, the creation of these sculptures has been placed variously between the eighth and fifth centuries BC.

The Mouriès stelae form the major single assemblage within what has been identified as an 'Alpilles group', which also includes the early stelae from Glanon (O. Coignard et al. 1998, 81). Similar images, however, are known well beyond the Alpilles, as at Saint-Blaise and Saint-Michel-de-Valbonne. Clearly the image of the horse, individually, collectively, or in association with a mounted warrior, was of central importance in this region from the Late Bronze Age, if not before.

The limestone lintel from Nages, near modern Nîmes, was found in 1901 near a spring below the oppidum of Castels. It is a substantial block, around 0.5 m in length, which has been worked with a high degree of technical sophistication (Py 1990, 821). On its front face is a frieze in bas-relief depicting two disembodied human heads alternating with two horses. The human heads are badly eroded, having lost their noses and much surface detail. Nonetheless, their elaborate hairstyles mark them out as members of the social elite. Their eyes appear to be closed or vacant, and where the mouth survives on one example, it is markedly downturned. They appear, in other words, to be dead. The accompanying horses, which bear traces of red ochre, are depicted realistically, one trotting and the other galloping. As at Roquepertuse, they move from right to left.

Although Arcelin and Rapin (2003, 212) date the Nages lintel to the first century BC on stylistic grounds, the iconographic parallels with Roquepertuse are very striking. Once again we see the close association of decapitated human heads and horses, placed together on a piece of monumental architecture. The image of the horse seems here to be bound up with ideas of death and high status. As at Roquepertuse, the carefully depicted movement of the horses suggests some conception of a journey. Together these images hint at the existence of an established set of religious ideas concerning the fate of the elite after death – ideas that, given the likely chronologies of the various sites discussed, must have developed over several centuries.

The association of human heads with horse imagery is by no means restricted to southern France. Lambrechts (1954, 44), for example, has suggested that the alternate heads and horses at Nages can be conceptually linked with human-headed horses seen on certain Celtic coin issues from further north in France. In Iron Age Britain, the Melsonby mounts, with their opposed human heads, discussed in the previous chapter, appear to have alternated with horse-head mounts as fixings on an iron-bound wooden bucket containing an elite cremation burial (Fitts et al. 1999). This decorative scheme, tying together ideas of death, the elite, the human head, and the horse, has strong resonances with much of what we see in the Iron Age sculpture of southern France (Ill. 4.21).

Indeed, Patrice Arcelin has suggested that the horse was regarded as a psychopomp (Arcelin et al. 1992, 218), or spirit guide, charged with leading the newly dead safely to the afterlife. Cross-culturally, psychopomps very often take animal forms, and these recurrently include horses, dogs, and birds of various kinds. Horses and dogs occupy a special place in many cosmologies because of their close links to the wild and their distinctive relationship to human societies. They tend to occupy a 'social' niche quite distinct from other domesticates, which were used primarily for food, or from fully wild animals. In southern France we see no obvious dog imagery, and birds we will come to later, but the horse was clearly a dominant cosmological symbol.

Another find, closely related to the Roquepertuse portico fragments, may provide some further insight. This unstratified lintel fragment, found in rescue excavations at the Villa Roma in Nîmes in 1992, bears two head-niches flanked by two apparently male human figures carved in low relief and displaying traces of red paint (Barbet 1992; Bessac 1992).

The right hands of these diminutive figures are held out to support the full-size human heads that would presumably have occupied the niches, while their left arms are raised high above their own heads. Combined with their curious, half-sitting posture, these arm gestures give a strong impression that the figures are in movement, perhaps even dancing. Most intriguingly, for present purposes, the opposing face of the lintel bears a further shallow relief showing two rearing horses, apparently in combat. The paired humans have previously been identified as likely psychopomps, and it is tempting to suggest that the paired horses occupy a similar role (Barbet 1992, fig. 2).

After horses, the next most common image at Roquepertuse is the carrion bird, perhaps a crow or raven. Given the paucity of burial evidence for the southern French Iron Age, it seems highly likely that the exposure of bodies was commonly used as a funerary rite, as it was in other parts of Europe (e.g., Carr and Knüsel 1997). Inevitably carrion birds would have haunted the locations where the dead were laid out, and would potentially have played a major role in the transformation of intact bodies into piles of disarticulated bones. This association did not go unrecognised during the Iron Age: for example, a pottery vessel from Numantia in Spain, dating to the first century BC, shows a dead warrior being consumed by birds (Almagro-Gorbea and Lorrio 2004, fig. 4). This iconographic association is expanded by literary references from Silius Italicus (*Punica*, 3.342–8) and Claudius Aelianus (*De Natura Animalium*, 10.22), writing in the late first century and second century AD respectively, who both describe the deliberate exposure of the bodies of high-ranking Celtiberians to carrion birds. Carrion birds also feature prominently in later Irish mythology, where their symbolism is associated primarily with death, war, and the underworld (Green 1986, 187–8), and they are frequently found buried in apparently ritual contexts in Iron Age Britain (Serjeantson and Morris 2011). All in all then, these birds seems a highly appropriate symbol for a monument associated with death and the transition to the next world.

Finally, we have the snake depicted on the underside of one of the Roquepertuse lintels. Again this is not an isolated image. A similarly sinuous serpent can be seen at Entremont on a pillar fragment built into the 'hypostyle' building. Cross-culturally, snakes have a wide range of symbolic associations, but links with healing and regeneration (due to their ability to shed their skin) are probably most common, and were certainly current in the ancient Mediterranean. Both birds and snakes seem also to have been recurrent victims of votive deposition. At the oppidum of Les Castels de Nages, during the first century BC, for example, a hand-built pottery urn containing the body of a grass snake, missing its head and tail, was set into a slab-built box. In another building on the same site, a similar urn, also set into a slab-built cist, contained the bones of a bird of unidentified species. Both deposits seem to have been associated with reflooring episodes in their respective buildings, again perhaps indicating a conceptual link between birds and snakes and renewal or transformation. The buried snake can be paralleled at the nearby oppidum of Ambrussum, in the early first century AD, where another snake was placed in an urn buried in the foundations of a wall (Py 1990, 805). An urn buried under a building of first-century BC date at Ensérune, also in Languedoc, also contained a bird of unidentified species (ibid.).

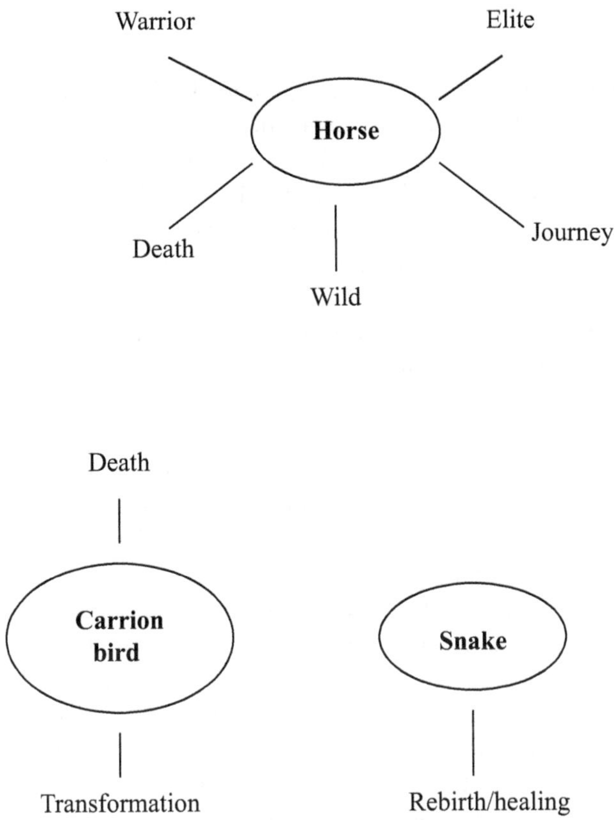

ILLUSTRATION 5.16. Simple diagrams sketching some of the conceptual associations relating to the various animals depicted on the Roquepertuse porticos. The horse, which is the dominant species represented throughout the iconography of the region, has a complex range of associations, of which only the most relevant are indicated here. Elsewhere, its associations with strength and fertility may have been equally (or more) important.

An association with the wild is obvious in the case of the snake and carrion bird, but less so for the horse, which occupies a more ambiguous classificatory position. However, the association of horse imagery with the ibex and deer on the Languedocien firebricks and the paucity of association with other domestic animals suggest that they were conceived as part of the wild. Horses may indeed have been taken directly from wild herds and broken for use as warrior mounts, rather than being bred on the Iron Age farms of the Midi. In any case, the Iron Age iconography of the region seems to focus almost exclusively on wild animals.

Given the likely symbolic associations of the horse, the carrion bird, and the snake (Ill. 5.16), it seems reasonable to suggest that the central theme of the Roquepertuse pillar decoration concerned death and the decay of the body, as part of a journey towards the afterlife. As such, the portico can be regarded, like the northern European caves discussed earlier, as a liminal structure. Passage through this transitional space seems to have symbolised the transformations associated with death.

Collecting skulls

However intriguing the symbolism of the animal decoration, the most striking feature of the Roquepertuse portico was of course the display of human remains. When it was found, lying face-down on Terrace 1, the three niches of Pillar C still contained human crania. These crania were coated and filled with clay that had apparently been used to hold them in place (Gérin-Ricard 1927, 23) in the absence of metal fixings. Burnt cranial fragments were also found in the vicinity of the other pillars (ibid.), and yet more have been recovered during the recent excavations (Boissinot 2003, 240). Clearly the portico had been in active use for the display of human remains at the time of its destruction, and there had been no attempt to remove them when the monument was broken up.

Although significant differences in size and shape between the various head cavities within the portico have been noted before (R. Coignard and Coignard 1991, 28), the implications of this variation have not been fully considered. The cavities within the lintels, for example, appear to have been generally smaller than those in the pillars and could probably have accommodated only crania with the mandible and cervical vertebrae removed (or else perhaps the heads of children). Certain of the pillar cavities, by contrast, were large enough to accommodate complete adult heads, with the upper cervical vertebrae still retained in the neck. Although no mandible fragments were found in association with the pillars, a fragment of an atlas vertebra was found with the cranium in the middle niche of Pillar C (Gérin-Ricard 1927, 23). This does not prove that the niche was occupied by a complete human head, because the atlas can remain attached to the cranium even after the mandible and other cervical vertebrae have fallen away (Christopher Knüsel pers. comm.). Nonetheless, it does show that at least some soft tissue remained attached, and this was not, therefore, some clean and fleshless ancestral skull.

Although a functional explanation for the disparity in the shape and sizes of the various niches has been suggested, based on the idea that the slightly smaller cavities would be less likely to compromise the stability of the lintel, this seems an improbable justification for such a major disparity in rite. In fact, the variation in the head cavities cannot be reduced to a simple contrast between the treatment of pillars and lintels. There was also considerable variation in the size, shape, and depths of cavities within the pillars themselves.

This variation would not be so problematic if the head niches were the late and rather barbarous additions to an existing monument that the original excavator believed them to be (Gérin-Ricard 1927, 40). Yet it is clear that at least some of the niches had been emphasised by incised lines or outlined in black paint or red ochre, showing them to be integral to the design of the monument (R. Coignard and Coignard 1991, 28), at least in its final painted form. There is no reason, therefore, to believe that any of the head cavities were carved after the completion of the portico.

This variation in the size and shape of head niches within a single monument is by no means restricted to Roquepertuse. The lintel fragment from Villa Roma in Nîmes, discussed earlier with reference to its figurative decoration, displays two very shallow head-niches, no more than 5 cm deep. One of these has a horizontal 'bar' cut across its otherwise flat inner surface (Barbet 1992). Clearly these two adjacent niches were designed to receive heads of rather different form, or displayed in rather different ways. These niches were, in other words, bespoke. They were carved not to hold 'generic' severed heads but specific ones.

The variations in design of the Roquepertuse head cavities suggest that they must have been subtly shaped to accommodate the particular requirements of a series of heads that were already available, prior to the creation of the monument. Either these heads had been processed in various different ways, or else some were much older and more decayed than others. In other words, the portico was not built to accommodate a fresh crop of heads obtained from a single battle, raid, or massacre, nor was it built in the hope or expectation of future additions. It was built to contain an existing, and rather mixed, collection, probably accumulated over a significant period of time.

If the Roquepertuse heads were not fresh battle trophies, then to whom did they originally belong? The problem may now be irresolvable. During the 1990s, the osteologist Eric Mahieu (1998, 65) reported that the museum box labelled as containing the cranial remains from Roquepertuse appeared instead to contain unrelated skull remains, possibly from the Languedocien oppidum of Ensérune. Rather depressingly, it appears that the crania from the original Roquepertuse excavations have now been lost. Our information on them, therefore, is restricted to the original examinations carried out after the early excavations. An initial analysis of the crania from the pillars, carried out in 1923 by Dr Berre, a medical practitioner from Marseille, suggested ages for three individuals of 30–45 (Gérin-Ricard 1927, 23). Subsequent examination in the early 1950s, by Robert Charles, revised the age of the youngest individual to around 25 and suggested that it might be female. That is pretty much all we have to go on.

There seems no way to tell whether these heads belonged to enemies or ancestors. Their sublimation within a larger decorative scheme might suggest a lack of respect for the identity of the individual head. However, we have no way of knowing how the clay that presumably coated these skulls was painted. Conceivably it may have re-created the individual facial features of the deceased. The placing and ordering of the heads may have followed a specific order indicating relationships of rank or genealogy. There are certainly parallels for similar displays of ancestral heads. According to early colonial accounts, for example, the lords of the Cocom Maya in the postclassic period underwent similar treatment after death; the heads were removed from their corpses, the backs of their skulls were removed, and their defleshed faces were plastered with stucco to re-create a lifelike appearance. The heads were then displayed and venerated (Serafin and Peraza Lope 2007, 234). Yet whether we choose to see the Roquepertuse heads as those of out-group or in-group, enemy or ancestor, the very fact the certain heads had been retained and curated over significant spans of time suggests that these heads held more value as individual objects than the undifferentiated trophies represented on the earlier Entremont head pillar and elsewhere.

The shock of the new

The monumentalisation of Roquepertuse was a major project carried out on what was already a significant and long-established sanctuary. The stone warriors were already at least five or six generations old, and there is little doubt that Roquepertuse would have been seen as an ancient and spiritually charged place. As we have seen, there are good grounds for believing that the monumental buildings at Roquepertuse formed an essentially religious complex, where formalised beliefs regarding the passage from life to death were given material form in the painted porticos with their embedded human heads. Yet the transformation of the sanctuary might also be seen as a political act, creating a place where religious experience was shaped and choreographed in quite specific ways. The resources involved were considerable: labour and materials were required for the building of the terraces, rampart, and buildings; skilled crafts-workers were needed for the carving, painting and assembly of the portico structure. This was a project that would have placed significant demands on the people of the valley, and perhaps even further afield.

We have to consider also where the various heads came from that were subsequently to be displayed in this newly monumentalised setting? We have already seen that they were probably acquired over a considerable period of time. Conceivably, they may have been displayed or stored at Roquepertuse itself in the earlier phases of the sanctuary's development. Alternatively, they may have been brought to the sanctuary for the assembly of the new monument by the various communities with a stake in the great project. The plastering and painting of the Roquepertuse skulls were intended to subsume them within the design of the porticos, blending previously disparate 'objects' within a unitary design. Yet the potentially lengthy curation and movement of these small numbers of heads suggest that they may have carried more importance as individual objects than the earlier heads depicted on the Entremont head pillar and the *bloc aux épis*. Perhaps they were the heads of venerated ancestors from the various communities, whose incorporation in the monument symbolised the sublimation of each group within a newer, larger social body? Or perhaps they were trophy heads, accumulated by different groups and brought together as a display of communal success?

The third century BC, as we will see in the next chapter, was a period of intense competition between indigenous societies in southern France, during which major territorial polities like the Saluvii were beginning to form. The creation of the monumental sanctuary lies at the opening of that century. Whatever their diverse origins, their assembly at Roquepertuse marked a powerful transformation of existing relics into a greater whole. If the heads did come from a range of different communities, this would have been a powerful symbol of a new cohesion – the enchainment of individual communities under the auspices of a powerful religious centre. Conceivably, then, the Roquepertuse portico structure, and the monumental sanctuary as a whole, may have been created as part of the wider regional process of social reorientation and polity building that characterised the second half of the Iron Age in southern France.

Iconoclasm

It is hard to be sure how long the Roquepertuse sanctuary lasted in its monumental form. Probably, though, it was no longer than a generation or so. For at some point around the middle of the third century BC, the oppidum seems to have gone through a further significant period of transformation (Boissinot 2004, 51). Although there is no obvious sign of burning, abandonment, or structural damage, there is damage of another kind. For the building rubble used to construct the next generation of buildings on the site contains numerous fragments of deliberately broken stone statues and pillars.

Before this, the dominant process at Roquepertuse appears to have been one of accumulation. There is no evidence that earlier statues were destroyed to make way for later ones. Despite careful sieving of the relevant deposits, no sculptural fragments have been found in earlier contexts. Indeed, the early warrior statues were most likely rehoused within or around the portico structure during the monumental phase of the sanctuary's use. Instead, all of the destruction for which we have evidence seems to have happened during the mid-third century BC, possibly during a single episode.

When it came, the destruction was violent and exceptionally thorough. Statues that had been present on the site for generations were broken up and, in some cases, pulverised into tiny fragments. The systematic destruction of these statues suggests a consciously iconoclastic act. Indeed, parts of the floor of the later Building 5, in the extramural settlement, seem to have been levelled using sculptural fragments ground almost into gravel (Boissinot 2004, 57). Even the two near-complete statues have been painstakingly reassembled from numerous fragments. It is surely significant, too, that not a single head survives from any of the statues. These must have been either removed from the site or singled out for complete obliteration. The portico structure too was cast down, the relic skulls and heads trapped under the weight of the fallen pillars.

If we construe a political as well as religious motive for the monumentalisation of the Roquepertuse, then it seems probable that its destruction was also a political as well as religious act. If the old statues, the decorated pillars, and the head relics had all come to symbolise a politico-religious authority with its centre at Roquepertuse, then it should come as no surprise that challenges to that authority might result in the destruction of the sanctuary and all its trappings. We shall explore further in Chapter 6, the regional power struggles that might underlie this event.

The afterlife of Roquepertuse

Any abandonment of the sanctuary was short-lived, and in the succeeding Period 12 (still within the mid-third century BC), the outer settlement was expanded and restructured. The main approach road was realigned and the entrance was further monumentalised by the construction of a formal stone stair. It is impossible to say whether any surviving statues continued to be displayed within the old sanctuary area, though it seems rather unlikely. In fact, there is little in the archaeological remains from this period to suggest any overt religious function for the site. The thoroughly scoured nature of the Terrace 1 deposits, however, means that we are unlikely ever to have a definitive answer on this point. On balance, it is perhaps more likely that the religious role of Roquepertuse had been usurped by this time, and it continued to exist only as a small defended settlement, serving a primarily domestic purpose.

The oppidum seems to have been finally, and violently, destroyed a couple of generations later, some time around 200 BC (Boissinot and Gantès 2000). A series of catapult projectiles (*boulets*) has been found in the final levels of the Period 12 settlement (Boissinot 2004, 56), which suggests a coordinated military assault. Like the statues, many of these boulets are of Coudoux limestone, and some may even have been worked from the broken remains of earlier sculptural fragments. Traditionally it has been thought that this destruction was the work of a Greek or Roman force (e.g., Gérin-Ricard 1927, 41), but there is no particular reason to assume that these weapons were entirely unavailable to indigenous groups. Other sites in the region, like the nearby oppida of La Teste-Nègre and Verduron, seem to have been destroyed around the same time, and we may be seeing the playing out of an intensive episode of internecine strife.

Although Roquepertuse was later reoccupied, it never seems subsequently to have formed more than an unenclosed farmstead. Nonetheless, deposition of a human cranium near the partially buried staircase suggests some residual votive activity during even this late period of the site's use (Boissinot and Gantès 2000, 267). A major religious centre since before the beginning of the Iron Age, it seems not to have survived the disruptions of the third century BC.

Heads and sanctuaries

Roquepertuse is by the far the best understood of the monumental sanctuaries that emerged during the middle centuries of the first millennium BC in southern France, but there were undoubtedly others housing stone monuments of comparable form (Ill. 5.17). Although superficially similar in design and conception, these often display subtle but important variations that suggest a range of practices relating to the curation and display of heads.

In Chapter 4, I mentioned the reused lintel from the Languedocien oppidum of La Ramasse (Garcia 1993, 303). Although I initially discussed its possible links with the simple engraved heads from sites like Saint-Michel-de-Valbonne, it is worth exploring a possible alternative explanation for the carvings on this stone. One face of the lintel bears a very simple engraving in the form of a head-shaped oval, some 30 cm high with a slightly pointed base. Off-centre within it is a small hole, some 3–5 cm in diameter. Although not a niche, in the sense of the others discussed here, it does appear that this was intended to enable the fixing of some object, rather than being a piece of decoration in its own right. It is certainly much more roughly finished than the stone on which it is carved, and thus may not have been intended to be seen. Garcia (1993, 308) suggests that the oval carving may have held some form of metal hoop that would in turn have held a human skull, attached to the lintel by a nail into the off-centre hole. It is hard to see quite how this would work in practice with something so bulky, although it may have been possible to attach a preserved face rather than a skull. It is worth noting that the 'nail hole' would have lain to the left side of the nose and would thus at least have prevented the nail from obscuring the main facial features of eyes, nose, and mouth. The context of this stone, in the body of the later rampart, shows that its primary use had come to an end by around 375 BC (309).

Two further blocks with head-shaped niches come from the oppidum of Cadenet and are now housed in the Musée de Cavaillon (Rolland 1964b, 559; Arcelin et al. 1992, no. 79; Hermary 2003, 525). The chronology of Cadenet remains unclear, although ongoing excavations may eventually put these early finds into some archaeological context (Isoardi 2007). The first piece constitutes the upper part of a pillar bearing two head-shaped niches.

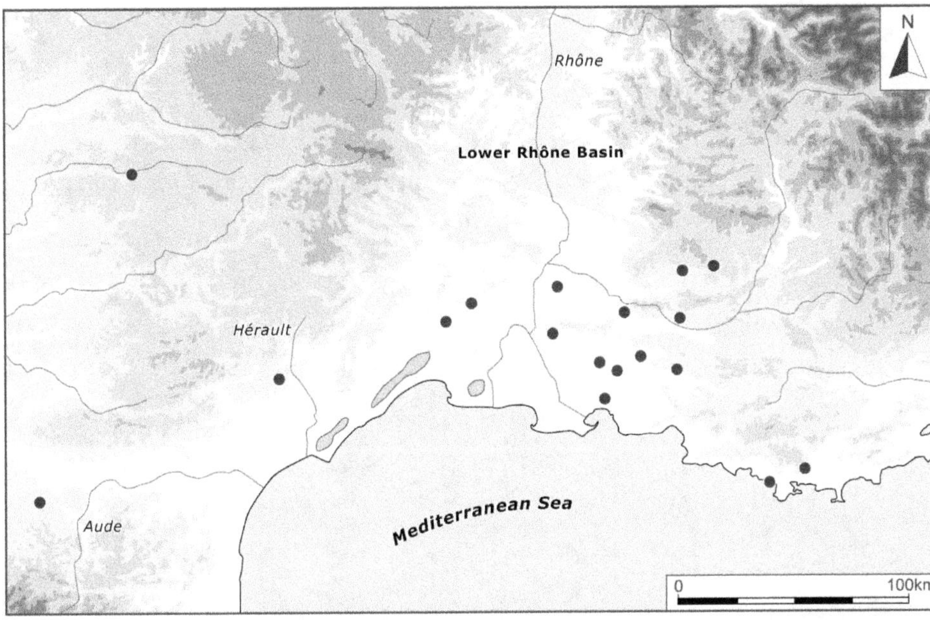

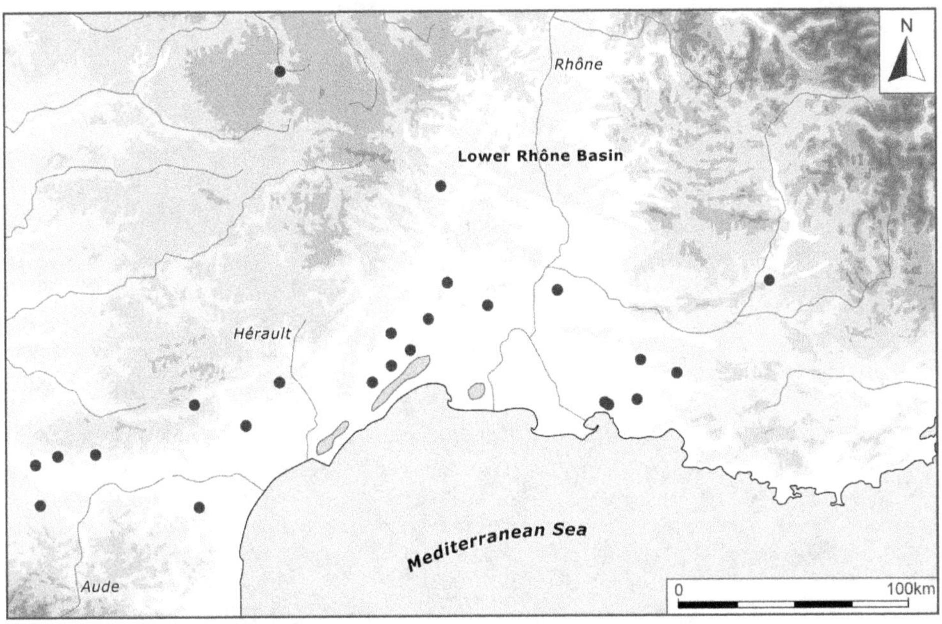

ILLUSTRATION 5.17. Maps showing the distribution of sculptural representations associated with the human head in southern France (*top*) and the occurrence of cranial remains on sites in the same region (*bottom*). (Drawn by Rachael Kershaw)

It has a projecting tenon on its upper surface, which would have secured a stone or timber lintel, demonstrating that it once formed part of a larger construction (Ill. 5.18). The second block is of similar dimensions and differs primarily in having just one head niche and lacking any sign of a tenon. Both are heavily weathered and appear to have stood exposed

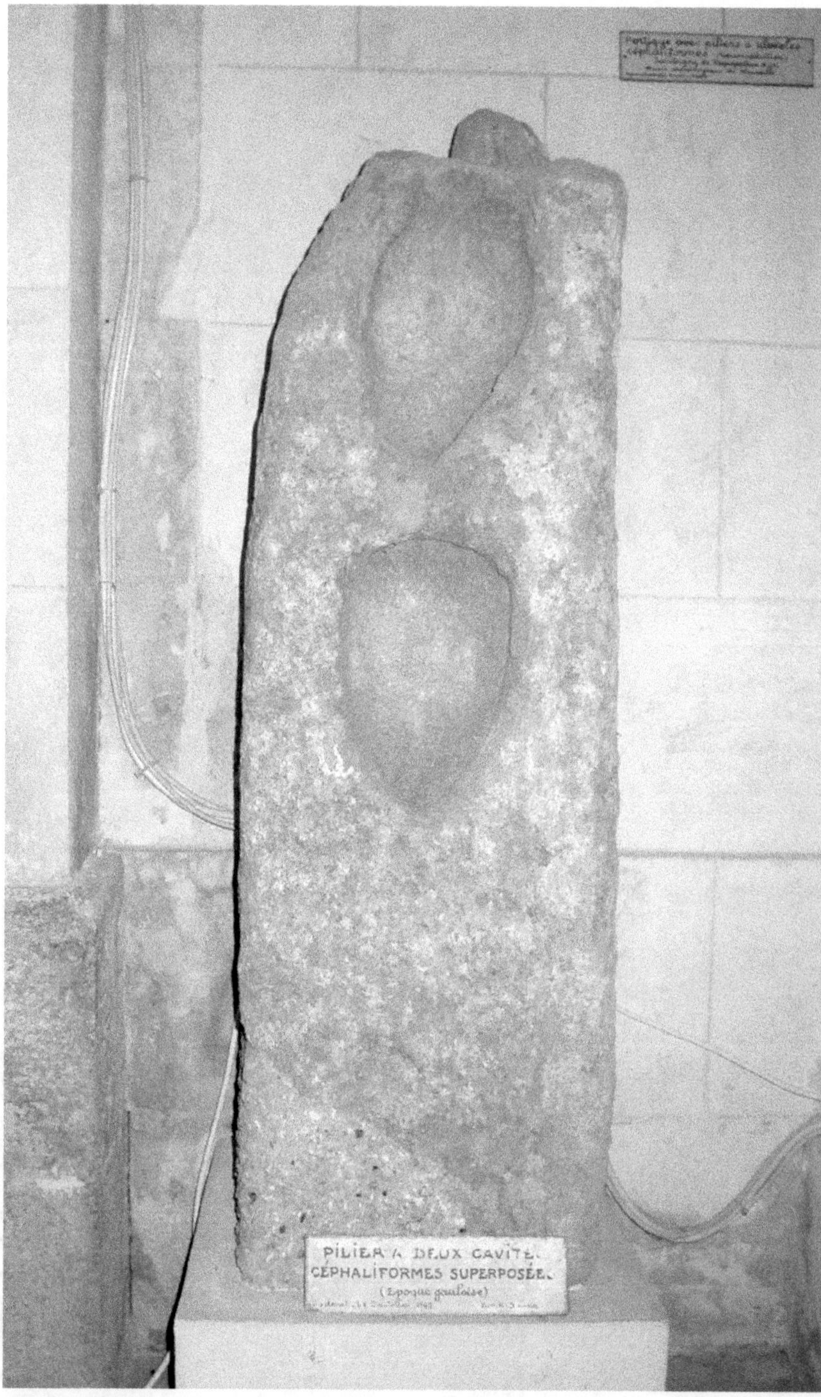

ILLUSTRATION 5.18. Eroded pillar from the oppidum of Cadenet with two shallow, head-shaped niches. (Photograph by author)

ILLUSTRATION 5.19. Detail showing part of a lintel from Glanon, elaborately carved in Hellenistic style and probably dating to the second century BC. It contains four head-niches, each roughly finished, with the distinctive knuckle-shaped boss, and each showing signs of an iron nail hammered downwards to fix some object into the stone, presumably a partial head or preserved face. (Photograph by author)

to the elements for a significant period. Thus there is no indication of the sort of painted decoration or other surface detail that we find at Roquepertuse. At only around 8 cm in maximum depth, the niches impose major restrictions on the size of object that could be displayed. If they contained human remains, then it seems likely that only the facial skin or front of the head could have been involved.

From the much larger assemblage of stones with head niches at Glanon, we see yet another variation on this theme. One extraordinary lintel from the site bears a striking mixture of Greek-inspired vegetal decoration and indigenous head display (Arcelin et al. 1992, fig. 21; Ill. 5.19). In keeping with its fine technical workmanship, the stone's four niches are carefully carved, though not absolutely identical in size and shape. Each has a distinct mound, or 'knuckle' of unworked stone, rising from its base. These occupy so much of the interior of the niche that it would have been impossible to accommodate whole heads or skulls. The remains of iron fixings, in the centre of each 'knuckle', nonetheless show that some object was secured within the niche. The most likely interpretation may be that the niches held preserved or reconstituted human faces, moulded across the 'knuckle' of stone. The recent discovery, from a cemetery in Reims, of an Early La Tène adult male skeleton with its face removed some time after burial, suggests that such practices were not unknown in Iron Age Gaul (Bonnabel and Boulestin 2008). This form of display seems

ILLUSTRATION 5.20. Head-shaped niche, on a pillar from Glanon, displaying the projecting 'knuckle' of stone so distinctive of this site. The pillar is around 1.5 m high and 0.5 m wide. The surface of the stone is smooth and finely worked with a chamfer down each side below the head niche; by contrast, the interior of the niche is rough and pitted. (Photograph by author)

characteristic of the Glanon assemblage in general and may represent a rather localised tradition. Two similar niches can be seen, for example, on a second, less complete and less ornate block from Glanon (Garcia 2004, 141), and on various other pillars from the site, although most examples lack the evidence for metal fixings (Ill. 5.20).

The formalisation of the ritual and iconography surrounding the human head is by no means restricted to southern France. A pottery vessel from Uxama, in central Spain, which probably dates to the second century BC, has painted decoration showing three severed heads in boxes interspersed with birdlike creatures (Ill. 5.21; Almagro-Gorbea and Lorrio 1993, 233–4). This association of heads with bird symbolism in Spain can be paralleled on the vessel from Arcobriga discussed in Chapter 4 (Ill. 4.13), while the apparent display of human

ILLUSTRATION 5.21. Motif, taken from a pottery vessel from Osma (Uxama) in Spain, showing three severed heads in boxes interspersed with birdlike creatures. (Drawn by Rachael Kershaw after Almagro-Gorbea and Lorrio 1993 fig.1, 15)

heads in boxes is reflected in a number of other iconographic and osteological contexts. A fragment of what appears to be a head moulded in relief, contained within some form of box or square framework, comes from Numantia, probably dating to around the first century BC (Taracena 1943, fig. 5), and similar examples from the necropolis of Carratiermes show human heads in boxes or under some form of structure (Saiz Rios 1992, 612). Both of these sites, like Uxama itself, are located in Soria, while Arcobriga lies in neighbouring Zaragoza, but the motif is echoed much further afield by the discovery of a box containing a trepanned human skull at the site of Garvâo, near Ourique in southern Portugal (Baixo Alentejo), which had apparently formed a foundation deposit (Beirâo et al. 1987).

In the absence of detailed chronologies for these sites, both French and Spanish, it is impossible to say how far their development might mirror the sequence at Roquepertuse. What seems clear, however, is that the display of human heads and faces, set within stone lintels and pillars, had become a dominant feature of the southern French sanctuaries. At Roquepertuse and Glanon, and by implication elsewhere, this distinctive practice was increasingly formalised within elaborate architectural settings. Although the inherent symbolism, within the Roquepertuse portico in particular, suggests a primary religious purpose, there is no escaping the important sociopolitical roles that such sanctuaries must also have acquired by the third century BC, if not earlier.

Return to Entremont

Some time after the destruction of the Roquepertuse portico, another monument was created that also incorporated a series of carved pillars. This once again depicts the twin symbols of the horse and the severed human head, though this time the heads are carved rather than real. The monument is important, firstly, for the additional depth of information it can give us on the religious imagery already seen at Roquepertuse and elsewhere; and, secondly, for the issues it raises concerning the relationship between the carved image of the severed head and its real, flesh and blood, equivalent.

An enduring iconography

The three limestone blocks that form the surviving remnants of this structure were discovered by walkers on the Entremont plateau in 1817 and were the first sculptures to be found on the site (indeed, it was these pieces that attracted the damning verdict from Prosper

Mérimée (1835, 141), quoted at the beginning of this chapter). They appear to have been found on the southeastern part of the site, some distance from the subsequent discoveries (Salviat 1993).

The blocks comprise three broken pillar fragments, probably deriving from a portico similar in conception to that at Roquepertuse. The Entremont blocks are much more abraded, and there is little surviving evidence for the painted decoration that would once have made the stones so vivid. Nonetheless, they bear a complex series of motifs carved in relief. Arcelin and Rapin regard these motifs as more technologically advanced renderings of the carved heads seen on the earlier monuments from Entremont (which I discussed in Chapter 4). Unlike their predecessors, they betray considerable Greek influence both in the use of relief and in the much more naturalistic modelling of facial features. Arcelin (2006, 157) dates them stylistically to the second century BC, though the depiction of a horseman on one of the pillars may suggest a third-century BC date (Arcelin and Rapin 2003, 212). In any case, they seem to postdate the construction of the Roquepertuse portico by a century or more.

Given their dimensions, and the close similarities in decorative schemes, it seems highly probable that the fragments derive from a single pillar (Salviat 1993, 219), which would have had an original height of at least 1.43 m. This being so, the marginal variations in their dimensions suggest that Salviat's '*pilastre* 31' was the lowermost, with 32 above it, and 33 on top (Ill. 5.22). I will discuss the decorative scheme based on this suggested rearrangement, referring to the individual stones as 'blocks' within a single pillar.

Like the Roquepertuse pillars, the Entremont one was decorated on three sides, with a fourth, blank face. What we might regard as the front face of the pillar (Ill. 5.22a) was divided into a series of at least four registers, each depicting a figurative scene. From the bottom, as suggested here, the lowermost register depicts a mounted warrior, riding bareback on a stallion. He wears a cuirass and holds a spear in his raised right hand, as if ready to strike. His left hand, meanwhile, holds the reins of his mount. The warrior also carries a long sword, sheathed in its scabbard. His horse, which appears to be trotting, has a round object suspended from its neck. This has been universally identified as a severed human head, recalling the descriptions of Livy, Strabo, and others (Chapter 2).

The register above this is extremely poorly preserved, but enough survives to show that it depicted a galloping horse heading, again, from left to right. Any trace of the rider has gone. Above this, however, is a better preserved image, perhaps depicting the same rider seen in the lowermost panel, or else a closely similar one. The surviving details parallel those of the lower image except that this time his right hand holds the spear much lower, and his mount appears to be rearing up or galloping. There is no trace of the suspended head.

The uppermost register is perhaps the hardest to reconstruct. The clearest part of the image represents the lower part of a standing human figure, facing to the right. He appears to hold a long, straight staff, which, in the context of the lower figures, is probably best interpreted as a spear. In front of him is a floating, disklike object. François Salviat (1993, 219), who provides the best descriptive catalogue of the Entremont assemblage, offers no direct interpretation for this image, preferring to quote what he calls Fernand Benoît's 'audacious explanation' that it represents 'the dead hero, naked, at the gates of Hades' – the mouth of Hell, once again. Whether we accept this or not, it is hard to escape the idea

ILLUSTRATION 5.22. Carved blocks from Entremont rearranged as elements of a single broken pillar. (Drawn by Rachael Kershaw after Salviat 1993, 214–19)

that these figurative scenes represent some form of narrative, and it seems likely that the figures each represent the same person; an elite, male, warrior horseman. This represents a new departure within the art of the region.

Intriguingly, all four of these carved figures are missing their heads. This could be put down simply to chance, as the stones are badly broken in places. However, for the lowermost horseman in particular, the head appears to have been deliberately chipped away from an otherwise more or less undamaged part of the stone. As at Roquepertuse, heads seem deliberately to have been obliterated.

The lateral face, to the left of the central panel (Ill. 5.22b), depicts a set of motifs somewhat more familiar from earlier works at Entremont and elsewhere. Although there is no formal division into registers, the motifs nonetheless comprise a vertical arrangement of highly stylised heads, carved in low relief. The lowermost head (following the reconstruction proposed here) is the most unusual and hardest to 'read'. Although the head is disembodied, its eyes are shown as open and the face appears to be alive. There are indications also of a heavy moustache below a strongly carved nose, and the swept-back, striated hair is clearly shown. Trouble arises with the interpretation of the weathered or defaced features above the head. These are usually interpreted as two projecting horns. If they are horns, however, they are very crudely delineated, and this seems out of keeping with the skilful and realistic rendering of the facial features. They also lean markedly to the left, and their overall shape does not really support this interpretation. A better idea of their form, if not necessarily their meaning, can be gained by examining the motifs that would have sat above them.

The equivalent face on the block above is very badly abraded, but there can be little doubt that the central motif represents another disembodied human head, the features of which have been wholly obscured. However, around this head weaves a broad, sinuous double-band, reminiscent of the swirling, vegetal motifs on contemporary La Tène art. It seems likely that the 'horns' on the lower figure represent the emergence of this vegetal matter from the living human head. The general impression is supported by the topmost block, which shows yet another disembodied head, this time with closed eyes, twice encircled by the vegetal band, which then extends upwards towards the missing summit of the pillar. This distinctive motif may represent plant tendrils (Arcelin and Rapin 2003, 194), though there have been other suggestions. Déchelette (1914, 1536), for example, suggested that it represents a serpent, or even the disarranged clothing of the decapitated individual.

The other lateral face of the pillar seems initially rather simpler (Ill. 5.22c). On the lowermost stone are two human heads, placed side-by-side and joined at the ear. Their eyes are puffy and closed and one of the mouths is markedly downturned; they appear to be dead. The block above depicts three more such heads, arranged vertically, each offset from the one below. The uppermost head may originally have joined with the lowest head on the block above, which again shows three. Taking the scheme overall then, we appear to have two conjoined heads at the base, with at least five above, set in irregular vertical order, as if floating upwards from the base.

On both of these side faces then, we have complex, vertical arrangements of disembodied heads. In both cases, specific relationships between them are suggested by their ordering and the physical links between them: in one case through the swirling vegetal bands that ascend the pillar, and in the other through the physical conjunction of the lower pair of heads. The nature of these relationships is hard to ascertain, though as at Roquepertuse it may involve concepts of genealogy or rank. As with the earlier carvings discussed in Chapter 4, the images reference ideas of fertility and growth.

The heads carved in relief at Entremont were by no means unique. Two limestone lintel fragments from Nîmes, found built into much later buildings in the remains of the Roman amphitheatre, combine to show seven hairless heads with large eyes and prominent noses (Py 1990, 821). Arcelin and Rapin (2003, 213) date these stylistically to the late second or first century BC. Michel Py has also commented on the stylistic links between the bas-relief heads from Nages and those at Entremont (1990, 826). Other heads in relief are probably rather earlier in date, including one built into a rampart at the oppidum of Bringasses, and another from the foundation of a rampart at Castellet à Fontvieille, close to Aix (Benoît 1955a, plate XXIV, 3). They also turn up in Spain, for example, at Puentedueme, La Coruna, and Cordoba (Lenerz-de-Wilde 1993).

Certain iconographic elements clearly link the Entremont pillar to the Roquepertuse portico. The repetitive display of heads, and their vertical arrangement, provides an obvious unifying symbolism, although other elements have changed. The animal imagery, so prominent at Roquepertuse, has slipped into the background at Entremont. The only animal depicted is the horse, and that has now become subordinate to its human rider. The carrion bird and serpent are apparently absent (although we should note the presence of a lintel bearing a relief serpent that was reused in the hypostyle building at Entremont). Most importantly, perhaps, is the relative positioning of the motifs. The central element at Roquepertuse, forming the frontal face of the pillars, was the severed human head. At Entremont, the heads (now carved rather than real) have moved to the lateral faces, making way on the central face for the new figurative human representations, which have no parallel in the earlier art. The elite warrior horseman, whether representing a particular mortal, a god, or a mythic figure, is by now the most prominent motif in the whole composition. This emergence of the individual is perhaps the most significant iconographic difference between the Entremont pillar and its predecessors.

Image and reality

Despite differences of detail, the Entremont pillar seems to derive much of its iconography from the same cosmological background as the portico from Roquepertuse. Yet one rather obvious difference remains unexplored. The heads in the Roquepertuse portico were real, whereas those at Entremont were carved in stone. In one sense, this may not seem too dramatic a change, since the Roquepertuse heads, being apparently plastered and painted, could be regarded as being as much about representation as reality. This, however, underestimates the importance of the materiality of the Roquepertuse heads, as physical objects of flesh, blood, and bone. Depending on their origins, the preparation of these heads may have begun with their violent removal on the battlefield, or with the controlled severing of the head from the corpse of a revered ancestor. Either way, these moments were highly charged and would have imbued the heads with great spiritual and emotional significance. Their subsequent preparation, and their apparently lengthy and varied histories before installation in the Roquepertuse portico, would have given added weight to their role as powerful ritual objects. Their very physicality, their weight, texture, appearance, and smell, would have formed a tangible link with the past. Frozen at the point of death, like the curated heads at the Sculptor's Cave or the Trou de Han, they would have embodied an immediacy of experience hard to replicate through a simple carving.

What then was the relationship between real heads, as displayed at Roquepertuse, Glanon, and elsewhere, and their iconographic representation on the Entremont pillar? It is clearly not a question of chronology, as the carved heads from Entremont, Saint-Michel-de-Valbonne, and elsewhere, discussed in Chapter 4, demonstrate that the representation of heads had a lengthy history, running parallel to the display of real heads. Both seem more or less equally abundant, and we have already seen how the carvings on the head pillar from Entremont actively mimic the niches within which real heads were displayed elsewhere. Furthermore, the plastered and painted heads from the Roquepertuse portico blur the line between image and reality, embedding and subsuming real human heads within a broader iconographic scheme.

Real and carved heads may even occasionally have appeared within the same monument, as in the case of the Entremont lintel (discussed in Chapter 4), where a single carved head occupied the same block as at least four rough and shallow head niches. Marcel Pobé and Jean Roubier (1961, 14) considered that the unusual (and probably erroneously identified) inclusion of a mouth on this head was 'so that it can refer to someone still living ... put there like a dummy egg in the nest to induce the presence of real heads of flesh and bone'. Once a real head had been obtained, they believed, 'the stone one is chiseled out and replaced'. Such practices are not unheard of in the ethnographic record: among the Nagas, for example, fake heads were sometimes displayed in newly constructed religious buildings until real ones became available (Jacobs 1990, 199). But while this hypothesis might (just about) work for this particular stone, head niches at Roquepertuse and elsewhere were clearly integral to the design of the monuments within which they were carved. More widely, we seem to see a fluid relationship between real and carved heads, with both carrying similar meanings, and with the preference for one or the other shifting between different times and places.

Conclusion: Fertility to efficacy

In Chapter 4, I examined the cosmological links between the human head and concepts of fertility and regeneration. This symbolic relationship, recurrent across societies in many different parts of the world, seems to have been particularly deep-rooted among the communities of Iron Age Europe. In southern France in particular, we saw how carved stones, showing carefully constructed groupings of anonymous human heads, were placed in special locations, as symbols of the fertility and fecundity of the small communities that inhabited the region. Nowhere, however, did we see images of recognisable individuals. If these were images of headhunting, then the headhunters themselves were notably absent from the scene.

In the case studies discussed here, links with fertility have been less overt. Instead, sites such as the Sculptor's Cave and the Trou de Han and the monuments from Roquepertuse and Glanon, emphasise a further symbolic dimension of the human head – its role as a symbol of liminality, inhabiting places where the worlds of the living and dead brush together. This does not mean that their deep cosmological associations with fertility had been usurped. Rather it highlights a parallel strand of symbolism that added further layers of meaning to the display and curation of heads. The skull racks, or *tzompantli*, of precolonial Mesoamerica, for example, in addition to their associations with fertility and regeneration, have been described as 'cosmic portals' to the underworld (Mendoza 2007,

418). The bleak, isolated interior of the Sculptor's Cave was a natural setting for the ritual dramas of death, decapitation, and display, where the living and the dead could commune, well away from the mundane world of day-to-day life. On a rather larger scale, the Trou de Han provided a space where the senses were assaulted by the cold and dark and by the power of its extraordinary underground river. These were places of ritual theatre, their potency reinforced by the physical presence of the dead. Curated heads, preserved with flesh intact, capturing the very moment of death, provided an apt metaphor for this curious underground meeting of worlds.

Despite the occasional ritual use of small caves and rock shelters, the Iron Age sanctuaries of southern France lacked the grand natural architecture provided by these north European cave systems. Here, however, we see the physical creation of monumental religious complexes that embody many of the same ideas. Porticos decorated with highly formalised iconographic schemes, and embellished with preserved and decorated human heads, gave entry to places of mediation between this world and the next. The relative informality of the early sanctuaries seems to have given way, at Roquepertuse, Glanon, and elsewhere, to the creation of monumental sanctuaries, involving large numbers of people in their construction. The inclusion of 'special' curated heads in the Roquepertuse portico hints at the coming together of previously disparate communities in larger and more cohesive units.

In all these developments, religion seems to have been the prime unifying force, through which new social relationships could be created. Rather than forming symbols of communal success, new monuments like the one at Roquepertuse are suggestive of the increasing prominence of particular individuals or groups. Certain heads were now sufficiently special to be retained, curated, decorated, and monumentalised, whereas others were presumably discarded or left to rot away. The emphasis had begun to shift away from the fertility and fecundity of the community as a whole, to the efficacy and potency of certain specific individuals or lineages. Clearly, the ideological implications of these new monuments did not go unchallenged, as the destruction of the Roquepertuse portico and the obliteration of its statues graphically testify.

This change of emphasis does not mean that the religious role of these monuments was simply a 'front' for deeper political agendas. The symbolism of the Roquepertuse portico reflects the concern of its builders with enduring religious issues: death, transformation, and the afterlife. But these increasingly formal religious monuments were linked to the development of more centralised polities, in which religious authority and political power were inevitably intertwined. In the next chapter, I will focus on the latter stages of indigenous development in the southern French Iron Age, where the religious concerns of the builders of Roquepertuse gave way to a much more pronounced emphasis on ideology and politics.

CHAPTER 6

From the dead to the living

The skulls of his enemies assured him that he was alive in the world of real things. He drank from skulls, he spat into skulls. Skulls formed the feet of his throne, the sides of his bed and the path that led to his bed-chamber. He knew the name of every skull in his Skull-House and held imaginary conversations with each in turn: the lesser enemies were piled on copper trays, but the great ones were wrapped in silk and kept in whitewashed baskets.

Chatwin 1982 (1980), 98

The two ritual acts of ancestor veneration and head-hunting mediate between the worlds of life and death, and provide the paths through which mana and efficacy flow from the dead to the living.

Sheppard et al. 2000, 35

My head belongs to the king

In 1893–4 a British surveying expedition, aboard the HMS *Penguin*, visited New Georgia in the Solomon Islands. Among the party was Lieutenant Boyle T. Somerville, a Royal Navy officer with ethnological, linguistic, and archaeological interests. During his eight months in New Georgia, Somerville compiled a wealth of information on the islands and their populations, which he subsequently published in the *Journal of the Royal Anthropological Institute* (1897). Among his other observations, Somerville recorded the impact of predatory headhunting on the local population in impressive detail. He states (1897, 399), for example, that raids on the island of Márova had reduced the population from around 500 in 1885, to around 100 at the time of his own visit, just eight years later. And Márova was only one of the islands that Somerville reports as 'played out' as far as headhunting potential was concerned at the time of his visit. The populations of other large islands, such as Wana wana, Kiso, and Tetipari, had been completely wiped out. Although inevitably based on secondhand testimony, there is little reason to doubt Somerville's basic point, that large-scale headhunting raids had had a devastating effect on the population of this New Georgia group of islands over a very short period.

Somerville (1897, 399) also describes one particular headhunting raid against the village of Bombatana on Ysabel Island in late 1893, during his own time in New Georgia. The raiding party, under the authority of the Roviana king Ingova, is said to have comprised some 500 warriors in 22 war canoes, as well as two English-built boats. The attack, which involved a sea journey of some 200 km, was seemingly motivated simply by a desire to replenish the stock of heads lost in a punitive raid by the British Navy two years earlier. Somerville's account, obtained from an English observer, includes numerous details on the disposition of the boats at sea and the precise tactics used during what was apparently a highly successful action. Again, there seems little reason to doubt its accuracy, at least in its essentials. Headhunting was not in this case a by-product of warfare; it was the primary reason for military action.

Recently, this turbulent period in the history of the New Georgia group has been subject to intensive anthropological and archaeological study. Shankar Aswani (2000a, 2000b), for example, has used the testimony of multiple oral histories, supported by the results of archaeological fieldwork, to describe indigenous predatory headhunting in Roviana in the period before intensive European contacts, that is, largely before 1830. For Aswani (2000c), the burgeoning of predatory headhunting was founded primarily on the political coalescence of various disparate groups into the chiefly polities of the Roviana coastal zone during the sixteenth and seventeenth centuries. It was with the emergence of stratified coastal elites that severed heads first took on a central significance, both in terms of the veneration of ancestral skulls, in societies where power hinged on secure claims to genealogical descent, and in the accompanying objectification of enemy heads as 'quantified political symbols' (Aswani 2000c, 44). Aswani conjectures that some form of smaller-scale headhunting probably existed in earlier periods, forming the matrix from which full-blown predatory headhunting later emerged, but the archaeological evidence is not strong enough yet in Roviana to provide such detailed insights into the period beyond the reach of oral histories.

For Aswani (2000c, 40), enemy heads were 'a medium to authenticate a chief's and his group's efficacious state and its ancestral endowment'. In this sense, Roviana trophy heads became a kind of capital, to be accumulated through warfare and stored as a measure, reflection, and symbol of efficacy. They confirmed the successful translation of ancestral efficacy to the living chief and, through him, to the community as a whole. Similarly, in Chapter 3 we saw how aspiring elites in East Sumba captured enemy heads as part of an 'ideology of encompassment' in which the appropriation of human remains symbolised the subjection of rival nobles and their territories (Hoskins 1996c, 225). In chiefdoms that were often only a few generations old and far from stable, like those of Roviana or East Sumba, the power of chiefs, even if it could be inherited, had to be continuously reinforced through action. Otherwise chiefly authority would simply collapse, or else a more capable leader would emerge from within the group. Successful headhunting legitimated chiefly authority and provided a means by which that success could be stored. In the case of Roviana, the development of the chiefly 'headhunting complex' was marked by the bringing together of previously segregated sacred and secular domains. Physically, this took the form of the incorporation into the settlement of shrines, which had previously lain isolated from any traces of habitation (Sheppard et al. 2000). Politically, it took the form of an integration of headhunting into an ideology of chiefly superiority.

Bruce Chatwin's (1982 [1980], 98) fictionalised description of King Gezo's palace in Dahomey conveys something of the significance that severed human heads held as symbols of social, spiritual, and political control in a much more centralised, state society. The historical Gezo, who ruled Dahomey from 1818 to 1858, did indeed flaunt the heads of his enemies much in the way that Chatwin describes. The Scottish explorer John Duncan (1847, 275) reports being shown between 2,000 and 3,000 skulls in Gezo's court, the most important being wrapped in fabrics and stored in brass pans, while lesser ones were kept in calabashes. Certain very prestigious heads were apparently 'ornamented with brass and riveted together with iron' (276). Remarkably, according to Duncan, Gezo's functionaries could apparently provide the names of all these individuals and recount details of their life histories. Further skulls were used to form paving in the royal apartments, and others embellished the facades of certain monumental buildings (Law 1989, 403). Even Gezo's throne, which still survives in the Musée Historique d'Abomey in Benin, was mounted on the skulls of enemy kings. Although unusually detailed, Duncan's report is by no means an isolated account: indigenous as well as colonial histories consistently stress the role of decapitation and the display of skulls in precolonial Dahomey and neighbouring states (Law 1989, 404). Although the skulls shown to Duncan were mostly those of outsiders, killed in battle, it was also within the king's power to decapitate anyone in his kingdom, with the sole exception of his chief minister, the Mingan (Adeyinka 1974, 545). Many were decapitated during various calendrical festivals such as the *huetanù* (literally 'yearly head business'), where judicial execution combined with religious ceremony (Law 1989, 402).

Although head-taking and display were far from uncommon in precolonial West Africa, the scale and flamboyance of Dahomean decapitation is nonetheless unusual. According to oral traditions, decapitation had at one time formed part of indigenous funerary practices, as it did in regions adjacent to Dahomey even into the twentieth century. In Porto-Novo, for example, crania were exhumed for secondary burial, while in Whydah, certain clans offered sacrifices to the skulls of their ancestors (Law 1989, 412). Traditionally, the prohibition of this practice in Dahomey was attributed to King Wegbaja, who ruled during the seventeenth century, most probably to undermine the ancestor cults that formed the ideological basis of clan power (ibid.). Yet although the local veneration of ancestral heads was abolished, decapitation by royal authority assumed even greater importance, both in warfare and in state-sponsored religious festivals. Thus, the potent religious and ideological symbol of the head became effectively a royal monopoly.

Royal ownership and display of heads, particularly those of rival rulers, acted as a political and ideological symbol on several levels. At its most basic, it demonstrated the death and abjection of Dahomey's enemies and the fearsome power of the state army. On another level, the appropriation of kingly heads prevented veneration by their descendants, disrupting the continuity of rival royal authorities. Furthermore, the retention and display of enemy heads in the royal palaces, particularly within the fabric of their walls and pavements, symbolised the incorporation of rival kingdoms into the expanding Dahomean state.

Despite their obvious differences, the nineteenth-century elites of Dahomey, East Sumba, and Roviana shared one thing in common. Drawing on long-established religious and cosmological associations, each group found the severed human head to be a highly potent and effective symbol of royal or chiefly power. It is this ideological dimension of headhunting that I want to explore in the remainder of this chapter, focusing on the emergence of increasingly complex societies in the last few centuries BC.

The Saluvii and their neighbours

Indications from the settlement record of southern France suggest increasing urbanisation and centralisation from the fourth to the second centuries BC (Garcia 2004, 80). The literary sources, too, hint at the formation of larger and more cohesive groupings on either side of the Rhône: the most important of these were the Volcae Arecomici in Languedoc, with their capital at Nîmes, and the Saluvii (or Salyens) in Provence (Pralon 1998). According to Strabo, the Saluvii were a confederation of 10 groups fielding a common army, which included cavalry as well as infantry (*Geographia*, 4.6.3). It is uncertain, however, exactly when the Saluvii first emerged as a recognisable political grouping. One clue comes from Strabo's remark that the Massaliote colony of Olbia was founded in response to Saluvian pressure on the coastal seaways (*Geographia*, 4.1.5). This seems to imply a relatively early origin for the Saluvii as a named people, because archaeological evidence suggests that Olbia was built around 330 BC (Bats 1990). Strabo was writing in the early first century AD, and his reference may simply be anachronistic, referring in general terms to aggression from the indigenous tribes of Provence. Nonetheless, it might suggest that if we want to understand the growth of indigenous polities in this region, we might do well to look to the fourth and third centuries BC.

Any understanding of the rise of the Saluvii is inevitably bound up with the question of Massaliote territorial control during these crucial centuries. The quantities of imported pottery and Hellenistic architectural borrowings at inland settlements like Saint-Blaise and Glanon led at one time to these sites being seen as Greek satellite towns within an extensive Massaliote territory, or *chora* (e.g., Ebel 1976). This view, however, has fallen sharply from favour in recent decades as archaeological evidence has accumulated. As Dietler (2005, 198) points out, for example, the Greek and Etruscan pottery from Saint Blaise is found alongside the full range of indigenous forms, which make up the vast bulk of the assemblage. Any Greek physical presence is likely to have been small and episodic. Similarly at Glanon, the enthusiastic assimilation of Hellenistic architectural forms by no means signalled a wholesale adoption of Greek culture. Most recent authors have thus taken the view that Massalia was essentially a sea power with only a small and heavily contested chora (e.g., Hermary et al. 1999, 113).

The foundation of a series of small oppida on the hills surrounding Massalia seems to represent an episode of indigenous expansion during the third century BC. The oppidum of Verduron, for example, dominates the coastal strip immediately west of Massalia and occupies high ground overlooking the Greek city just a few kilometres away (Bernard 2000): indeed, Verduron is now in the suburbs of modern Marseille. It is highly improbable that this native stronghold would have been tolerated deep in the heart of the Massaliote chora. More likely, as the excavator Loup Bernard suggests, its foundation represents an act of aggression directly threatening Massaliote interests and penning the inhabitants within the city. After a short period of occupation, perhaps a generation or less, Verduron and other similar settlements on the hills surrounding Massalia were destroyed around 200 BC; perhaps in the same campaigns that brought about the final destruction of Roquepertuse. These sites give an insight into what must have been a complex and shifting pattern of territorial control.

Further inland, larger foundations perhaps with a stronger mercantile role, such as at Arles and Glanon, show a striking mix of indigenous and Greek cultural forms. At Glanon, for example, several monumental buildings of Phase 1b, which dates c. 200–125 BC, incorporate Hellenistic architectural elements (Roth Congès 2004, 32). Even at sites like Verduron, imported Greek pottery was plentiful, suggesting close economic (if not cultural) links. It would be a mistake, therefore, to see this period as a straightforward conflict between Greeks and barbarians, regardless of how the Massaliotes and later the Romans chose to perceive it. Massalia had been a fixture in the political geography of the region for several centuries and cultural habits, such as wine drinking, were so embedded in indigenous culture that they are unlikely to have been regarded as anything other than a traditional practice. We can speculate that Provence, in the fourth and third centuries, was a contested landscape at a number of levels, economic, cultural, and political, with numerous factional interests and complex webs of alliance and obligation.

It was from this fluid background that the Saluvii emerge into the written records in the second century as a politically dominant confederation of 10 constituent peoples. The creation of the major oppidum of Entremont around 180–170 BC, on the site of the earlier sanctuary, and its massive enlargement a generation or so later, provide compelling archaeological evidence for the consolidation of Saluvian power. By the 120s they had become sufficiently strong to threaten the existence of Massalia, forcing the Greeks to call in military assistance from Rome (Ebel 1976, 64).

At the time of the Roman invasion of 123 BC, for which the literary sources provide reasonably detailed coverage, the Saluvii seem to have been led by a king named Teutomalius (Livy, *Periochae*, 61), implying centralized political control. It is also clear from the written sources, however, that the Saluvii were backed in their opposition to Massalia by a number of northern allies, notably the Arverni under their king Bituitus. This group, based in the Massif Central, had emerged as the most powerful of the Gaulish kingdoms under Luerius, the father of Bituitus, and they seem to have been the dominant partners in an extensive alliance that incorporated numerous other powerful groups, such as the Allobroges (Ill. 6.1). Ebel (1976, 66) suggests that the Arverni were behind the Saluvian decision to challenge Massalia as part of a broader policy of expansion of their political and economic interests. At one extreme, we could see the confederation of the Saluvii as an Arvernian creation; a client kingdom intended to extend the interests of what has sometimes been described as the 'Arvernian Empire' (e.g., Hawkes 1958; Scheers 1981, 18). This is certainly the implication of a reference in Strabo, which describes the Arverni during the reigns of Luernius and Bituitus, as 'masters of the tribes' over a wide area of Gaul, and specifically 'as far as … the boundaries of the Massaliotes' (Strabo, *Geographia*, 4.2.3). Whatever the truth of this claim, there is no doubt that the Saluvian elite did not operate entirely independently and were bound up in broader relationships of clientship and the wider power politics of the region.

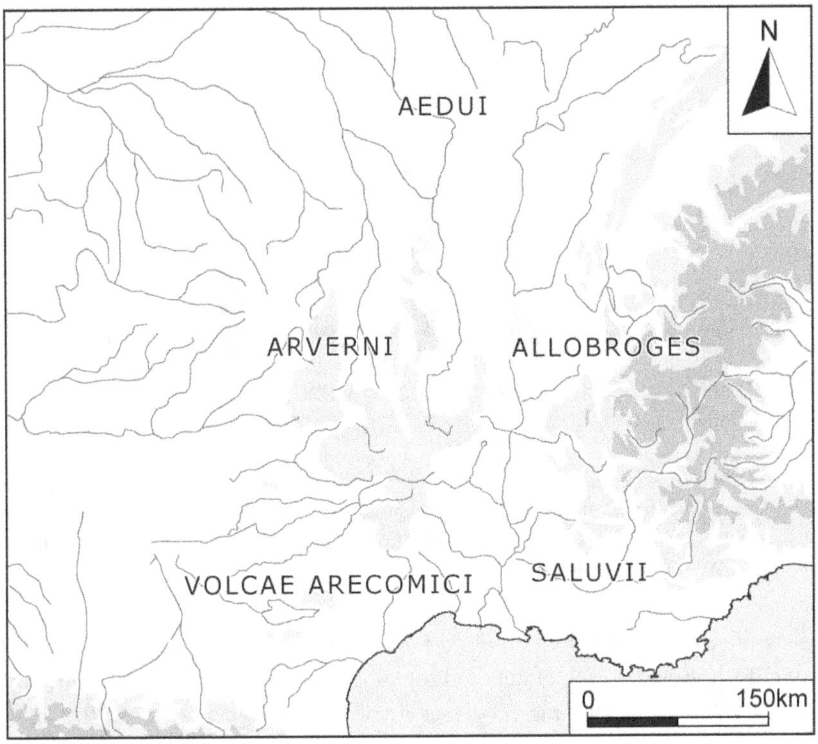

ILLUSTRATION 6.1. Map showing the approximate locations during the late second century BC of the principal tribes of southern and central France mentioned in Chapter 6. (Drawn by Rachael Kershaw)

Heads and houses

The frequent appearance of cranial fragments on southern French settlement sites of the Late Bronze Age and Early Iron Age has already been discussed in Chapter 4. Not only did this pattern continue through the remainder of the Iron Age, but it also intensified from the fourth century BC, reaching a marked peak in the third century (Ill. 6.2). As before, the best evidence comes from Languedoc, west of the Rhone, where Bernard Dedet and Martine Schwaller (1990) have quantified all human remains found on early settlement excavations. Similar finds have, however, been made frequently in Provence, for example, in the oppida of Saint-Pierre and l'Île de Martigues, where they were apparently mixed with domestic refuse in and around the houses (Chausserie-Laprée 2005, 227). Boissinot (2003, 240) also records skull fragments under or adjacent to house walls at Roquepertuse. Many of the most relevant Provençal sites, however, have not been fully published, and there has been no thorough cataloguing for this region to set alongside Dedet and Schwaller's exhaustive compendium of the Languedocien material. For this reason, only the latter material has been used in Illustration 6.2.

As in the earlier period, the overwhelming majority (89 percent) of disarticulated human bones recovered from Languedocien settlements comprise cranial or mandible fragments. These are usually highly fragmented, which suggests that they had been curated or left

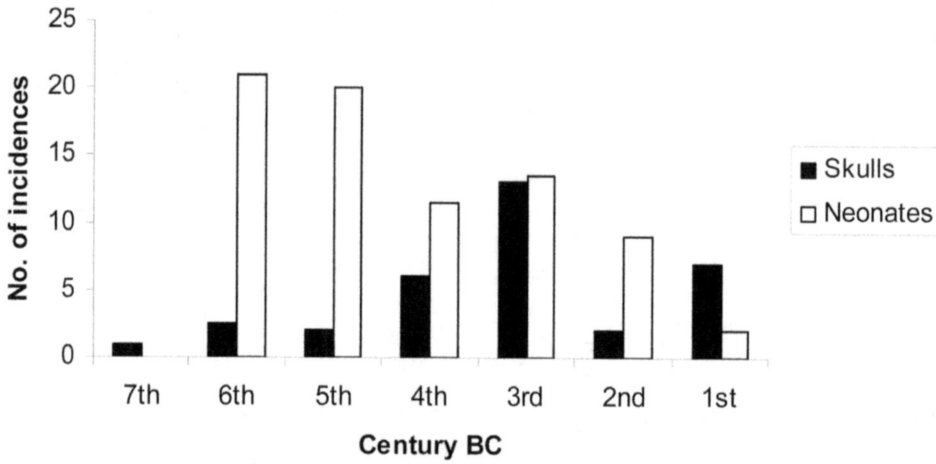

ILLUSTRATION 6.2. Graph showing the prevalence of skull deposition and neonatal burial on Iron Age settlement sites in Languedoc (based on data from Dedet and Schwaller 1990 with additions). There are numerous uncertainties in the dating, making this a rather crude analysis, but it does appear to indicate the existence of certain patterns in the data.

exposed for a significant period before entering the deposits in which they were later discovered. Generally, however, a lack of detailed osteological analysis prevents any attempt to reconstruct the precise taphonomic processes involved. As always, we have the additional problem of distinguishing the heads of enemies from those of ancestors. Although both may be represented, the iconographic evidence from Entremont and elsewhere provides fairly unambiguous evidence for the importance of predatory headhunting during this period, and I will proceed on the assumption that this was the primary means by which human skull fragments ended up on Iron Age settlements (Table 6.1) (Armit 2006b, 2010b).

As well as the data on cranial and mandible fragments, Illustration 6.2 also shows the numbers of neonatal or infant burials, commonly found under house floors or in wall foundations, from the same range of Languedocien settlements. These are of considerable interest in their own right, since inhumation of the dead is otherwise absent in the southern French Iron Age. They are usually interpreted as representing an alternative mortuary rite for individuals too young to merit whatever funerary treatments were applied to the rest of the population, though it is not impossible that in some cases they represent acts of sacrifice or dedicatory rituals (e.g., Py 1990, 802). They are less common in Provence, though they have been found at major sites, including Saint-Blaise, Entremont (Chausserie-Laprée 2005, 232), and Roquepertuse, which has numerous neonatal burials under the walls of its houses (Boissinot 2004, 52). For present purposes, however, the primary importance of these burials is that they can act as a control for the deposition of disarticulated remains. Illustration 6.2 shows a significant decrease in neonatal burial over the course of the Iron Age, in complete contrast to the pattern reflected in the cranial and mandible remains. At the very least, this marked divergence demonstrates that these trends are not simply a reflection of the relative intensity of past excavation or the numbers of sites inhabited in any given period (Armit 2010b).

TABLE 6.1. *Principal Iron Age sites yielding evidence for headhunting in Provence and Languedoc*

Site Name	Region	Statues with severed heads	Heads carved or in relief	Head niches in pillars and lintels	Skull fragments
Entremont	P	X	X	X	X
La Courtine à Ollioules	P	X			
Bringasses	P		X		
Badasset, Vernègues	P		X		
Le Pied de L'Aygues, Rustrel	P		X		
Le Puech, Noves	P		X		
Castellet Fontvieille	P		X		
Saint-Pierre, Martigues	P		X	X	X
Saint-Michel-de-Valbonne	P		X	X?	
Buffe Arnaud	P				X
Constantine	P			X	
Saint-Blaise	P			X	
Cadenet	P			X	
Roquepertuse	P			X	X
Glanon	P			X	X
La Cloche	P				X
Île de Martigues	P				X
Bouriège	MP	X			X
Rodez	MP		X		
Les Castels, Nages	L		X		
Nîmes	L		X	X	
La Ramasse	L			X	X
Pech Maho	L				X
Puech de Lascours	L				X
L'Agréable	L				X
Carsac	L				X
Vié-Cioutat	L				X
Camp Redon	L				X
Ensérune	L				X
La Lagaste	L				X
La Rallongue	L				X
Tonnerre I	L				X
Roque-de-Viou	L				X
La Liquière	L				X
Ambrussum	L				X
Lattes	L				X
Cayla de Mailhac	L				X

Note: This table excludes sites that have produced statuary without explicit iconography of severed heads (e.g., the heads carved on pillar capitals at Glanum, Apt, Rustrel, and Avignon). Regions: P = Provence; L = Languedoc; MP = Midi-Pyrénées.

Source: Based on data from Hermary 2003, Arcelin et al. 1992, and Arcelin and Rapin 2003 with additions.

Given that these patterns do seem to reflect the intensity of deposition through the period, can we then relate the varying intensity of head deposition to the wider historical developments underway in southern France at this time? The increasing frequency of head deposition from the fourth century, and its peak in the third century BC, certainly coincides with the trend towards greater urbanization and the growth of larger polities. As we have seen in the previous chapter, this is also the period during which many of the most important developments at Roquepertuse occur. Since the sanctuary lay only 14 km west of Entremont, it seems highly likely that its shifting fortunes were firmly keyed in to the broader social and political currents that culminated in the establishment of Saluvian power. The remarkable monumentalisation of the sanctuary around 300 BC seems to reflect a coming together of social groups and a new configuration of religious expression. The subsequent sacking of Roquepertuse around the middle of the third century marks a dramatic change in the fortunes of the religious authority with which it was identified. Other sites in the region also display signs of periodic disruption, notably the wave of destruction around 200 BC, which affected many sites in Provence, including Verduron (Garcia 2004, 91). The intensification of head deposition thus occupies a period of political and military competition during which the major indigenous polities were created: the Saluvii in Provence and the Volcae Arecomici in Languedoc.

As well as the frequency of skull deposition, the nature of the practice also changes around this time. Before the third century BC, skull fragments were generally found in house floors or in occupation or midden deposits, as at the seventh–sixth-century BC site of La Liquière discussed in Chapter 4. These bones appear to derive from heads displayed or stored in private dwellings. During the third century BC, by contrast, the display of heads seems to become more public. The most important Languedocien assemblage from this period comes from Yves Solier's excavations at the oppidum of Pech Maho in Aude from the 1950s to the 1970s (Belarte et al. 2011). Among the human bone assemblage were five crania found at the base of a squat stone pillar just inside one of the main entrances to the oppidum. According to Solier, the pillar was 'intended for the exhibition of trophy heads' (cited in Dedet and Schwaller 1990, 148). Other public spaces, such as roads and passageways yielded further cranial fragments, some with holes bored to enable suspension or display. Yet another, from House 58A, had what appears to have been a nail hole through its right parietal, suggesting that it had been attached to a wall or post (Solier 1979, 118, no. 140). In at least one case, the preservation of vertebrae suggested the display or deposition of a fleshed head rather than simply a cranium (Dedet and Schwaller 1990. 148). Both the frequency and the visibility of bone deposition thus suggest that headhunting acquired increasing prominence during the third century BC.

During the succeeding second century BC, skull fragments are found with far less frequency. Yet, as we have seen, this was precisely the period in which the military power of the Saluvii in particular had become sufficiently strong to threaten the survival of Massalia (Hodge 1999, 101). If the deposition of crania and mandibles on settlement sites really does reflect the intensity of predatory headhunting, then it would seem that the practice was especially prominent during the unstable years leading to the establishment and consolidation of Saluvian power. Headhunting was thus particularly associated with wars of internal domination, where certain communities were absorbed or subjugated by others in the creation of larger polities. By the same token, headhunting was seemingly much less important later, during the second century BC, when the military energies of a consolidated Saluvian confederacy were directed externally, against Massalia and Rome.

Finally, it is worth noting that even though the Roman invasion destroyed Saluvian power and signalled the end of indigenous autonomy in southern France, the presence of skulls on settlement sites once again increases in the first century BC. Although the Roman military was nominally in control of the region at this time, early efforts at provincial organisation were impeded by major disruptions, including the invasion by the Cimbri and Teutones in the final decade of the second century, and internal revolts by the Allobroges, Saluvii, and others in the first half of the first century BC (Ebel 1976, 78). Headhunting, it appears, intensified once more in the fragmented social conditions that followed the collapse of indigenous power structures.

The osteological data have highlighted some intriguing patterns that can be mapped against wider sociopolitical developments in the region. The detailed interpretation of this patterning, however, depends on the assumption that the osteological material can act as a proxy for the intensity of headhunting through time. In order to explore this assumption in more detail, and give some more depth to these ideas, it is essential to consider another strand of evidence: the iconography of headhunting at Entremont.

Entremont

The oppidum of Entremont occupies a limestone plateau overlooking the modern city of Aix-en-Provence. It dominates the valley of the river Arc, which formed a major east-west route linking Iberia and Italy and is well placed to oversee one of the major north-south routes linking Massalia with the Alpine passes. The site initially came to prominence with the discovery of the first sculptural fragments in 1817, but major excavations did not begin until the work of Fernand Benoît in the aftermath of World War II, during which Entremont had been a base for occupying German forces. Although he recovered a substantial artefactual assemblage, and the spectacular warrior statues that form the basis of this discussion, Benoît's work was heavily criticized in some quarters for its lack of stratigraphic rigour (e.g., Wheeler 1958), and problems remain in understanding the context of certain key finds.

In recent years, however, research at Entremont has been led by Patrice Arcelin whose work has greatly clarified the site sequence and deepened our understanding of activity within the oppidum. The sequence begins with the emplacement of a sanctuary somewhere on the plateau during the Early Iron Age. This, however, is known only from the various sculptured fragments incorporated within the fabric of buildings in the second-century oppidum and from a deposit of several hundred small pieces of sheet bronze (including folded strips, small plaques, and pierced disks) and glass beads found near the summit (Arcelin et al. 1992, 233). As we have seen in Chapter 4, the dating of the early carved stones is imprecise, but they could suggest a foundation as early as 700–400 BC. As such, it was broadly contemporary with the early sanctuary at Roquepertuse, 14 km to the west, and Glanon, around 60 km northwest in the Alpilles.

The character of the site changed dramatically with the foundation of the oppidum around 180–170 BC (Arcelin 2006, 125). This 'Upper Town' occupied the southern extremity of the plateau and was enclosed by a stone wall that defined a roughly square interior around a hectare in extent. This enclosing wall was relatively modest in size, being only around 1.5 m thick, but its defensive capability was enhanced by the provision of projecting rectangular bastions along the two most exposed sides. Elsewhere the ground fell away sharply, precluding the need for additional protection. The interior was subdivided by a regular grid pattern of streets, defining *îlots* made up of long rows of small rectangular buildings. Extrapolating from the best-preserved parts, there were probably around 200 of these buildings within the interior. These were closely packed, leaving no room for any substantial open space, and there is no obvious sign of any major public buildings or distinct activity areas. Indeed, the rows of near-identical, single-roomed buildings are more reminiscent of barrack blocks than a developed urban centre. The intervening streets were around 3 m wide, enough to allow access by wheeled vehicles, but the frequent use of the roads as cooking areas would have made access cramped and difficult.

Within a generation or so, around 150 BC, the original oppidum was enveloped within a new defensive system that more than trebled the extent of the enclosed area (Garcia 2004, 112). The new wall was also far more impressive than its predecessor, around 3.5 m wide and possibly standing to around 6–7 m high, with regular projecting bastions. Although a much smaller proportion of this Lower Town has been excavated, it is nonetheless clear that it was quite different in character to its predecessor. While still laid out in regular *îlots*, individual buildings were larger, with multiple rooms, stone stairs to access upper storeys, and occasional wine and olive presses. Recent geophysical survey shows an extensive system of streets and houses extending across the unexcavated area (Armit, Gaffney, and Hayes 2012). The refurbishment of the oppidum represented both a major investment of resources and a significant growth in the population. It was an ostentatious display of wealth and power.

Entremont has long been identified with the fortified capital captured by the Roman consul Sextius Calvinus in his final defeat of the Saluvii in 123 BC (Diodorus Siculus, *Bibliotheca Historica*, 34.23). The identification is based both on the scale of the second-century fortifications and on the proximity of the site to Aquiae Sextiae (Aix-en-Provence), the garrison town founded by Rome to control the region after annexation and eventually the administrative centre of the province. The extraordinary quality of the surviving stone statuary underscores the importance of Entremont and provides further support for its identification as a place of political and religious importance. Entremont does indeed appear to have suffered a major period of destruction in the 120s BC, probably at the hands of the Romans, although it was subsequently reoccupied until around 110–90 BC.

Entremont was by no means the only important centre in Saluvian territory, however, and it was far from the largest. An important pre-Roman centre is known to have existed, for example, at Arles, on the left bank of the Rhone. Although its layout can only dimly be discerned through sporadic and piecemeal excavations conducted in the course of modern redevelopment, it is nonetheless clear that the Iron Age settlement at Arles was both substantially larger and substantially older than that at Entremont (Arcelin 2000, fig. 1), as well as being sited on the major regional river route. Glanon too, was a long-established centre in Saluvian territory with monumental defensive works dating to the late Iron Age. As effectively a 'new-build' development of the second century BC, albeit on the site of a long-established sanctuary, Entremont seems to mark a deliberate statement of intent by its founders to establish a power base away from the traditional centres of political and economic power in the region.

The warriors

In 1877 a farmer on the Entremont plateau was busily digging out holes to plant almond trees, when he came upon a most unusual block of limestone. Although heavily battered and broken round its edges, the stone clearly bore relief carvings of four human heads, clustered together with their eyes closed and mouths downturned (Ill. 6.3). The carving was acquired by the antiquarian Alfred d'Aubergue, who later passed it to the museum in Aix. This was not the first stone carving to be found on the plateau, since the carved stone pillar discussed in Chapter 5 had been found by walkers some sixty years earlier. It was, however, the first element of the most celebrated series of statuary from Entremont: the collection of near-life-size figures representing warriors seated cross-legged and bearing trophy heads.

It was several generations later, however, before the importance of the site became fully apparent. During World War II, the Entremont plateau was occupied by a German garrison, which constructed a number of military buildings on the site. In 1943, in the course of these works, a cache of statue fragments was discovered, including numerous delicately carved heads. After successfully campaigning to retain these extraordinary finds in Aix, Fernand Benoît, who was at that time director of the Archaeological District of Provence, was able, in the immediate aftermath of the German withdrawal, to mount a series of large-scale excavations on the site. These recovered many more statue fragments and revealed much of what we know today about the archaeology of the Entremont plateau.

The warrior statues are very fragmentary, and none can be reassembled in anything approaching its complete original form. However, because individual elements of the same size and shape occur repeatedly in the assemblage (e.g., torsos, warrior heads, trophy heads), the general picture is quite clear (Ill. 6.4). There are eight definite torsos and at least one more represented by a shoulder fragment, which suggests a minimum of nine original statues. To these torsos belong five heads, depicted with open eyes, though the similarities in style, material, scale, and pose are such that it is impossible to match them up with any confidence.

None of the statues were found in their primary context. All were heavily (and deliberately) fragmented, and most were found within a late resurfacing of the principal roadway through the oppidum (Arcelin and Rapin 2003, 208). This part of the site is dominated by an unusually large building, sometimes referred to as the 'hypostyle', which was built during the overall enlargement of the oppidum around 150 BC. The building was set against the outer face of the former rampart that had enclosed the Upper Town at Entremont and occupied the space between two of its projecting bastions. The hypostyle incorporated a number of old carved stones within its fabric, including the head pillar discussed in Chapter 4, which formed its threshold. In front of the building, the narrow roadway

ILLUSTRATION 6.3. Photograph showing the 'nest' of heads from Entremont. One block, bearing four of the trophy heads, was found in 1877, while the conjoining fragment was not discovered until 1946. (Photograph © C. Durand, Centre Camille-Jullian/CNRS)

opened out, providing a rare patch of open space. Overall, this was a remarkable structure that clearly played a significant role within the life of the Saluvian capital. The road surface into which the fragmented statues were placed seems to have been laid down in the 120s BC, and the statues may thus have been broken up during the Roman destruction of the site.

The warrior statues were all carved from a soft limestone, and their unweathered condition suggests that they could not have been displayed in the open air for any significant time. Salviat (1993, 166) identifies the limestone as deriving from the plateau of Célony, close to Entremont, in an area where quarrying has continued into modern times. The stone seems to have been specially selected to allow the carving of fine detail using simple tools. Other sources of harder, more durable limestone were used for other carvings on the site, such as the pillars and lintels (ibid.), suggesting a detailed understanding of the properties of the various available stone sources. Traces of paint adhering to certain of the fragments confirm that they would have been highly coloured when originally displayed.

The cross-legged posture of the warrior statues is widespread in the iconography of the area and mirrors what we have already seen from Roquepertuse and Glanon. As has frequently been pointed out, this was a position commonly associated with the Gauls by

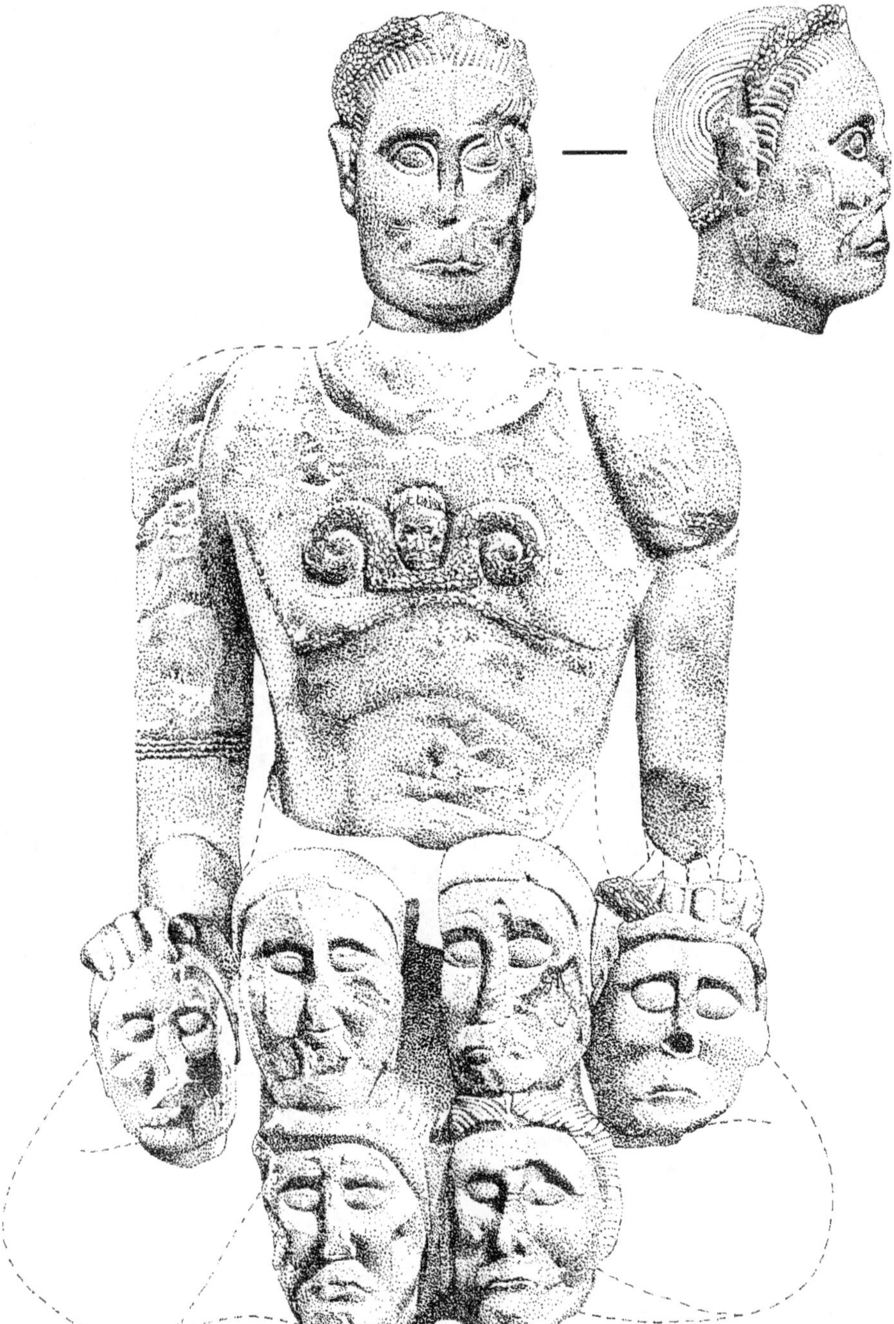

ILLUSTRATION 6.4. Drawing showing one of the Entremont warrior statues reconstructed from fragments (Salviat 1993, nos. 3, 8, 22, and 23). It depicts a 'warrior', cradling in his lap a nest of severed human heads. It is not certain that the warrior head and torso all belong to the same sculpture, because the nine or more warrior statues are remarkably similar in size, equipment, clothing, and pose. (Drawn by Libby Mulqueeny)

ILLUSTRATION 6.5. Two warrior heads from Entremont with elongated facial features and elaborately styled hair. Their wide-open eyes are carved very differently from the closed or puffed eyes of the trophy heads. (Photograph © C. Durand, Centre Camille-Jullian/CNRS)

classical writers (e.g., Strabo, *Geographia*, 4.4.3), and one that appears throughout the iconography of the European Iron Age, as with the bronze deity from Bouray (V. Megaw, 1970, 142) and the antlered god depicted on the Gundestrup cauldron (Bergquist and Taylor 1987, 13).

Clean-shaven and dressed in their elaborate armour, the Entremont figures are a far cry from the hairy moustachioed Celtic warriors familiar from the literary descriptions (Ill. 6.5). Indeed, the tightly curled hair of at least two of the statues recalls fashions

ILLUSTRATION 6.5. *(continued)*

represented in Greek iconography, including second-century BC coinage from nearby Massalia (Salviat 1993, 170). Two others have elaborate coiffures, featuring a raised band, rather like a diadem, framing the temple, while the fifth wears a distinctive, close-fitting helmet with cheek and neck guards (170–8).

There is a distinct 'family resemblance' between these warrior heads. The faces are markedly elongated, and uniformly clean-shaven, with gaunt, hollow cheeks, narrow mouths, and rather stylized ears. Their eyes are wide open, the eyelids strongly emphasized by a heavy encircling band of flesh, as if to differentiate them all the more completely from the dead eyes of the trophy heads. Despite their marked similarities, however, great care has been taken to individualise each head, most obviously through hairstyle and headgear.

The eight surviving torsos have, in most cases, been assembled from smaller fragments. Where enough surface detail survives, each seems to show the warrior dressed in a cuirass protecting the wearer from the chest to the midthigh. A flap covers the shoulders and upper chest, secured by an elaborate metal brooch decorated, in at least one case, with a disembodied human head (as with the hair, each brooch is quite individual in design). A sword in its scabbard hangs from a belt on the right of each figure, one displaying a distinctive trilobate pommel (see Salviat 1993, fig. 267d). The best-preserved cuirass (torso 6) is stippled with drill holes, as if to depict chain mail (Ill. 6.6), though it has also been suggested (less convincingly) that it might indicate some form of leather or animal skin (Benoît 1955a). A similar torso from the broadly contemporary site of Fox-Amphoux in Var has a more realistic depiction of chain mail, suggesting that this may also have been the intention in the Entremont figure (Salviat 1993, 181). Whatever the material, the flexible cuirass, held fast by a brooch on the chest, was quite different from the stiffened plates worn by the earlier figures from Roquepertuse and Glanon. Where the relevant portions survive, each warrior wears a heavy torc and an armlet on the upper right arm. Finger rings and elaborate heavy bracelets are also represented.

The arms of each warrior were held tight against the body as far down as the elbow, with the forearms held out to the front. Each left hand rested on a severed trophy head, placed on the knee. Five trophy heads survive as recognizable fragments, each clasped from above by a warrior's hand. In addition to these, the well-known 'nest' of six heads demonstrates that multiple trophies were sometimes present (Ill. 6.3). In their right hands, several warriors held iron objects that had been inserted into specially drilled holes. Only corroded stumps of iron survive in the drill holes, but they seem originally to have been long, narrow objects like sceptres or wands. They were almost certainly not swords since, as we have seen, these are depicted as sheathed in their scabbards.

As with the warrior heads, the carvers responsible for the trophy heads have gone to considerable lengths to individualise them. Two of the trophy heads in particular are highly distinctive. The first is a highly fragmented head, with a high forehead and carefully dressed hair, which has been split down the middle with tremendous force. Unlike the others, the eyes are depicted in the same style as the warrior heads, with the eyelids emphasized, although here the eyes are half-closed. In any case, the presence of a left hand, clasping the head by its hair, leaves no room for doubt that this is a severed trophy head. The soft features have given the impression that this may be intended to represent a youth or child (e.g., Salviat 1993, 205). A lock of hair passes through the fingers of the warrior's hand between the thumb and forefinger. The second head is very different, with a flat face that, despite several mutilations, seems to have had a moustache. The hair is pulled back from the low forehead in parallel striations very much in the style characteristic of classical descriptions and images of the Gauls (e.g., Rankin 1987). Again the treatment of the eyes is unusual, showing them as closed and puffed up (Ill. 6.7).

The working of the statues gives some clues as to how they were displayed. Although their backs are plain and unworked, suggesting that they may have been displayed against a wall, the backs of their heads are elaborately carved. Their hair in particular is intricately

ILLUSTRATION 6.6. The best-preserved warrior torso from Entremont, depicted wearing chainmail. (Photograph © C. Durand, Centre Camille-Jullian/CNRS)

detailed. This odd combination may suggest that the figures were displayed on wooden chairs or thrones, with the backs of their heads visible. Such an interpretation would also account for the lack of any evidence for plinths. Conceivably, they may even have been paraded around on wooden litters.

Unlike the earlier statues, parallels for the Entremont warriors appear to be rather limited in distribution. Fragments of a seated warrior statue were found in destruction

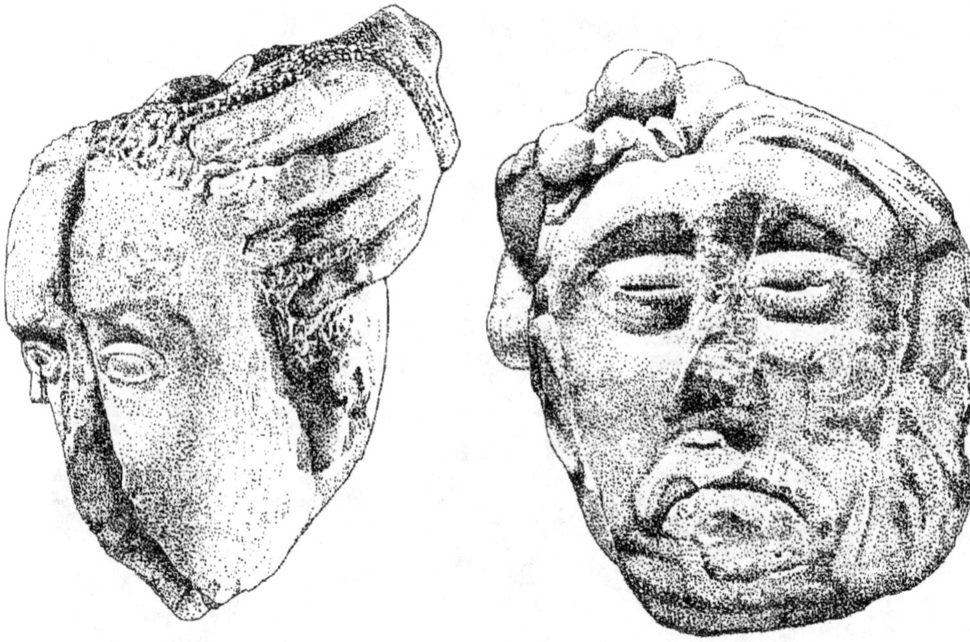

ILLUSTRATION 6.7. Reconstruction showing two sculptural fragments from Entremont depicting severed heads grasped by hands (Salviat 1993, nos. 3, 8, 22, and 23). The hair and facial features of the head on the right are reminiscent of the archetypal image of the Gauls familiar from classical literary descriptions and images. (Drawn by Libby Mulqueeny)

layers dating to around 50 BC at the oppidum of La Cloche close to Massalia. The statue itself, however, has been dated to around 300 BC based on the distinctive ring of Münsingen-Rain type clearly carved on one finger of its right hand (Arcelin and Rapin 2003, 209). Sculptural fragments found during nineteenth-century excavations at the oppidum of La Courtine, Ollioules, Var, include a life-sized left arm and hand clasping a much-eroded human head (Brun 1999, 543; Arcelin 2004a, 79, fig. 10), and a fragment from Saint-Pierre de Martigues may have come from the back of a slightly larger-than-life warrior statue (Arcelin and Rapin 2003, 207). Elsewhere in the historical territory of the Saluvii, only a male arm with a bracelet and a veiled female head from Mont-Garou indicate the presence of this rather exclusive sculptural tradition.

Well outside of Saluvian territory, and far to the west of the main distribution of the southern statuary, another statue offers one of the closest parallels for the Entremont warriors. The Bouriège statue from Aveyron in Midi-Pyrénées endured a long and unhappy process of discovery, loss, and rediscovery. It was originally found in three fragments sometime around 1920, in deep pits dug through alluvial deposits at a site known rather prosaically as Devant la Ville, near Bouriège south of Limoux. This valley location lies below the Iron Age oppidum of La Carla Bouriège. Having been reburied, the pieces were again exhumed in the 1950s and displayed in the nearby village. The statue seems to have been found in an area yielding occupation of the fifth to second century BC (Arcelin and Rapin 2003, 215).

ILLUSTRATION 6.8. Sculpture from Le Carla, Bouriège. (Drawn by Rachael Kershaw)

The Bouriège sculpture is small and squat, the surviving portion being around 55 cm high, and lacks the technical proficiency of the Entremont statues (Ill. 6.8). It depicts a stocky, cross-legged figure with a long beard, wearing a tunic and a bracelet on his right wrist. In his lap he cradles a severed human head, clean-shaven, with vacant eye sockets and swollen lips fixed in a grimace, depicting death with a 'horrible realism' (Barruol et al. 1961, 50). Although the surface is much-abraded and little detail is preserved, there is no indication of weaponry. The hair of the severed head is shown in tresses, one of which seems to fall between the right thumb and forefinger of the seated figure, recalling the similar instance at Entremont. Although Arcelin and Rapin (2003, 215) suggest that the head is held in a position suggesting 'respect' and is thus not a war trophy, there seems little reason to separate the basic iconographic statement of this piece from that of the Entremont warriors.

The main problem relating to the Bouriège statue concerns its dating. Sometime between the first and second excavations, the head seems to have disappeared and the reconstruction offered by Barruol and colleagues (1961, 47), and reproduced in subsequent works (e.g., Arcelin and Rapin 2003), is based on local recollections that the statue wore a 'type of bicorn hat, like that of Napoleon' (quoted in Barruol et al. 1961, 46). This seems a rather flimsy basis for the speculation that the figure wore a helmet similar to those on the early busts from Saint Anastasie and the Warrior of Grézan. The fifth-century BC date suggested on this basis thus seems poorly grounded. This is especially so since the Bouriège statue appears to have little in common with the clean-lined hieratic figures of Roquepertuse and Glanon. Even aside from stylistic considerations, his flowing beard, lack of the distinctive tunic and back protection, and lack of a bracelet on his left upper arm all set him apart from the other known examples. By the same token, however, he by no means meets the Entremont dress code, lacking the decorative buckle on his chest; the sword, belt, and distinctive cuirass; and the clean-shaven face. Although he does not wear a bracelet in the usual upper-right-arm position of the Entremont statues, he does wear one on his right forearm, and, of course, the favoured Entremont accessory of the severed human head is present. All in all, there is more to link him with Entremont than with the earlier statues, although he clearly belongs to a rather different (and rather poorly understood) cultural context, geographically distant from the Saluvian heartlands of Provence.

Even further afield, several sites in Galicia have produced carved heads similar to those at Entremont, including the oppida of Allariz (Benoît 1955a, plate XIV, 4) and Armeá (Gonzalez-Ruibal 2004, 137–8). The latter in particular includes examples where the dead, puffed-up eyes exactly mirror the treatment of the severed heads at Entremont and probably date to the same period. A stone block built into the wall of La Tour San Magin at Tarragone also bears heads in relief (Benoît 1955a, plate XXIII, 1), while a sculptural fragment in the Musée de Madrid, from Cerro de los Santos, Albacete, depicts a human head grasped by a hand, strikingly similar to the Entremont statues (plate XVIII, 2). Perhaps the closest Iberian parallel for the Entremont sculptures, both stylistically and thematically, however, is the fragmentary stone column from the Catalan site of Sant Martí Sarroca. This fragment, which probably dates to the later part of the Iron Age and may be broadly contemporary with the Entremont warriors, depicts three superimposed human heads, carved in relief, with closed eyes represented by deep straight lines (Lenerz-de-Wilde 1993, 243–4). Each head is similarly rendered, with striated hair and a stylized triangular nose, and each is disembodied, separated by irregular gaps. The pillar also bears the fragmentary remains of a standing male figure, depicted at more or less the same scale as the heads. When complete, the heads would have formed a group ranged vertically down the left side of the central figure. Conceivably, there could have been a matching range of heads down the right side. In either case, the thematic links with Entremont are clear.

Less well–understood than the warrior statues from Entremont is a series of fragments of three-dimensional sculptures of cavalrymen. These figures are rather less than life-size, and even more fragmented than the warrior statues, at least partly due to their innate fragility. They are also a little later than the warrior statues, dating to the final quarter of the third century (Rapin 2004, 19), but still predate the foundation of the oppidum. Once again there are Iberian links in both the nature of the imagery and the techniques of execution, though the Iberian material is rather earlier in date (e.g., Aranegui Gascó 2004, 106). Too little can be reconstructed of these equestrian figures to know if they continued the iconographic themes embodied in the earlier warrior statues. They do suggest, however, that the tradition of monumental sculpture of the Saluvian elite carried through the third century.

The women of Entremont

Much less attention has been paid to another group of equally fragmented statues from this same collection. These are a series of seated female figures carved of the same soft limestone and to the same scale as the warriors (Salviat 1976). The three definite female heads each have the same wide-open eyes as the male warriors and are rendered in the same style, although the faces are a little less narrow and elongated than those of the men (Ill. 6.9). Although there are slight differences in facial shape between each of the women, there has been no obvious attempt to distinguish them individually through their costume. Since each wears a long veil, pulled across the head to cover the hair, there was no possibility of distinguishing them through their coiffure, as was done for the men. In two cases the veil is lifted just high enough to reveal elaborate earrings.

From various reassembled fragments it is apparent that the women wore ankle-length dresses covered by a short cloak and a long veil that terminated in a fringe low on the back. Much of the clothing was decorated with geometric grid patterns. Unlike the warriors, these women seem to have rested on low seats, with their knees drawn up together in front of them and their arms held tight against their bodies. Also unlike the warriors, the backs of the female statues are carved in considerable detailed, suggesting that they were displayed in the round: they may also be associated with some fragmentary carved feet on plinths. Despite the profound stylistic similarities then, the women of Entremont were apparently displayed quite differently from the men.

Representations of women are exceptionally rare in the southern French Iron Age, as they are elsewhere in Europe. Indeed, only two examples are known from other sites in the region. One fragment from the oppidum of Saint-Pierre preserves part of a fringed veil, suggesting that it derived from a statue of a veiled woman, slightly larger than life-size, similar to those from Entremont (Chausserie-Laprée 2005, 228). Another much-eroded head from Le Mont Garou, Sanary, wears a veil and an ostentatious earring much like the Entremont women (Arcelin 2004a, 79, fig. 10). A much abraded head from La Courtine à Ollioules, probably represents a final example (78). All four of these sites have also produced male warrior statues in the Entremont style, suggesting that the production and display of male and female statues in this period were thoroughly intertwined.

It is tempting to relate the female depictions at Entremont and elsewhere to the slightly earlier tradition of female statuary along the Mediterranean coast of southeast Iberia, which reached its apogee in the first half of the fourth century BC with a series of extraordinary carvings such as the Dame d'Elche, Dame de Cabezo Lucero (Aranegui Gascó 2004), and the Gran Dama Oferente from Cerro de los Santos (Blázquez 1990). Each of these female carvings is richly dressed with huge ear ornaments that recall the more modest, but nonetheless prominent, earrings of the Entremont women. The veil of the Dame de Cabezo Lucero also resembles the costume of the French statues, while slightly later

ILLUSTRATION 6.9. Head belonging to one of the female statues from Entremont. (Drawn by Dan Bashford)

examples, such as the abraded female head Cerro de los Santos, are even closer in general appearance (Aranegui Gascó 2004, fig. 13).

Arcelin and Congès (2004, 10) have suggested that the seated women at Entremont, along with possible children (for whom there is actually little evidence), may have been placed as 'adorants' around the men. These posthumous family groups would thus have reflected the central figure's status as a husband and father as well as a warrior. This interpretation, however, perhaps rather underestimates the importance of these female figures. It has always been assumed that the female statues were not directly associated with the trophy heads. Given the degree of fragmentation of the Entremont statues, there is actually no evidence one way or the other for this belief. However, for present purposes I will proceed on the assumption that the severed heads, along with the unidentified iron objects, were essentially male attributes. There is, however, another small group of carved stone objects that may be related to the Entremont women. Three carvings representing small metal vases, or situlae, some 20–30 cm high, were found in the same vicinity as the statues (Salviat 1993, 198). One has a representation of a fine sieve across its opening, suggesting that it was used to strain wine. The consumption of wine was a widespread high-status activity in the region and is well attested at Entremont through the remains of imported amphorae and bronze ladles. The situlae appear to have been broken off from larger statues, but they do not appear to fit within the compositional scheme of the warrior statues. They may, however, be associated with the female figures.

As well as being a high-status activity, the consumption of wine was also highly ritualized and lay at the centre of elite ideology in Iron Age Europe (B. Arnold 1999; Dietler 2005), as witnessed by the inclusion of prestigious vessels associated with the serving and drinking of wine in richly furnished grave assemblages (e.g., Biel 1981). Bronze wine-serving vessels often appear to have female associations. In the situla art of north Italy and Slovenia, for example, veiled women are depicted serving from vessels similar to those depicted at Entremont (e.g., Turk 2005). These women have sometimes been interpreted simply as servants, but the symbolically charged nature of situla art imagery and the context of these vessels as funerary offerings suggest a more important role for these women. The largest and most exceptional of all such vessels was the Vix krater, found in a rich female grave below the Mont Lassois hillfort in Burgundy and dating to the end of the sixth century BC (Chaume 2001). Chris Knüsel (2002) has argued that the artefacts in the Vix grave denote the woman's role as a ritualist, with the krater forming the centrepiece of her ritual paraphernalia. Given this broad European Iron Age background, it is tempting to see the carved stone vessels as attributes attached to the female statues at Entremont, in much the same way that the trophy heads were attached to the male warriors.

Naming the dead

The detail of the clothing and jewellery suggested to Arcelin and Rapin (2003, 209) that the Entremont statues, both men and women, were sculpted between around 300 and 250 BC. As well as predating the foundation of the oppidum by at least 70 years, this dating implies a chronological gap of more than a century between the Entremont statues and their predecessors at Roquepertuse, Glanon, and elsewhere. Although the Entremont warriors share certain elements with this earlier group, most obviously their general posture and scale, there are also notable differences in dress and accoutrements. Where the earlier statues wore tunics with distinctive back protection, the Entremont figures wear flexible cuirasses, secured by a decorative brooch, sometimes in the form of a head. Where the earlier series always (where sufficiently well preserved) wore a bracelet around their left arm, the Entremont figures never do. Instead, they seem uniformly to have worn bracelets on their right arms. Where the earlier statues sat on stone plinths and had elaborately detailed backs, the Entremont ones had no plinths and had no detailing on their backs, which suggests quite different modes of display. Perhaps most crucially, the earlier statues have no obvious associations with severed heads. The Entremont series clearly references the earlier statues and draw much of their iconographic power from that earlier religiously imbued tradition. But they represent, at the same time, something radically new. For the first time the motif of the severed head and the image of the individual warrior were drawn together in a single composition.

During the fourth and especially the third centuries BC, the processes were played out that led ultimately to the emergence of the Saluvian confederacy as a dominant political and military power in the region. This period was marked by intense and violent competition between social groups seen, for example, in the final destruction of such centres as Roquepertuse and Verduron around 200 BC. Following the foundation of Entremont, the mid-second century BC represents the high water mark of Saluvian power. As in Dahomey and Roviana more than two thousand years later, the symbolism of the severed human head played a significant role in these political transformations.

In a nonliterate society, images carry an enhanced importance, forming an instantly accessible and powerful medium to convey ideological messages. Given their sociopolitical context in the early third century BC, and their location on an already-ancient and symbolically important sanctuary, it is no surprise that the Entremont warrior statues were intended to communicate certain powerful ideas. So what were these ideas, and who were the statues meant to represent? Even when they were first seen, during the third century, different observers would probably have made different readings; the age, gender, ethnicity, and social class of the observer would all have affected individual understandings of the meaning and power of the statues. Yet the more detail that was put into these statues, the more the range of potential meanings and readings was constrained. And these were very detailed statues.

Perhaps most striking is the tension in the statues between individualisation and standardisation; they are the same, but different. Unlike earlier sculptures, each warrior is individually portrayed in terms of hairstyle and facial features and in the detail of his armour and clothing. These variations would have served unambiguously to identify the subject to the contemporary observer. Yet, they all share the same pose, the same facial shape, the same style of armour and clothing, and they are all the same size. We are clearly intended to view them as either a group or a series. There is little to suggest that they were intended to depict a pantheon of gods, since we would presumably expect multiple deities to have highly varied attributes, such as the animal images carved on figures like the statues from La Coutarel, Tarn (Gruat 2004), and Euffigneix, Haute-Marne (V. Megaw 1970, no. 226), or at least a wider range of physical appearance. It is much more likely that they represent individual members of the Saluvian elite, either living or remembered.

There are numerous possible interpretations of the Entremont statues, but I want to explore just two. One possibility is that the carvings represent successive members of a Saluvian dynasty of the early third century BC, perhaps war leaders, or kings responsible for the unification of the Saluvii. If so, were they carved piecemeal over a number of years, as each leader reigned, or else as a consciously defined series following the consolidation of the dynasty. Given the stylistic unity of the statues, the latter seems more likely, because we would surely expect more variation if they were carved over eight or nine successive reigns (however short and turbulent). Perhaps then we may be seeing the conscious creation of a heroic lineage – ideological mythmaking in action.

The severed heads held by these seated warriors are just as much individually conceived as the warriors in whose laps they nestle. Are we looking here at the faces of specific individuals, defeated in the process of Saluvian polity building? Or are they perhaps emblematic – caricatures of particular tribes or peoples subdued or absorbed by the Saluvii? Given the expansionist nature of the Saluvian polity, the taking and curation of heads could be read as a metaphor for the assimilation of previously independent groups. As for Gezo in nineteenth-century Dahomey, turning one's rivals' heads into objects, whether for display, vilification, or ceremonial use, was a powerful and potentially lasting statement of control.

This interpretation assumes that the Entremont statues were conceived as a series, like successive issues of the same coin. Yet this may not be so. Perhaps, instead, they were conceived as a group. They may, for example, represent individual heroes of some great Saluvian victory. But perhaps more intriguing is the possibility that they relate to the component groups that came together to create the Saluvian confederacy. Strabo, for example, tells us that there were ten separate groups within the confederacy (*Geographia*, 4.6.3), though this precise figure may not of course be accurate or constant. Did the individual statues then represent the various societies that made up the Saluvian state – real leaders or personifications of tribal deities differing in key details, yet in their essentials the same? Were certain aspects of the statues, such as the shared facial shape, intended to subtly stress the familial relationships between these societies and distinguish them from the enemies, whose round (and decapitated) heads sat in their laps?

Neither of these two competing interpretations accounts particularly satisfactorily for the presence of the female statues. In the first interpretation we might see the women as representing occasional female dynasts who ruled at various times in the evolution of Saluvian power. In that case, however, we might expect them to be more analogous to their male counterparts. As we have seen, that was not the case, and the women were clearly displayed rather differently. In the second interpretation, we might suggest that certain of the component groups were personified by female figures, though the same objection applies. Perhaps it is mistaken to assume that the female figures form part of the same series as the men. They may instead relate to some quite different aspect of Saluvian ideology, perhaps associated with female ritual power. Alternatively, female secular power may have been sufficiently common to have produced its own norms and conventions of representation that differed from those applied to males.

There is no real evidence at present to allow us to choose between these sorts of interpretations. But any alternative explanation has to account for this peculiar tension between similarity and difference, as well as the many other detailed features of the carvings. However these images are interpreted, it is clear that the specific identities of both the trophy heads and the warrior were important and were intended to be understood by contemporary observers. As such, they reflect a very specific series of ideas associated with the emergence of the Saluvian confederacy and the power of its elite. Not only were these ideas sufficiently important to lead to the initial creation of the statues; they were also sufficiently enduring to ensure that the statues were still prominently displayed, more than a century later, at the very heart of the Saluvian capital. And the most powerful component of these messages was the decapitated human head.

The 'museum' of heads

As we have seen, the Entremont warrior statues were carved in the third century BC, but we have no idea of the context of their original display. Indeed, there are no archaeological deposits or buildings from the site that can be dated to before 175 BC. Only with the creation of the Lower Town around 150 BC do we begin to get any sense of how and where the sculptures were placed; this, of course, need not tell us anything about their original display.

Central to the final location of the Entremont statuary is the hypostyle building that occupied a privileged location close to the heart of the oppidum. In front of this building, the main road widened to form a rare open area within the packed grid of streets and buildings. The hypostyle itself was formed of reused blocks and pillars from much earlier structures, including the head pillar discussed in Chapter 4 and other stones bearing niches for the insertion of heads. It appears that the warrior statues were probably displayed in roadside structures close to the hypostyle, but the evidence from the old excavations is unclear (Congès 2004, 63).

So why were all these disparate and already ancient stones brought together in and around the hypostyle? This 'museum' of carvings, some probably at least three centuries old, seems unlikely to have been collected together simply for aesthetic reasons. More likely the authorities at Entremont were explicitly stressing their genealogical links to a past that included the long-established sanctuary and the heroic third-century figures represented by the warrior statues. As a 'new-build' capital of the second century, it was perhaps especially important to use the power of the past to legitimate control in the present.

Surprisingly perhaps, there is remarkably little evidence for headhunting or its representation within the main period of settlement at Entremont. There are some carvings that appear to date to the second-century period of the Upper and Lower Towns, but these are rather different in character. The carved pillar fragments discussed in Chapter 5 include a mounted warrior with a severed head dangling from his horse's neck (Ill. 5.22), and these have been dated to the second century BC (Arcelin 2006, 157). There are also relief fragments showing cloaked human figures bearing animal sacrifices, which Arcelin and Rapin (2003, 212) have dated to the same broad period, but iconography related to the severed head is otherwise limited to a small crescent-shaped phalera bearing a single disembodied head (Benoît 1955a, plate V, 11). The stark presentation of headhunting in the earlier warrior statues, however, was not matched by anything produced while the oppidum was occupied.

Iconoclasm and transformation

In the preceding chapter we saw how the monumental sanctuary at Roquepertuse and its associated warrior statues were destroyed sometime around 250 BC. The destruction was so thorough as to suggest a deliberately iconoclastic act, probably associated with the overthrow or removal of those in control of the sanctuary. Just over a century later, exactly the same fate was to befall the Saluvian warrior statues at Entremont.

The Roman military campaigns of the 120s BC effectively destroyed the Saluvii as an independent force. The taking of Entremont (probably in 123 BC) is marked by the presence of *boulets*, stone projectiles thrown from Roman siege engines, in destruction deposits in the Lower Town (Congès 2004, 63). After the defeat, at least some of the

Saluvian leaders fled to the territory of the Allobroges where they were given sanctuary and support (Livy, *Per.* 61). The following year, however, a Roman army under the consul Ahenobarbus pursued the war among the Allobroges with a much larger force, and the campaign was continued by Domitius in 121 BC (Ebel 1976, 70–1). An account in Appian (*Roman History*, 4, fragment) describes a joint embassy to Domitius on behalf of the Arverni, Allobroges, and the deposed Saluvian leaders, showing the web of alliance that existed at this period. Under the consul Q. Fabius Maximus, however, the Roman army won a major military victory over these combined forces in Gaulish territory in 121 BC, which seems to have settled things for good. The surrender of the Arvernian king Bituitus shortly thereafter sealed the Roman victory and ushered in the period of Roman control over southern France.

Entremont was not, however, abandoned at this time. According to Diodorus Siculus, some 900 Saluvians who had supported the Massaliote cause were spared (*Bibliotheca Historica*, 34.23), and it may have been this group that continued to occupy the oppidum after the defeat of the Saluvii by Rome. The leader of this pro-Roman faction was a Saluvian with the Greek-sounding name Craton, in sharp contrast to the deposed Teutomalius with his quintessentially Celtic name. The renewed occupation of Entremont is marked by a transformation in the way in which power was materialised. Several factors suggest a conscious attempt to break with the past and dissociate the new Saluvian leadership from their deposed predecessors. The first concerns the treatment of the warrior statues, and the second involves the treatment of real human remains.

The Entremont warrior statues, like their predecessors at Roquepertuse, were comprehensively destroyed. All were decapitated, arms and legs were severed, and torsos fragmented. The destruction seems to have been frenzied and deliberate. Subsequently, the fragments were incorporated into the makeup of the newly widened road that passed in front of the hypostyle (Congès 2004). Given the heavy symbolic charge of these images, and their likely association with the anti-Roman leaders of the Saluvii, it is hard not to see this treatment as a deliberate attempt to erase any links to the past and to mark a new beginning for the community. Not only were these embodiments of past authority destroyed, however; they were also debased by disposing of their fragmented remains in the road makeup, where the population of Entremont would tread over them on a daily basis. This treatment recalls the practice in Roviana of burying the skulls along tracks to be trampled underfoot as a form of final insult (Aswani 2000c, 68).

Although the stone images of headhunting warriors were no longer on view, an equally remarkable exhibition of heads was created. A series of human crania pierced by substantial iron nails found in the final destruction layer of the roadway suggests that this reoccupation of Entremont under its new regime was marked by a display of real human heads attached to the façade of the hypostyle building. The most recent analysis suggests that at least 15 skulls are represented, although there may of course have been many more. The material is generally too fragmented for age or sex attributions, although all were adults or adolescents. Most intriguing is the recent identification of numerous perimortem injuries from bladed weapons, and cut marks, particularly visible on the mandibles. Clearly a number of these individuals had suffered violent deaths, and their remains may have been deliberately defleshed. The enormous iron nails pierced right through the frontal bones of the skulls, presumably attaching them to a wooden post, although one example appears to have been suspended from a cord (Courtaud et al. 2010, 327). Such massive nails could have been inserted only while the skull was fresh, as a dry skull would have shattered with the impact. The combination of violent death and the apparently rather irreverent form of display suggests that these were most probably trophy skulls, displayed as a form of disparagement. Did they perhaps belong to the deposed elite of Entremont – the anti-Roman Saluvian leaders who claimed descent and legitimacy from the individuals depicted in the warrior statues?

Although these nailed skulls form a dramatic ideological statement, associated with a very late stage in the development of Entremont, there are other human remains from the site. Parts of at least two individuals, including an articulated arm, were recovered from excavations in 1986–7 in the Upper Town (Square 1, *îlot* 33) (Mahieu 1998; Courtaud et al. 2010). Mahieu regards these as fragments of casualties from the final destruction of the site, though it is equally possible that they might represent earlier deposits or human remains curated within the buildings.

Equally prominent displays of nailed human skulls can be found on other Saluvian sites of the same period, including various deposits from Glanon (e.g., Roth Congès 2004, fig. 14). At the oppidum of Buffe Arnaud, north of Entremont in the foothills of the Alps, numerous crushed fragments of human crania from at least two (possibly many more) adults were found in the entrance passage under a monumental gate (Duday 1995). These seem to have been burnt during the destruction of the oppidum, and the excavators not unreasonably suggest that they are the remains of heads displayed in the covered entranceway to the oppidum (Garcia and Bernard 1995, 122). As at Entremont, no mandibles or cervical vertebrae were found. Crania pierced with iron nails or spikes have also been found at Glanon (Arcelin et al. 1992, 215), and in Languedoc at the oppidum of Pech Maho (Dedet and Schwaller 1990, 147–9). Further afield, there are also several Iberian sites with similarly treated remains, including crania pierced by large nails at Catalonian oppida like Puig de Sant Andreu and Puig Castellar (Rovira-Hortalà 1999, 19, 25).

More dramatic still is the evidence from the oppidum of La Cloche on the north flank of the Nerthe hills northwest of Massalia. The rugged summit of La Cloche was visited from at least the sixth century BC for the deposition of small offerings, including twisted and pierced strips and plaques of sheet bronze similar to those at Entremont, though again there is no trace of accompanying settlement activity. Later the hilltop sanctuary seems to have housed at least one seated warrior figure carved in more or less identical style to those of Entremont. The details of the hair and weaponry suggest a date similar to the Entremont warriors, and the distinctive snake-shaped finger ring of La Tène B2 Münsingen-Rain style narrows its date to the decades around 300 BC (Arcelin and Congès 2004, 11). The statue was found in a destruction level dating to around 50 BC, suggesting that it had not suffered destruction during the Roman annexation of the area in the 120s; by the time of its destruction it was more than two centuries old, and nothing is known of its original context or location on the site. Only much later, at the end of the second century BC, the oppidum was created, with a grid plan of streets and small back-to-back buildings familiar from Entremont and elsewhere (Chabot 2004). The foundation of this

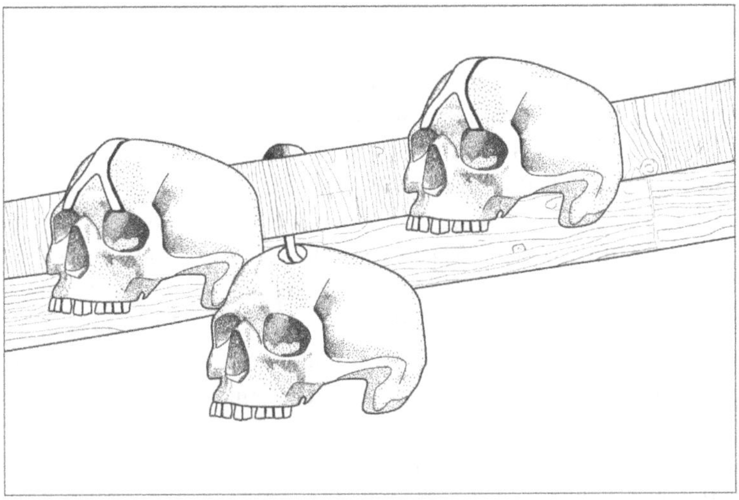

ILLUSTRATION 6.10. At least one of the crania found at the entrance to La Cloche was held in an iron clamp that had apparently been fixed to the gate structure of the oppidum. Other were fixed using iron hooks. (Drawn by Rachael Kershaw after L. Chabot)

defended indigenous settlement within Roman-controlled southern France suggests that direct Roman intervention in the region remained slight at this time.

Within the entrance to the oppidum, the excavator Louis Chabot found two human crania. One was perforated by a large square-sectioned iron nail, while the other was held tight in an extraordinary iron 'armature' or clamp, which extended from the temple to the back of the head and was secured in both eye sockets (Ill. 6.10). Both were adult males, aged 30–40 and 55–60 (Mahieu 1998, 64) and, like the skulls at Buffe Arnaud, they appear to have been originally displayed over the entrance to the oppidum. Other nails in the vicinity, similar to those used on the surviving cranium, suggested that further skulls had once been present.

Aside from these crania, several other groups of human remains were found at La Cloche. These included a more or less complete skeleton of a woman of around 30–40, seemingly crushed within a collapsed building, and a stray head of another adult of 40+ found in one of the streets (Mahieu 1998, 64). These have been explained as massacre victims killed during the violent destruction of the oppidum, but the apparent disarticulation of the woman's bones, which showed no evidence for rodent or carnivore marks, do not suggest casual disposal of a battle casualty. The woman also appears to have been wearing a heavy iron ankle ring that is particularly interesting as it is virtually identical to two ankle rings found in the grave of another woman buried at foot of the hillside below the oppidum of Le Vallon du Fou, not far to the west. Jean Chausserie-Laprée (2005, 41) has interpreted the heavy iron ankle rings as badges of slave status, citing parallels from the La Tène site of Sanzeno in northern Italy. Since formal inhumations are virtually unknown in Iron Age Provence, it is hard not to make some connection between these two skeletons and to suggest that they may represent the structured deposition of the bodies of low-status women.

Terror and abasement

The nailed skulls from the last days of Entremont, the displays at Buffe Arnaud and La Cloche, seem to move us into a new realm of meaning. Though far from the scale of the Mesoamerican *tzompantli* discussed in Chapter 3, the elaborate head clamp from La Cloche might legitimately be described as representing a 'technology of terror' (Mendoza 2007, 401). The iron-clad skull suspended above the heads of all those passing in and out of La Cloche seems to signal a fairly unambiguous message of debasement and humiliation. The overwhelming dominance of adult males, often fairly old ones, suggests that these individuals were elders of some form, perhaps political leaders rather than necessarily warriors. Here we see the display of human heads apparently shorn of any lingering religious significance, demonstrating simply the power of the headhunter and the destruction of his enemies.

If we look overall at the chronological development of headhunting iconography in Provence and the osteological reality of headhunting practice in Languedoc, similar trajectories are quite evident (Ill. 6.2). Both are most prominent in the period of polity building that culminates in the third century BC. The graphic displays of the Entremont warrior statues and the increasingly prominent head deposits on Languedocien oppida together serve to highlight the potency of headhunting chiefly as a strategy in the practice and performance of power, especially when that power was potentially unstable and had to be visually reinforced for the subject population. Once some internal stability had been achieved, however, by the second century BC, both the reality and depiction of headhunting subside. The warrior statues remained on display but now as a conscious archaism, and there are few skull deposits from settlement sites of this period. Change came again with the arrival of Roman authority at the tail end of the second century. Saluvian power collapsed, and in the absence of any strong control by the occupying power, indigenous communities returned to a situation of internecine competition and aggression. Once again, most vividly at La Cloche, the human head emerges as a symbol, this time of power and aggression, with few obvious remnants of its earlier religious and cosmological associations.

It was during this period that the Greek writer Poseidonius visited Massalia and composed his descriptions of the Gauls. We have seen in Chapter 2 that Poseidonius claims to have seen the severed heads of enemies displayed above the entrances to Gaulish homes. If this is true, and there is no real reason to doubt it, then he probably came across these displays in the vicinity of Massalia. It is not beyond the bounds of possibility that he visited La Cloche itself, where the skulls date to more or less the right period, but there are numerous other oppida around Massalia where he could have encountered trophy heads at first hand. The important point, however, is that Poseidonius experienced an indigenous Gaulish society stripped of its independence and suffering under a barely consolidated military occupation. Although they have been used as a template for the understanding of 'Celtic' headhunting, his descriptions capture no more than a snapshot of one specific, and traumatic, moment in the development of Saluvian society. The Gauls he encountered did indeed collect the heads of their enemies, but that does not mean that their actions, beliefs, and attitudes in relation to this practice were identical to those of

ILLUSTRATION 6.11. Motif from an Armorican coin dating to the early first century BC showing a large head in profile, apparently set on a pillar or support of some kind, suggesting it is intended to represent a decapitated head. Several smaller heads are attached to it by chains or cords. Motifs of this sort are not uncommon in Armorican coinage. (Drawn by Rachael Kershaw after Gruel 1993, 306)

their ancestors; the elite warriors depicted in stone at Entremont, the guardians of the fourth-century Roquepertuse sanctuary, or the stone carvers of the earlier Entremont head pillar.

Coins and heads

The rich stone sculpture of southern France is not matched in the north, but there is nonetheless some iconographic evidence relating to the human head. Indigenous Gallic coinage, which emerges from the late fourth century BC, blends imported conventions with fantastical native imagery, including enigmatic depictions of disembodied human heads. The heads depicted on the obverse of Celtic coinage initially followed Greek and Macedonian prototypes, although it is entirely possible that a pre-existing enthusiasm for head imagery contributed to the uptake of these alien objects. Later elaborations developed this motif in decidedly un-Greek directions, including human-headed horses and early first-century BC Venetic coins showing a large head in profile, with smaller heads attached to it by chains or cords (Ill. 6.11).

Such regional iconography merits more detailed study, but for now I want to focus on an apparently more straightforward image that appears on silver coins of Dubnoreix (Dumnorix), an Aeduan aristocrat killed by Caesar's cavalry in 54 BC. This coin issue shows a central figure, presumably Dubnoreix himself, carrying in his right hand a boar-headed carnyx and in his left hand a severed head (Ill. 6.12). The severed head is given

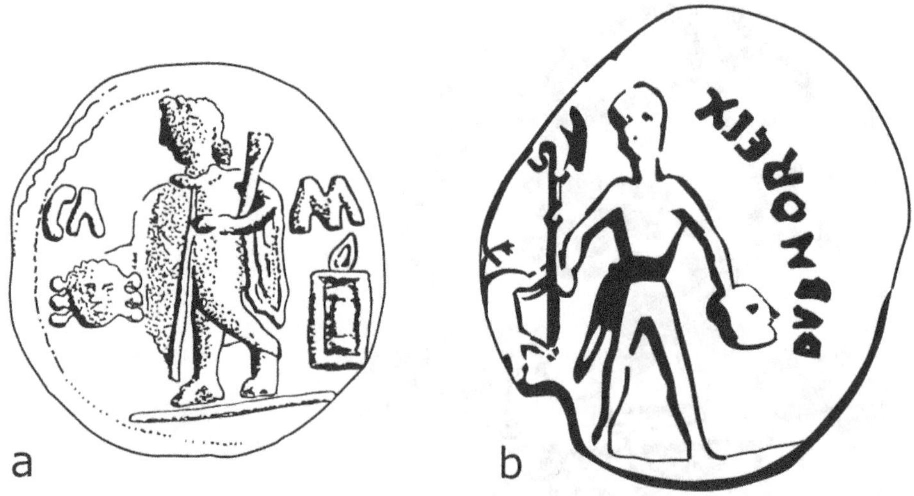

ILLUSTRATION 6.12. (a) A first-century AD coin of Cunobelinus from southeast England. The imagery here probably draws on the myth of Perseus and Medusa, yet the appeal to indigenous traditions remains striking (after Creighton 2000, 129). (b) Silver coin showing an Aeduan warrior (probably Dubnoreix) carrying a boar-headed carnyx and a severed head (after Olmsted 2001, plate 52). It dates to the period of the Gallic War (50s AD). (Drawn by Rachael Kershaw)

extra emphasis through its size. According to the descriptions in Caesar's *Gallic War*, Dumnorix was a charismatic and powerful anti-Roman leader who appealed to the independence and autonomy of the Gallic tribes, placing himself in opposition to those like his brother Diviciacus who were more sympathetic to Roman influence (*Gallic War*, 1.19). It is possible that the image on the coin was intended to reinforce his Celtic credentials. Although such motifs are far from common, we do see something very similar on coins from southeast England in the first century AD (Ill. 6.12). John Creighton (2000, 129) has suggested that such coins refer to the imported myth of Perseus, who beheads the gorgon Medusa. This may indeed be the case, but the very adoption of a 'foreign' myth involving decapitation would seem to suggest an indigenous interest in this practice and a belief that this was a highly appropriate scene through which to represent kingly authority. Clearly, the depiction of power through the motif of the severed head was not restricted to Entremont and the Saluvii.

The northern sanctuaries

Disproportionate numbers of cranial fragments are commonly encountered on a whole range of Iron Age secular and religious centres far beyond southern France. At La Tène itself, in Switzerland, crania considerably outweigh other bones among the large assemblage of human remains, and decapitated skeletons are also present (Jud 2007). Further crania and cervical vertebrae, many showing clear signs of decapitation, also dominate the human bone assemblages from the oppidum of Manching in southern Germany (Lange 1983), the sprawling unenclosed Basel Gasworks settlement in Switzerland (Trancik Petitpierre 1996), and numerous other Late Iron Age sites. In none of these cases are we dealing exclusively with heads, however, and the body treatments represented are clearly complex. It is impossible here to examine every case, but it is worth looking at one group of sites in some more detail.

Excavations over the past 20 to 30 years have shown that a series of northern religious sites, such as the sanctuaries of Gournay-sur-Aronde and Ribemont-sur-Ancre, were gathering places for large numbers of people. They were thus particularly significant places in landscapes that did not otherwise contain the sort of nucleated population centres seen along the Mediterranean littoral. Ideological concerns, in communities of increasing size and complexity, were here played out in overtly religious rather than urban settings. And osteological remains from a number of these sanctuaries suggest that the human head was as potent a symbol here as it was in the south.

The Middle La Tène sanctuary at Gournay-sur-Aronde, Oise, in northern France appears to have been the focus for elaborate religious observances centred on a sequence of temple buildings set within a large quadrangular enclosure (Brunaux 2003b). The bulk of the activity seems to have involved animal sacrifice and the display of enormous quantities of deliberately bent and broken weaponry, probably obtained as spoils from military operations. This practice accords with Caesar's observation from the first century BC that the Gauls offer war booty to the gods in *loci consecrati* – sacred spaces protected by religious law (*Gallic War*, 6.17). Similar displays of military trophies within a religious setting were also common in the Mediterranean world and widely practised, for example, by the Greek city-states.

Although not present in large numbers, there were also human remains at Gournay, including concentrations of crania and cervical vertebrae clustered around the entrance to the enclosure. Some of these showed clear signs of decapitation and enlargement of the *foramen magnum*, presumably to enable heads to be displayed on stakes or poles. Of the 12 individuals identified, at least 3 were female, suggesting to the excavators that the dead at Gournay were not simply enemy warriors but perhaps represented the victims of human sacrifice (Brunaux and Malagoli 2003, 25–6). Given the scale of the military conflicts suggested by sites like Gournay, we might also consider the possibility that the heads belonged to a ruling family from some defeated group, as may be the case at Entremont. There are plainly many possibilities, but it is at least clear that the curation and display of decapitated heads was being practised at major centres in northern France at broadly the same period that the Entremont statues were being carved.

Although it is the best known, Gournay is only one of a number of major religious centres now known in this part of northern France. In each case, religious performance seems to have centred on sacred spaces defined by square or rectangular enclosures. Deposits of metalwork (especially weaponry) dominate, and there is usually evidence for feasting or animal sacrifice: human remains are a persistent but numerically small component and are often dominated by cranial fragments. There is, however, one sanctuary where human remains take centre stage.

Some 50 km north of Gournay, Ribemont-sur-Ancre occupies the edge of a plateau overlooking a broad stretch of the Ancre Valley, from its upper reaches to the east, to its confluence with the river Somme to the west (Fercoq du Leslay 2000, 113). The Iron Age remains cover around three hectares of previously cultivated land and focus on a group of ditched enclosures (Ill. 6.13). In the northern part of the site is a large square enclosure, outwardly quite similar to Gournay, and it was this which formed the focus of the early excavations.

Curiously, the central part of the square enclosure at Ribemont seems to have been left empty and the excavators have interpreted certain irregular features in that area as tree throws. This has suggested to Jean-Louis Brunaux (2003a) that the complex may have had at its heart a sacred grove of the kind described in classical accounts of Celtic religious practice. Attached to the square enclosure is a larger trapezoidal enclosure extending to the south and east, which has been excavated only in a few limited areas. Within this is a third, oval enclosure with a length of projecting ditch that may once have linked it to the square enclosure. As we shall see, each enclosure is intimately associated with the curation and display of human remains, but in each case this takes on a quite different character, suggesting a complex and interrelated set of mortuary activities. Although the complex has been heavily disturbed by later Gallo-Roman constructions and twentieth-century military trenches, enough remains to enable the piecing together of a remarkable series of activities. It is simplest to break these down by outlining the evidence from each of the three enclosures in turn before discussing what beliefs and motivations might have underlain them.

The first hints of the extraordinary practices carried out at Ribemont came in 1982 with the excavation of a macabre little building tucked into the northeast corner of the square enclosure (Cadoux 1984). This structure, identified as an 'ossuary', takes the form of a square with sides around 1.6 m long, built entirely of human long bones with occasional fragments of iron weaponry and horse bones. Within its cramped interior was a single, deep central pit containing at least 200 litres of cremated human and animal bone. All in all, the ossuary appears to have contained the bones of around 300 adult men and some 50 horses. The human bones had clearly been obtained from corpses that, although heavily decomposed, still retained some articulation of the bones. It appears, for example, that femora were selected with the *os coxae* (hip bones) still attached and were placed such that the *os coxae* formed an inner wall for the ossuary (Fercoq du Leslay 1996, 204). Other articulated bones, including hands and feet, were also intermingled in the construction. Heads, however, were entirely absent.

Renewed excavations, led by Jean-Louis Brunaux in the 1990s, added a second ossuary. This had unfortunately been badly damaged by a First World War trench, but enough survived to show that it had been of similar construction and with slightly larger dimensions. At least one further ossuary has since been identified, and others may well have existed around the periphery of the enclosure.

The immediate source for these body parts seems to have been several extensive concentrations of articulated human remains displayed at various points in and around the square enclosure and the adjoining trapezoidal enclosure. In every case, these remains appear to have comprised men of fighting age, many with perimortem weapon trauma (Fercoq du Leslay 1996, 202). Once again, no heads were found, and several skeletons retained cervical vertebrae showing cut marks associated with decapitation (Brunaux 2003a, 67). Mixed in with these bones were the corroded remnants of weaponry, which enable the excavators to date the deposits to a relatively brief episode at the beginning of La Tène C1, around

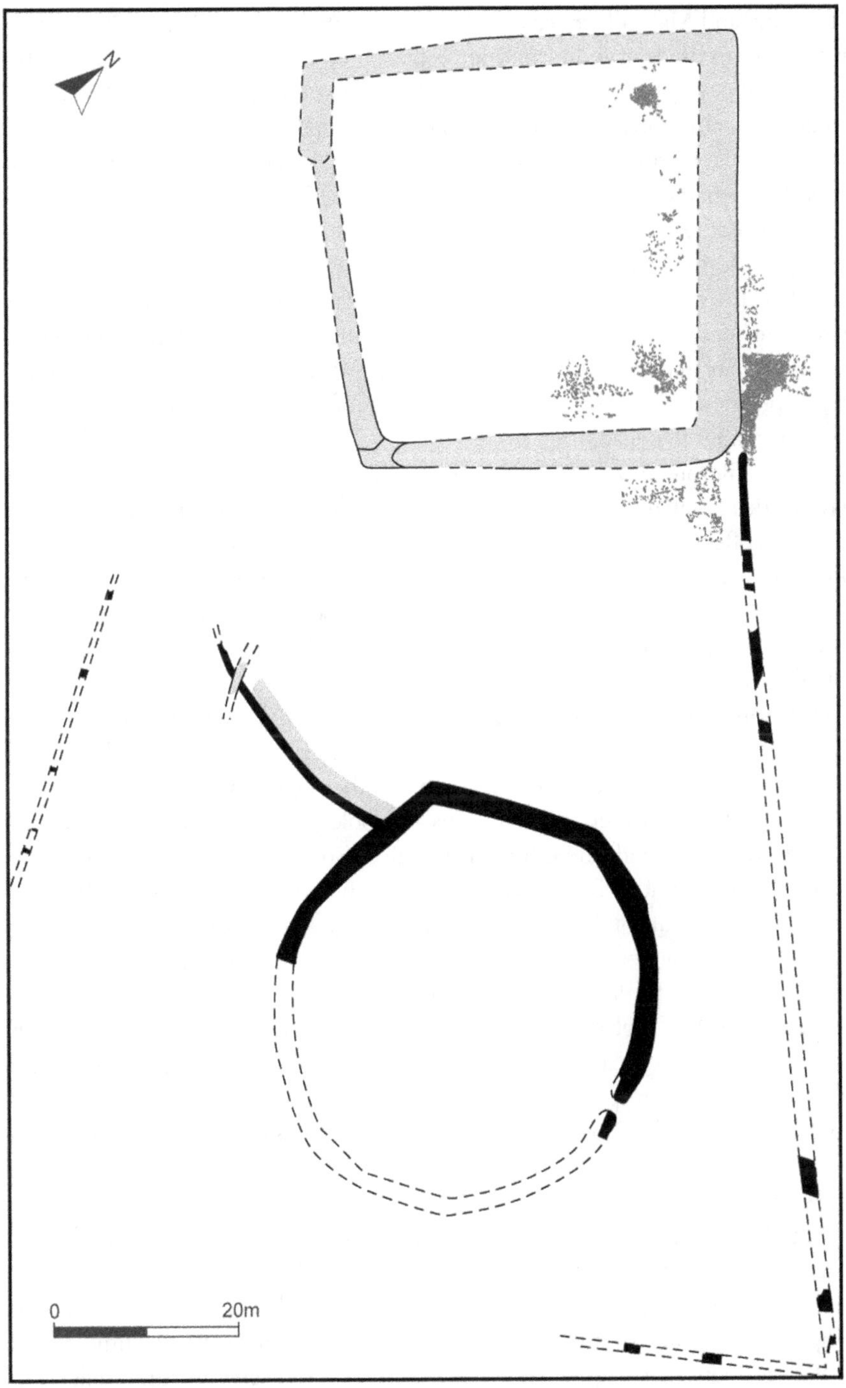

ILLUSTRATION 6.13. Three interconnected enclosures that formed the focus for Middle Iron Age activity at Ribemont-sur-Ancre. The square enclosure at the north end had an open interior with 'ossuaries' of human bones around the periphery. The oval enclosure contained exposed corpses of warriors and was eventually marked by a ring of sandstone stelae. Other dense deposits of human remains occupied parts of the larger, trapezoidal enclosure. (Drawn by Rachael Kershaw after Brunaux 2003a, 65)

260 BC (Brunaux and Malagoli 2003, 34). From the limited excavated areas, it is already clear that several hundreds of individuals are represented. Painstaking excavation of the various concentrations of bones and metalwork shows that individual human bodies had been exposed in great numbers and allowed to decay *in situ*. Brunaux has suggested that the headless warriors, complete with weaponry, may have stood upright on timber racks or platforms. As the bodies decayed, bones and body parts fell to the ground, where they formed intermingled heaps of partially articulated bone. It has been suggested by Brunaux that these displays formed the reservoir of bodies from which the ossuaries were built. If that is true, then the ossuaries cannot be a great deal later than the original displays of corpses, since the bones used to build them were still partially articulated. It seems reasonable then to see the ossuaries and the displays of decaying bodies as parts of a single complex mortuary practice that was used to treat the bodies of several hundreds of warriors – a vast processing area for the dead.

The dominant interpretation of these extraordinary displays is that, like the metalwork at Gournay, the decapitated human bodies represent battle trophies, the mortal remains of defeated warriors offered up to the gods. The recovery of Armorican gold coins among the human bones has led Brunaux (2003a, 67) to go further and to suggest that these men were a defeated force of 'foreign' warriors – victims of a large-scale regional conflict between local Belgic tribes and enemies from the far west. Brunaux believes, however, that he has also found dead of the victorious side.

A little way to the south of the square enclosure was a near-circular enclosure of similar size that seems to have been surrounded by a continuous timber screen. The floor of the enclosure was dug down to solid rock, which was then set with paving to receive the remains of around 60 adult men. On the basis of typological differences in the shields and spearheads associated with these human remains, Brunaux suggests that these men had belonged to the victorious Belgic army – heroes whose corpses were respectfully set aside to be defleshed by carrion birds, while the high plank walls excluded dogs, rats, and spectators. After a lengthy period during which the bodies were allowed to decompose, the enclosure was eventually dismantled, and the course of the infilled ditch marked with around 50 squat sandstone stelae. For Brunaux, this was intended to create a permanent memorial enclosure to the dead heroes, within which their living descendants would come to feast and pay their respects (Brunaux and Malagoli 2003, 37).

It is debatable whether the Ribemont complex should necessarily be seen as the monumental aftermath of a single great battle between the regional powers of northeastern and northwestern France. The complexity of the various deposits, together with the relatively limited extent of excavation and analysis carried out to date urges caution. Nonetheless, there are certain key points that we can take from the work done so far. Firstly, this was a place where the dead were processed on a vast scale within what seems to have been an overtly religious context, and over what appears to have been a quite restricted time period. Secondly, the mortuary rites conducted at Ribemont were restricted to adult males, and the presence of so much weaponry suggests strongly that these individuals were primarily represented in death as warriors. Thirdly, the variety of treatments suggests that these warriors were carefully divided into various categories for whom specific treatments were deemed appropriate; certain of these treatments, notably the display of standing, fleshed cadavers, seem unlikely to have been intended as respectful. For Brunaux, this was essentially a division between in-group and out-group, but other possibilities could be considered. Finally, the nature and scale of all this activity suggest a strongly militarised and hierarchical society, much as we have seen emerging in southern France at around the same period.

One of the most puzzling aspects of the Ribemont complex is, of course, the absence of heads. Brunaux (2006, 107) estimates that at least 500 crania must have been removed from those individuals excavated so far, and the original number may have been considerably greater. Indeed, of the thousands of human bones recovered from Ribemont, only three cranial fragments appear to have been found. Although poorly contexted, these come from near the supposed entrance to the square enclosure (Fercoq du Leslay 1996, 204), suggesting that they may have been fixed to the entrance structure, as was the case at Gournay and at southern French sites such as La Cloche. What then has happened to the many hundreds of missing heads from Ribemont?

One possibility is that heads were removed for special treatment or display and we may simply not yet be seeing those parts of the complex within which heads were processed. A second possibility, however, is that the heads of these dead warriors were removed by their killers on the battlefield, as personal trophies, echoing the written accounts by Livy and others (see Chapter 2). In this scenario, we would seem to have an overt division of control over the body, where the head belongs to the individual victorious warrior, while the remainder passes to the corporate group for separate display and treatment. Decapitated heads would presumably be dispersed to the various home bases of the head-takers for display, curation, and eventual disposal. If so, this would suggest that control over the fate of enemy bodies and enemy souls was far from a chiefly monopoly – hardly suggestive of a highly developed central authority. However, this sort of interplay between individual warriors and the broader corporate group may be appropriate in a situation where corporate power is only just beginning to emerge. In northern France, we see no major centres of population and no great monuments to individual leaders. Indeed, it is only with the appearance of sanctuaries like Gournay and Ribemont in the Middle La Tène period that we see strong evidence for large numbers of people coming together in any form. The overtly religious nature of these complexes may suggest that sacred spaces mediated the interaction of social groups who otherwise maintained a certain social distance from each other, as may have been the case in the earlier part of the southern French Iron Age. While the displays at these sanctuaries were indeed spectacular, real power may have been more fragmented and locally based. The division of bodies and body parts between the corporate group and the individual may have stood for a wider division between regional and local authorities.

Conclusion: Heads and power

Aggression between highly competitive communities within Provence seems to have led to the eventual consolidation of power in the form of the Saluvian confederacy. From the statuary at Entremont and elsewhere, it appears that this grouping was presided over by elite individuals and lineages whose position was founded on a blend of military and religious authority. They distinguished themselves in life through impressive and costly dress, weaponry, coiffure, and jewellery. In death, they were memorialized in stone. Although elite males were presented primarily as warriors, they were not depicted in the midst of battle, but sitting in its aftermath, calm, secure, controlled, and reposed. The primary signifier of the successful conflict is, of course, the severed trophy head lying, singly or in groups, within the grasp of the seated warrior.

The centrality of the human head as a cosmological motif remained constant throughout the long sequence of statuary at Entremont. The strategies applied to the manipulation of this motif, however, had by now changed, as society itself had changed. The earlier Iron Age sanctuaries where the head motif is found were not associated with large secular settlements. The later Entremont statues, however, were displayed within a major (perhaps *the* major) centre of Saluvian power. It is this unification of secular and religious authority that is embodied in the warrior statues. The emergent Saluvian elite had by now begun to promote an ideological vision of itself that encompassed all the traditional cosmological and religious associations bound up in the symbol of the severed head.

As among the nineteenth-century elites of Dahomey, East Sumba, and Roviana, the severed human head had emerged as a central symbol of political power. In each case, the head already carried a potent bundle of associations, with fertility, reproduction, efficacy, and communion with the ancestors. An additional layer of symbolism now accrued, however, related to the subjection and assimilation of outsiders. The head nests of the Entremont warriors, the skull trees of Sumba, the temple offerings in Roviana, and the vast arrays of enemy heads held in Gezo's chambers in the Dahomean capital all served to embody the reduction of enemy bodies to chiefly possessions. As a symbol of this 'ideology of encompassment', the severed human head was uniquely powerful.

In contrast to the third-century BC period when the statues were being carved, the relative paucity of indications of headhunting in southern France for the greater part of the second century BC marks a fairly dramatic change. By now the Iron Age polities of the region seem to have become more internally secure. Groups like the Saluvii, rather than being focussed on internecine fighting, were increasingly drawn into power politics on a much larger scale, involving major regional powers such as Massalia and the expansionist Arverni. As part of this reorientation, headhunting became less prominent both in iconography and in public ritual, reinforcing the suggestion that the practice was peculiarly appropriate for the earlier period of polity building, with its progressive assimilation or 'encompassment' of neighbouring tribes.

The dramatic revival of headhunting during and after the collapse of Saluvian autonomy from the 120s BC reflects a further shift. Some of the most obvious regional manifestations of headhunting date to this period, including the crudely nailed skulls from Entremont and Glanon. Yet, by now, little remains of the overt cosmological or religious symbolism of earlier periods. Instead we see, for example, the grotesque iron skull clamps from La Cloche, something rather closer to the brutalistic trophy display familiar from much later periods.

The similarities in the ideological uses to which severed heads were put in so many diverse times and places are striking. What each of our examples shares is a historical context in which new relations of dominance were being established. In each case aggressive new elites were engaged in the ongoing subjection and absorption of rivals. In such situations, control and manipulation of the human body provided an attractive means by which aspirant elites could stake their claims to social power. Severed heads of enemies could be displayed and curated, giving material form to power relationships that may often have been insecure or transient. Turning one's rival's heads into objects, whether for display, vilification, or ceremonial use, was a powerful and potentially lasting statement of control.

Although heads have formed the focus of this study, it is clear from the sanctuaries of northern France that manipulations of the human body could be much more complex than the simple removal and curation of the head. Near-complete bodies at Ribemont were displayed in a grimly theatrical series of displays, while at Gournay and elsewhere we see the structured use of animal and human body parts in various combinations and configurations. In the final chapter, I will move away from southern France and consider some broader issues relating to ritual violence and the treatment of human and animal bodies.

CHAPTER 7

Gods and monsters

All war propaganda, I went on, proceeded on the assumption that you must degrade the enemy into something bestial, infidel, cancerous, and so on. Or, alternatively, your fighters must transform themselves into surrogate beasts – in which case men become their legitimate prey.

Chatwin 1987, 248

My soul does not live in my body. It lives in the leopard. It is not in me now. It visits me in sleep. I meet it in dreams.... If anything happened to my leopard in the day, my soul would come and tell me. I would get the same wounds.

Attributed to Sangtam Naga 'were-leopard', from private notes of W. G. Archer, 1940s, cited by Jacobs 1990, 85–6

The boy below the floor

The coastal sand dunes of the Hebridean island of South Uist are prone to aggressive tidal erosion. In the early 1980s, a group of stone buildings became exposed at Hornish Point, in the northern part of the island (Barber 2003). Under the floor of one small roundhouse was a cluster of four pits with quite extraordinary contents (Ill. 7.1). Unlike the usual rubbish pits found on such sites, these were narrow, only around 0.4 m across, but up to 0.8 m deep (James and McCullagh 2003, 90). Their shape suggested that they might have held timber posts, but examination of their contents told a different story. First, a deposit of stained sand filled the lower half of each pit. How it got there is not clear, but it was probably deliberately placed. Next, a layer of bones was deposited in each pit in what the excavator described as a 'compacted, jumbled mass' (Barber et al. 1989, 704). Part of one pit was lined with stones set on edge. Finally, the pits were sealed with darker soil, perhaps domestic midden found lying around the settlement. These complex and deliberate fills are puzzling in themselves, but it is the composition of the bone assemblage that makes Hornish Point exceptional.

Although the pits produced a mixture of human and animal bones, it soon became apparent that all the human remains belonged to a single individual. This was, most probably, a boy of around 12 years old, although sexing is imprecise for individuals of that age (Lee 2003). Recent AMS dates have shown that he died around 170 BC–AD 20, and there appears to be no significant difference in date between his death and that of the animals

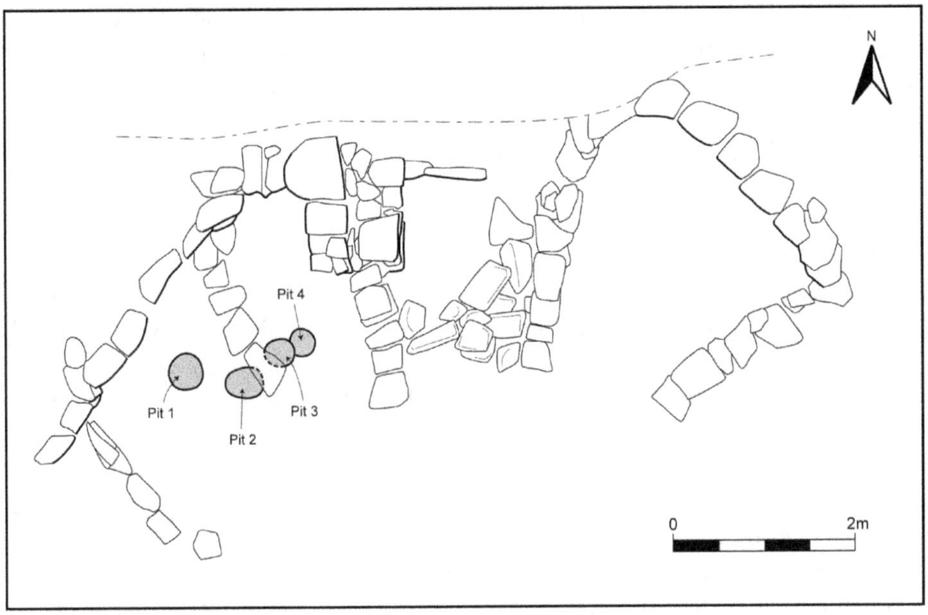

ILLUSTRATION 7.1. Simplified plan showing the pits under the floor of the fragmentary roundhouse at Hornish Point. (Drawn by Rachael Kershaw after Barber 2003, 89)

buried with him (Tucker and Armit 2009). In one pit were his cranium, parts of both arms, hands and the upper trunk; a second pit held most of the pelvis and parts of the left leg; another held much of the right leg and more of the pelvis, while the final pit contained just parts of the left foot. Other small bones were distributed seemingly randomly between the pits.

The general lack of cut marks on the body shows that most of the soft tissue had already gone before the bones were divided up. There were, however, two deep cuts to the lumbar vertebrae which suggested that the boy had received a massive wound to his back: this would certainly have been fatal. Although the excavators suggested that these injuries may have been caused during the division of the body, to enable its deposition in separate pits (Barber et al. 1989), the elements affected by these cuts were actually found in the *same* pit; furthermore, this would hardly have been the most stubborn part of the body to break up, and other more robust articulations, such as the pelvis, seem to have been pulled apart without the need for cutting tools (cf. Tucker 2010). It seems highly likely, therefore, that the Hornish boy was deliberately killed, being struck from behind by an assailant using a sharp bladed weapon with considerable force. It is possible that this untimely death, perhaps the result of a homicide or ambush, had singled out the Hornish boy for this highly unusual treatment. Equally, however, he may have been killed specifically in order to be buried under the roundhouse. This may, in other words, be a case of human sacrifice to provide a foundation deposit for the new building. Foundation sacrifice, usually of captured enemies or slaves, is not uncommon in the ethnographic literature of the nineteenth and early twentieth centuries. Among many other examples, we have already encountered the sacrifice of slaves by Northwest Coast communities to inaugurate carved house poles

(Donald 1997, 171–2), and the killing of outsiders to provide heads for the consecration of log gongs among the Sangtam Nagas (Jacobs 1990, 120). The killing of a young boy to sanctify a new roundhouse is not, therefore, a wholly improbable scenario.

Intriguingly, Strabo records a gruesome practice of the continental Gauls, who allegedly sacrificed human victims by stabbing them in the back (*Geographia*, 4.4.5). In his account, which probably derives from Poseidonius, the death throes of the victim were apparently used in divination. Tempting as it is to associate this practice with the evidence from Hornish Point, it would be rather a stretch to assume some unified divinatory practice across such a broad area, especially given the lack of any comparable osteological evidence in intermediate areas. In any case, the marks on the Hornish bones were apparently 'diagonal chops or cuts' (Lee 2003, 141), which might suggest slashing blows with a sword or axe rather than stabbing. Nonetheless, a ritual killing of some kind would seem the most likely cause of death for the unfortunate Hornish boy.

Murdered or not, the death and burial of the Hornish boy was a complex and drawn out affair. The condition of the bones retrieved from the various pits seems to provide rather contradictory evidence for the treatment of the body. On the one hand, the corpse had clearly become so decomposed that it was possible to break it up into very small pieces without the use of cutting tools. Yet the occurrence of the epiphyses of certain long bones in their correct anatomical position shows that not all the soft tissue had disappeared (Lee 2003). At first glance this might seem like a clear case of excarnation, where the body is exposed to the elements to enable decomposition, before being retrieved for secondary burial. Bits and pieces in varying states of decomposition might be collected and removed to some appropriate place for final deposition. Indeed, excarnation has been a popular explanation for the general absence of formal burial sites in the British Iron Age as a whole (Carr and Knüsel 1997). In this case, however, it does not quite work.

When bodies are exposed to the elements, the first parts to become detached are generally the small bones of the hands and feet. Over time, these and other fragile elements are dispersed, fragmented, and removed, by birds and mammalian predators and through general exposure to the elements. In communal excarnation cemeteries, the bones of multiple individuals become inextricably mixed. In such circumstances, it is more or less inconceivable that the remains of a single adolescent could be recovered and removed intact. Yet the Hornish pits contained large numbers of even the tiniest bones of the hands and feet. Furthermore, the good surface condition of the bones suggests that the corpse was kept safe from the ravages of animals, birds, wind, and weather (Tucker 2010). Although some bones are missing, this can probably be explained by the circumstances of the discovery. The excavators mention, for example, that the pits were 'not preserved in their entirety' and that they lay in part of the site that had been badly scoured by tidal erosion. Perhaps crucially, the excavators also mention three similar-looking pits in a nearby part of the same building that were not excavated (James and McCullagh 2003, 90, 92). It is quite possible, therefore, that all the bones of the skeleton were originally deposited. What seems certain, however, is that this boy's body had been allowed to decompose under controlled and protected conditions, and when the time came to bury him, the bones were carefully collected and handled.

Three of the Hornish pits also contained the bones of young cattle and sheep, each slaughtered at prime meat-bearing age. This in itself was highly unusual, since Iron Age cattle in the Hebrides were either slaughtered for meat shortly after birth or kept for dairying until they were much older (Halstead 2003). Sheep similarly, were either killed in their first year or retained for breeding, in which case they survived to a much greater age. The animals in the pits, therefore, had been deliberately removed from the normal subsistence regime, selected for the quality of their meat. All had been filleted and butchered, and in some cases the bones had been broken to extract marrow (James and McCullagh 2003, 89–92). One had even been chewed by a dog. Even if the animals were eaten on separate occasions, there would have been a significant amount of meat to be consumed, suggesting communal feasting rather than a private meal. Once consumed, the remains of each animal were carefully collected, sorted, and individually placed within the pits; even small splinters of bone were deposited. Thus, two pits each contained a single young bovid, while another contained two sheep, all mixed with body parts from the Hornish boy.

So what was going on at Hornish Point? First, it seems that we have a deliberate killing of a young individual, perhaps as a sacrifice associated with the construction of a new building. There followed a protracted period, during which the body was allowed to decompose almost fully. This must have taken place in a protected environment, where the body was kept safe from predators, wind, and weather. Perhaps it was wrapped and stored within a building, possibly in the smoky roof space where it would have kept dry and out of reach of vermin. Possibly it may have been buried for later retrieval, though this seems improbable, given the less controlled conditions. Both possibilities have also been suggested for certain sites in Wessex where similarly well-preserved but disarticulated human bones have been observed (Wilson 1981, 174). In either case, it must have taken several months, at an absolute minimum, for the body to decay to the point where robust elements like the pelvis could easily be broken up without the use of cutting tools. Mike Parker Pearson has suggested that Late Bronze Age bodies buried at Cladh Hallan, also in South Uist, may have been curated as mummies for several centuries before their eventual disposal (Parker Pearson et al. 2005). While the correspondence in AMS dates between the human and animal remains from Hornish Point shows that this was not the case here, they could nonetheless allow for a gap of several decades between death and burial, if his body was deliberately preserved by drying or smoking.

At some appropriate time, perhaps when the new building works were under way, the body was retrieved with extreme care. Even the tiny finger and toe bones were collected up, and the long bones were handled with sufficient delicacy that the decaying epiphyses remained in place. Prime cattle and sheep were selected for slaughter, and one or more feasts were arranged, perhaps involving a fairly substantial number of participants. The feast took place, one imagines, with considerable ceremony and finished with the painstaking retrieval of the bones of each animal. Carefully regrouped, these were then placed, each in a single pit, along with the remains of the boy. Finally, the pits were sealed and the new building constructed above.

Human bodies, animal bodies

Strange and complex as the deposits at Hornish Point undoubtedly are, there is nothing to suggest that the site itself had any specialist religious role. Indeed, everything else from the excavations is suggestive of domestic occupation on a long-lived settlement site. As we saw with the Cnip wheelhouse discussed in Chapter 1, domestic sites in Iron Age Scotland frequently produce evidence for the deposition of human remains. Here the dead accompanied the living, embedded in the physical fabric of their dwellings. Bodies, partial or complete, had meanings for the community that were important in assuring that life progressed as it should, and that the house and its inhabitants were adequately protected from supernatural threats. The diverse ways in which human remains were deployed suggests that such threats were themselves perceived as many and varied. And the frequent occurrence of animal remains in such deposits suggests that the boundary lines between animal and human were weak and fluid.

As we have seen, the animals in the Hornish pits had been butchered and consumed in a fairly conventional way, while the human bones had been subjected to no such treatment. Were the animal bodies in the Hornish pits simply food debris, mementoes of a funeral feast, or did they relate in some more significant way to the body of the boy himself? Did the age of the animals, not so young as those normally consumed for meat, but not as old as those that died at the end of their useful lives as dairy cows or breeding ewes, reflect the age status of the boy, between childhood and adulthood? Did the consumption of these animals in some way stand for the consumption of the boy himself, a form of surrogate cannibalism? Certainly the final treatment of the animal remains, carefully collected and sorted, reflects the treatment given to the boy.

The mixing of animal and human remains occurs at other sites in the region. The Cnip skull discussed in Chapter 1, for example, was accompanied by a small fragment of unidentified animal cranium. At the roundhouse of Whitegate, in Caithness, a chamber within the walls held a dense deposit of fragmented animal bones, on top of which had been placed a human cranium and articulated arm bones (Heald et al. 2007; Tucker 2010). Another human skull, found in a pit within a roundhouse at Kintradwell, Caithness, was apparently accompanied by the skull of a horse (Tucker 2010).

More detail comes from a recent excavation at Sloc Sabhaidh in North Uist, where a remarkable series of deposits occupied part of the roundhouse floor (Rennell and McHardy 2008). At the base was the upper stone of a rotary quern, its central perforation blocked by a small stone. The stone could easily be removed to give access to the hole, which perhaps acted as a conduit between the inhabitants of the roundhouse and the world below. Later, access was blocked by a deposit of burnt material containing a good deal of charcoal and several iron objects. Next, a low stone kerb was built around the area containing the quern, into which was deposited a mass of cremated animal remains, burnt bones, and stone. Over this material, perhaps when it was still smouldering, was placed a group of semiarticulated animal remains. Finally the whole deposit was sealed by the placement of a single human mandible (Rennell and McHardy 2008, 15; Ill. 7.2).

The Sloc Sabhaidh deposit dates a little later than the one at Hornish Point, around 220–390 AD (Tucker and Armit 2009), but it lies within the same broad tradition. Again the date of the human and animal bone is almost indistinguishable, which suggests no long-term curation, although the disarticulated nature of the human bone suggests that an extended period may once again have passed between death and deposition. The solitary human bone, perhaps importantly a part of the head, seems to have acted as a token in a deposit made up overwhelmingly of animal remains. Other pits contained solely animal

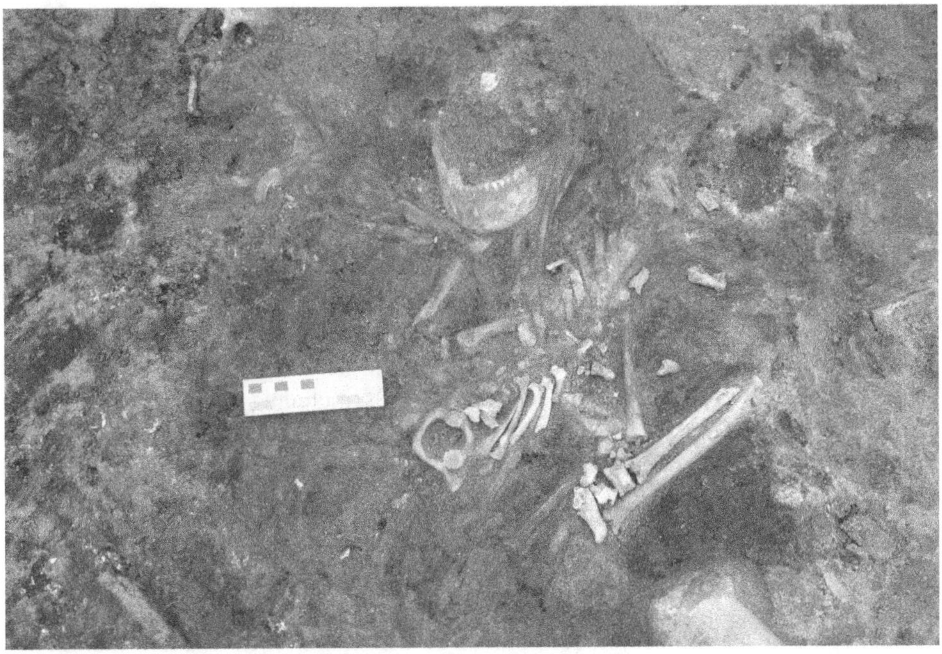

ILLUSTRATION 7.2. Deposit within the floor at Sloc Sabhaidh. (Photograph by Tom Dawson © SCAPE Trust)

remains, some semiarticulated (Rennell and McHardy 2008). Did the human bone give added power to this particular deposit? Did it complete some supernatural 'recipe' required for the pit to fulfil its (presumably ritualistic) function?

Although deposits containing human remains are reasonably common in the Atlantic Scottish Iron Age, deposits containing purely animal remains are even more widespread. At Cnip for example, as well as the human cranial deposits, there were numerous animal deposits, including bird's heads and cattle vertebrae built into the walls and sheep burials associated with thresholds (Armit 2006a). The archetypal site for such deposits is the Sollas wheelhouse in North Uist, where hundreds of animals in varying states (cremated, disarticulated, complete) were deposited within pits inside the building during the course of its occupation (E. Campbell 1991).

Similar deposits across Atlantic Scotland suggest a shared cosmology, where the boundaries between humans and animals were fluid and could be transcended in death. The deployment of various animal species (usually domesticates but occasionally also deer) in structured deposits within domestic locations suggests that different animals were perceived as having diverse meanings or powers. It is possible that they symbolised various deities or spirit beings, but the diversity of practice does not suggest the widespread recognition of clearly defined gods with specific rites of worship. As with human remains, animals could be deposited as complete bodies, stray bones, cremated bone, or partial articulated elements. They could be old, young, or mature individuals. Everything suggests that the same set of principles was applied to the deposition of humans and animals. Does this mean that these were special animals that had been accorded an enhanced status on a par with humans? Or does it mean that the people whose bodies were deposited on domestic sites were regarded as less than fully human – perhaps outsiders, slaves, or transgressors?

Mixed deposits containing both animals and humans are also found throughout the Iron Age of Wessex. The grain pits at Danebury hillfort are among the best-known examples, containing a range of human bodies and body parts analogous to the deposits found in Atlantic Scotland (e.g., Cunliffe 1995; Craig et al. 2005). Frequently these remains form parts of elaborately structured deposits made over periods of months or even years (Hill 1995). Animal remains are a frequent component of such assemblages, and again these reflect the variable condition of the human bones, being variously complete, partial, or disarticulated. Wessex and Atlantic Scotland share a range of soil types conducive to the preservation of bone. In many intervening areas, including virtually all of southern Scotland, unburnt bone simply does not survive to the levels necessary to preserve such complex deposits. The underlying cosmologies may potentially, therefore, be considerably more widespread than the various 'islands' of preservation currently suggest.

People and animals in southern France

Cosmological relationships between animals and humans, such as those that existed within the relatively fragmented and nonhierarchical societies of Atlantic Scotland, can also be discerned within the more complex social structures of the southern French Iron Age. Much less attention has been paid to animal deposits in this region, but examples are beginning to emerge, particularly from recent work on sites around Martigues in Provence.

The incidence of animal deposits varies greatly over time and space, even within fairly small geographical areas. Around Martigues, for example, sites such as L'Arquet and Saint-Blaise yield only a small number of deposits concentrated in particular buildings, whereas houses from L'Île de Martigues produced many tens of examples, and Saint-Pierre de Martigues many hundreds (Chausserie-Laprée 2005, 232). Deposits from the latter site date from the fifth through to the first centuries BC and are mostly contained in small pits around 0.2–0.4 m in diameter and around the same depth.

In some instances wild animals have been chosen, as at the oppidum of Saint-Blaise, where the remains of a fox cub were found buried under the threshold of a house. The cranium, jaw, and hind-quarters of this unfortunate animal had been carefully pierced with bronze needles in what was clearly a propitiatory or apotropaic offering (Chausserie-Laprée 2005, 230). We have already seen in Chapter 5, the burials of wild birds and snakes, often placed in urns, at sites such as Les Castels de Nages, Ambrussum, and Ensérune, in Languedoc (Py 1990, 805). Overall, however, wild animals are in a minority. Around 75 percent of the pits from Saint-Pierre de Martigues contained sheep or goats, and another 20 percent held pigs, although larger domesticates, like cattle or horses, were extremely rare (Chausserie-Laprée 2005, 232). The proportions were thus quite different from those found in apparently unstructured deposits of food debris and show a deliberate selection of certain species. The presence of ashes and charcoal, as well as indications of butchery and burning on certain bones, shows that some had been cooked and partly eaten before burial, though this was not always the case, and many were buried whole and unbutchered, suggesting sacrifice rather than feasting. Occasionally animal deposits were accompanied by small objects, such as arrowheads, pins, brooches, and polished bronze disks, of the sort found as offerings at religious sites, such as hilltop sanctuaries and caves. The focus on smaller animals may reflect the domestic nature of the deposits and the limited resources of the households involved. This is further supported by the fact that more than 80 percent of these animals were less than three months old with consequently limited investment in their upkeep. Pits containing neonatal or very young pigs and sheep, for example, were found at the base of many house walls at Roquepertuse (Boissinot 2004, 52). Often, the deposits were contained within rough handmade ceramic vessels, quite unlike the wheel-thrown fine wares common in the region at this time, again suggesting a relatively modest level of investment. This is not to say that there were never any more substantial offerings. In one deposit at Saint-Pierre, for example, nine sheep skeletons and three pigs, all complete or near complete, were found.

In virtually all cases, animal deposits occurred within private houses, avoiding the communal streets and open spaces. As in Atlantic Scotland, these animal deposits seem to result from essentially small-scale domestic rituals associated with particular households. In this respect they resemble the earlier deposits of human heads, at sites like La Liquière, discussed in Chapter 4. Unlike human heads, however, animal deposits remain limited to the domestic sphere throughout the Iron Age. Even in the earlier Iron Age there is no evidence for the deliberate combination of animal and human remains in single deposits, as we see in Atlantic Scotland at Hornish Point, Whitegate, Sloc Sabhaidh, and elsewhere. The deliberate combination is also seen fairly widely in northern France, for example, in the ossuaries of human and horse bone at Ribemont, discussed in the previous chapter. Here in southern France, however, the cosmological boundaries between human and animal may have been more rigid.

Inhabiting the leopard

Cosmological relationships between humans and animals have been equally significant in more recent societies. In Naga cosmology, for example, humans are perceived to have multiple souls, certain of which can move beyond the physical body of the individual and inhabit the body of an animal. The testimony of a Naga 'were-leopard' (cited by Jacobs 1990, 85–6, and quoted at the beginning of this chapter) shows the degree to which the human and animal realms were perceptually intertwined. Many Naga groups believed that particularly close relationships existed between humans and large predatory cats; the tigers and leopards that headed the natural food chain. Origin myths related the shared ancestry of humans and tigers in particular (85). Warriors who took the heads of these animals underwent the same purification rituals required by those who captured human heads. Monkeys too were regarded as closely related to humans, to the extent that they could be perceived as 'a low clan' by the Konyak Nagas (125). Their skulls, sewn into headdresses, bags, and clothing, could serve to symbolise the status of successful headhunters. This close affiliation prompted unsuccessful attempts by the British colonial government to persuade the Nagas to take monkey heads in place of human ones.

Human souls could also inhabit other animals such as the mithan, a large indigenous ox that played an important role in Naga economy and belief. Clans within Naga society identified themselves with certain species of animals and birds, and a series of taboos existed in relation to their consumption. Totemism is common ethnographically, and it would be no surprise to find it in Iron Age Europe. When we first hear the names of tribes in Atlantic Scotland, in the writings of Ptolemy in the second century AD, we hear of a tribe in Argyll named the Epidii (horses), others in Sutherland called the Caereni (sheep) and Lugi (ravens), while the Latin name Orcas (for Orkney) refers to island of the young pigs (Armit 2005, 70). Parker Pearson (1999) has suggested that subgroups within Iron Age East Yorkshire were identified with the boar or the sheep and that joints of meat from the appropriate species were placed with them in the grave. Against such a background, it is easy to see how the deposition and physical manipulation of animal body parts might relate closely to equivalent treatment of humans.

This distinction between Atlantic Scotland and southern France in the fluidity of human or animal identities may relate to the wider characteristics of communities in these two very different regions. Atlantic Scotland was occupied almost exclusively by individual households occupying monumental stone roundhouses, notably the broch towers that seem to have flourished in the last three or four centuries BC (Armit 2003). These were usually set some distance from their neighbours and project a sense of isolation and self-sufficiency, even if some cooperation between groups must have taken place at certain points in the agricultural cycle. There is no archaeological evidence for trade or interregional contacts on any significant scale, and contacts with nonlocal groups would have been sporadic and limited. In tightly bounded societies with strong internal ties and few external contacts, outsiders will often be regarded with fear and mistrust. Their manners, appearance, and language may be alien to daily experience; more a part of the untamed outside world than the familiar domesticated environment of the village, household, or family. Their humanity may be hard to gauge, particularly where clear boundary lines are not always drawn between humans, animals, and spirit beings. In such circumstances, outsiders may very well have been conceptualised as wild or semihuman. Such ideas can be found in latter-day island folklore, where dangerous, shape-shifting, supernatural beings like the seal/human selkies give expression to deep-rooted anxieties over the dubious humanity of strangers and outsiders who appear from the sea. Here the distinctions between human and animal bodies blur almost completely.

The communities of southern France, even at the beginning of the Iron Age, lived in significantly larger groups and were exposed to a much wider range of social interactions on a regular basis. The area lies on the intersection of long-established trade routes that link communities characterised by very different ethnic, linguistic, and social backgrounds. The presence of imported pottery shows the receptiveness of local communities to alien products and ideas. Outsiders may still be dangerous, and life was still lived mainly amid the security of the kin group, but a much larger social world was regularly experienced and increasingly embraced. In such circumstances, it seems that the human-animal divide was more clearly conceptualised and reflected in an absence of hybridity in depositional practice.

Fighters into beasts: The Tarasque of Noves

Before we leave the issue of hybridity, it is worth considering one last, rather problematic object from the southern French Iron Age. The creature known as the Tarasque of Noves, now housed in the Musée Calvet in Avignon, is an extraordinary stone carving depicting a ferocious animal, sitting in an unnatural, humanlike position. Although usually identified as either a wolf or a lion, it appears to draw on the artistic or mythological imagination more than any physical reality (Ill. 7.3). Rather appropriately, it takes its name from a chimerical river monster of medieval legend, said to haunt the Rhône and its surrounding area. The sculpture is carved in the round at more or less life-size and has a slightly cartoonish appearance, with exaggerated teeth and claws, and an erect phallus. Its teeth clasp a somewhat diminutive human arm (probably originally part of a complete figure), and gripped under each of its front paws is a bearded human head. The two severed heads are shown with dead, empty eyes, and both appear to wear skullcaps similar to those visible on several of the trophy heads associated with the Entremont warrior statues. Indeed, the whole presentation is remarkably reminiscent of the Entremont figures, representing the aftermath of a violent confrontation. Here, however, the aristocratic Saluvian warriors of Entremont are replaced in the composition by a ravenous wild beast.

The Tarasque was acquired in 1849, having been found 'on a pile of stones, 2.5 m deep in "the field ... behind the church at Noves"' (Arcelin 2004b, 41–2). The surrounding area contains evidence for a long-lived Iron Age and Gallo-Roman settlement. Although some reservations have been expressed over its Iron Age credentials (e.g., Duval and Heude 1983), these seem to have little evidential basis. The sculpture has no convincing Roman or medieval parallels, and since the Tarasque was found long before the Entremont warrior statues, it can hardly be a fake consciously referencing them. Frustratingly, the closest iconographic parallel for the Tarasque is also unprovenanced and undated. The 'Linsdorf monster', reputedly (but probably not) from Alsace in eastern France, depicts an even more cartoonish beast, once again clutching two severed human heads (Stead 1985; Ill. 7.4).

Opinions on its date have generally varied between the third and first centuries BC on the basis of various stylistic and iconographic features (e.g., V. Megaw 1970). Pottery found with the Tarasque, however, would seem to date its deposition to the period around 100–50 BC, the second generation or so after the Roman annexation of Provence. That has led to recent studies taking the view that the statue itself dates to the immediately postconquest period (Arcelin 2004c). Given the exceptionally long use-lives of the other Iron Age carvings considered in earlier chapters, however, the statue may actually have been carved rather earlier.

One important recent observation concerns the striking similarity between the technical rendition of the flowing beards on the severed heads gripped by the Tarasque and the equally sinuous manes of the horses in the equestrian statues at Entremont (O. Coignard and Coignard 2004, 79). These are so close stylistically that they may even have been carved by the same group of crafts-workers. These particular Entremont statues are dated typologically to the end of the third century BC, slightly later than the warrior statues but predating the foundation of the oppidum. A similar date for the Tarasque seems plausible,

ILLUSTRATION 7.3. Two views of the Tarasque of Noves. The detailing on the back (*on facing page*) appears to represent a shaggy mane, suggesting that the statue is intended to represent a lion. The overall effect is hardly naturalistic, however, and it may be a deliberate composite of wolf, dog, and lion. The statue stands around 1.1 m high. (Photograph by author)

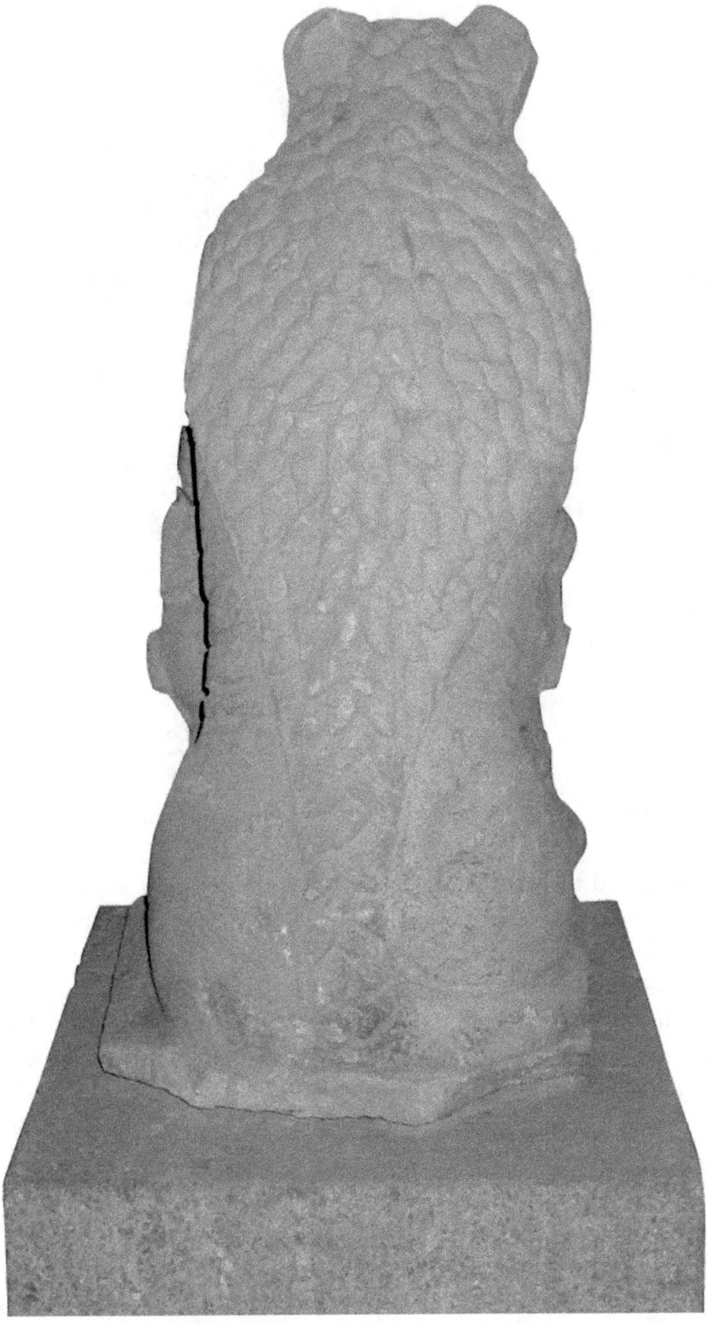

ILLUSTRATION 7.3. *(continued)*

ILLUSTRATION 7.4. Two views of the 'Linsdorf monster', an undated and effectively unprovenanced object with striking parallels to the Tarasque of Noves. The two trophy heads in its paws are clean-shaven but otherwise mirror those of the Tarasque, with their prominent dead eyes. The statue is rather smaller, at around 0.55 m high. (Drawn by Rachael Kershaw after Stead 1985, plates 1 and 3)

allowing us to bracket its creation between the end of the third century BC and the middle of the first.

Patrice Arcelin has suggested that the Tarasque might have come from a necropolis, but in fact there is no real evidence as to its original context or purpose. Certain aspects of the sculpture, however, suggest that it may have been an object of veneration in much the same way as the earlier Entremont statues. One of its most prominent features is a smooth hollow directly in front of the phallus and between the two severed heads. The surface treatment here contrasts with the rest of the carving, which is heavily detailed with stylised representations of the animal's shaggy coat. Although its purpose can only be conjectured, it is interesting that this smoothed area forms a perfectly sized and shaped receptacle for a real human head. Indeed, when a skull is placed within this niche, the phallus appears to emerge directly from the top of the head (as can be seen in the photograph in O. Coignard and Coignard 2004, fig. 9, where they do just that), recalling the cosmological links between heads and fertility discussed in Chapter 4. Intriguingly, the Linsdorf monster is also equipped with a space between the two carved heads for the insertion of a real human head or some similarly sized object (ibid., plate 2).

One further observation made by Roland and Olivier Coignard relates to the detailed depiction of the severed heads. In their study of the morphology of the Tarasque, they point out that there is a very sharp line dividing the front of each head (carved in great detail, showing face, beard, and skullcap) from the back, which is plain and smooth, as if showing the fleshless skull. There is also a marked change in level between the front and back, which appears to suggest that the face has been placed over this bare skull (O. Coignard and Coignard 2004, 63–4, figs. 4 and 5). We have seen in previous chapters how faces may have been preserved or reconstituted for display in pillars at Roquepertuse and Glanon. Indeed, the 'knuckles' of stone within the niches at Glanon seem designed for just such a purpose (Ills. 5.19 and 5.20), and it appears that the sculptors of the Tarasque may be making explicit reference to that same practice. This rather changes the implied narrative of the statue, however, since the heads could now be seen not as recent victims of the Tarasque but rather as trophies or relics of individuals (whether enemies or ancestors) who had already been dead for some time.

If we assume, as seems probable, that the Tarasque is a genuine product of the Late Iron Age, it is clear that it was intended to make specific reference to the slightly earlier Entremont warrior statues or at least to invoke the iconographic tradition of which they were a part. Here, however, the image of the headhunter at repose in the aftermath of violent conflict is transposed to the image of the predator at rest with the remains of its victims. The multiple trophy heads of the Entremont warriors are reflected in the two heads held by the Tarasque, which occupy more or less the same position within the overall composition – in this case, clamped between front and rear paw rather than between hand and knee. For at least some of those who viewed this object, certain ideas and associations would undoubtedly have arisen relating to the established ideological expression of Saluvian military power.

As well as these indigenous resonances, the Tarasque also bears influences from broader Mediterranean traditions. The motif of the predator carrying human body parts in its mouth has innumerable parallels in Roman and Etruscan art, from funerary contexts dating at least as far back as the fifth century BC. The same theme is recurrent in the broadly contemporary situla art of northern Italy and Slovenia (Turk 2005) and extends northwards into the La Tène art of Central and Western Europe (Duceppe-Lammarre 2004). Indeed, the motif of the 'devouring beast' extends both geographically and chronologically over a remarkably wide area, even turning up in the form of a monumental, life-size lioness from Cramond in central Scotland in the second or third century AD (Hunter and Collard 1997). Closer to home, the Tarasque also evokes a series of life-size lion carvings (some clasping human heads) found in southern Iberia (Aranegui Gascó 2004b) and at the Early Iron Age oppidum of Mailhac in Languedoc (Gailledrat and Bessac 2000, 302). On one level, therefore, it seems to mark a translation of indigenous cultural expression into a more conventional classical form. None of these other images, however, displays the anthropomorphism of the Tarasque, seen in both its humanlike sitting position and its mirroring of the earlier warrior figures.

In the Roman world, the predator motif would usually be taken to symbolise the triumph of death over the living, making it highly appropriate for a funerary context, and some such interpretation may plausibly explain the appearance of the Tarasque. There may, however, have been more to it than this. The Tarasque is later than the Entremont warrior statues; it may date to the end of the third century BC, or perhaps later still. By this time, the Saluvian polity was entering its period of greatest power and stability. Rather than celebrating individual members of the Saluvian elite, as the Entremont statuary had done, the Tarasque seems to embody a more abstract ideal; perhaps a corporate Saluvian identity formed in opposition to the Greek cultural centre of Massalia. As in the quote from Bruce Chatwin's *Songlines* (1987, 248), that opens this chapter, the unified components of the Saluvian confederacy may by now have conceptualised themselves as a vengeful beast, with humans as their 'legitimate prey'.

Medicine, muti, and the spiritual anatomy

The discussion in this chapter has moved some way from the consideration of heads and headhunting. The ritualised violence seen in the killing and dismemberment of the Hornish boy, and implicit in the iconography of the Tarasque, suggests a broader concern with the symbolism of the human body that extends beyond any simple fetishisation of the head. The blending of human and animal, seen in the mixing of bones and bodily attributes, calls into question the conceptualisation of humanity relative to other species in Iron Age cosmologies. The highly varied evidence from Atlantic Scotland also forces us to address the complexity of body treatments in the Iron Age. Although heads were still privileged as objects of curation and deposition, a range of other body parts was quite deliberately deployed in structured deposits of one kind or another. The southern French obsession with the head seems more diluted here, submerged within a broader range of cosmological expression.

In this respect, the communities of Atlantic Scotland are perhaps more representative of the broader European picture. In southern England, for example, many combinations of body parts turn up in the Danebury pits (Cunliffe 1995). In northern France we have already seen the range of mortuary rites practiced at Ribemont-sur-Ancre and elsewhere. At other northern French sites, most notably Acy-Romance in the Ardennes, we see the burial of the complete bodies of young men in a squatting position similar to that of the Entremont warrior figures (Lambot 2006, 180–3). What then might such variety of depositional practice mean?

In the sanctuaries of northern France, we see the manipulation of human bodies on an enormous scale. Distinct categories of people were defined for separate treatment. Various parts of the body were removed, reassembled, and displayed in a vast corporeal choreography of death, power, and religion. The display and processing of human bodies followed regular and recurrent patterns. By contrast, in Atlantic Scotland, communities operated on a smaller canvas, expressing similar sets of ideas through the treatment of the dead in a more intimate, domestic context. The blending of humans, animals, and objects according to varying depositional recipes, here displays a seemingly improvisational approach to the material expression of ideas about people and their place in the world. Depositions of human and animal remains seem not to have been regular and predictable events, repeated time and again to the same script. Rather they seem to have been irregular and expedient, drawing on a body of shared cosmological knowledge but applying that knowledge creatively to each new circumstance. Among a hugely diverse range of deposits in Orkney, for example, we find articulated hands deposited in the broch village of Gurness, severed legs in a boundary ditch at St Boniface, and a severed arm under a wall at Bu broch, as well as the expected range of cranial fragments and isolated long bones (Armit and Ginn 2007).

The relationship between creativity in religious expression and the shared cosmological knowledge that underlies it is analogous to that between artistic creativity and established structures of thought. Creative tension exists in traditional societies between 'a shared understanding of cultural structures' and 'individual intuition' (Layton 1991, 143) – a tension, in other words, between structure and agency. Individual artists can create fresh images and explore new metaphors and new ways of representing deeply embedded cultural ideas; these underlying ideas, however, may be very resilient and resistant to change, even if the accumulated actions of artists over time might eventually begin to impact on the way the world is understood and represented.

Who then were the 'artists'? Who made the decisions over which body parts to place in these various pits, walls, and floors? What factors influenced the selection of a calf, a lamb, or a human being for deposition? Why was it sometimes an arm, sometimes a head, or sometimes just a fragment of bone that was selected? The formality and drawn-out nature of some of these ritualised events (Hornish being the prime example) might suggest the presence of ritual specialists or 'occult technicians' (Gell 1992, 49). The fragmented nature of society in Iron Age Atlantic Scotland, however, and the sheer diversity of ritual practice, is not suggestive of any organised priesthood or codified religion. It is perhaps more likely that certain individuals or groups were recognised as having greater ritual knowledge than their peers, perhaps by virtue of age, descent, or life experience – shamans, sorcerers, or elders, for example. Certain small-scale ritual performances may, however, have been undertaken entirely without recourse even to semispecialist ritualists. Among the Nagas, for example, individual households carried out most domestic rituals independently, deferring to village shamans only for more serious or solemn matters (Jacobs 1990). Even violent rituals involving the killing of humans need not imply the active involvement of specialists. Among Native American communities of the Northwest coast, wealthy individuals killed their own slaves for ritualised deposition at the inauguration of new house poles and other significant events, without the intercession of dedicated ritualists (Donald 1997).

Displays and deposits of human and animal remains were presumably intended to bring benefits to the household or wider community through mediation with the supernatural. In certain cases, they may have been intended to bribe, cajole, or placate ancestors, gods, or spirits of various forms. This seems especially likely where remains were removed from circulation, for example, through burial in house floors and pits or under thresholds. It is worth noting, however, that deposition of this kind is only rarely documented in ethnographic accounts of more recent societies. Much more common is the circulation of objects or body parts within the world of the living.

Across much of North America, indigenous peoples curated small collections of sacred objects within skins or hide bags known as medicine bundles. The precise role and importance of medicine bundles varied between communities, but among the Mandan, for example, they formed a major focus of ritual life (Seeman 1988). Individuals within Mandan society assembled their own medicine bundles, which comprised collections of apparently ordinary objects that were nonetheless invested with considerable symbolic and spiritual significance (Bowers 1950). More important still were bundles held collectively by the tribe, containing objects associated with ancestral spirit beings or representing tribal creation myths. One Mandan tribal medicine bundle, for example, contains two curated human skulls believed to belong to semidivine bird-men with powers to bring rain. The detailed symbolism of such collections could never be unravelled from the objects alone, but the recurrence of certain object types, such as human and animal skulls, hides, and pipes, would nonetheless provide some clues as to the shared cosmologies underlying them. It is interesting in this context to recall the deposit from the Cnip wheelhouse discussed in Chapter 1. Here we had an adult male cranium, an unusual smooth stone, two fragments of pottery, one of them decorated, a piece of animal skull, and a few other stones, all laid together within a scoop in the sand. During excavation, my focus, as the excavator, was on the human skull and, to a lesser extent, the other artefacts; the other natural, unmodified objects seemed initially to be random and perhaps fortuitous inclusions. But in retrospect, it may have been the collection itself, and the relationship between these objects that was most important, even if we cannot now grasp what the nature of these connections was.

Perhaps the closest parallels among present-day traditional societies for the sorts of deposits encountered in Iron Age Britain come from the practice of muti in southern Africa. The term muti describes types of traditional medicine that draw upon a range of natural substances to treat illness of various kinds. However, the same belief systems are also employed more generally to induce wealth, fertility, or success. Among the substances commonly used in muti are human and animal body parts, and it is here that we start to see some parallels with the types of evidence we find in the Iron Age.

In some parts of southern Africa, muti killings, carried out to retrieve human body parts, are endemic (e.g., Bhootra and Weiss 2006). These killings are distinct from religious sacrifice in that the purpose is to retrieve specific body parts from the victim that are held to provide benefits to the patient or client (depending on the nature of the commission). Muti is held to be effective, not through the intervention of any specific god or supernatural being, but because of the cosmological connections between the substances used and the intended effects. Judgement as to exactly what substances these might be is mediated by ritualists, in this case traditional healers who draw upon a shared cosmological system yet exercise considerable interpretative autonomy. The parallels between modern muti and the Iron Age deposition of human and animal body parts are hardly exact, but the parallels are nonetheless striking. In both cases we see the selection and harvesting of specific body parts, animal and human, for specific purposes – signs of an underlying 'spiritual anatomy', where the various parts of the body have specific symbolic value (Leach 1977, 174).

Conclusion: Fragmentation and identity

Modern muti operates in the service of individuals with very specific objectives in terms of their personal ambition, health, and affluence. The deposits of human and animal remains that characterise the Iron Age of Atlantic Scotland and southern England seem intended, by contrast, to benefit the corporate group. Thus, we find them in communal locations, inside houses where extended families spent the bulk of their existence, eating, sleeping, and working together. If Iron Age deposits were intended to have effects similar to those of modern muti rituals, then these were probably intended to benefit the family as a whole. Indeed, the evidence for any form of individual, as opposed to communal, identity in this period is problematic.

Archaeologists have recently become interested in the idea of 'personhood' in the prehistoric past – the ways in which personal identities were conceived and expressed, often through the deployment of material objects (Fowler 2004; Borić and Robb 2008). It has been suggested that Iron Age communities are unlikely to have shared modern Western concepts of the individual as an autonomous, bounded entity, or the Cartesian separation of mind and body. Anthropologists, including Marilyn Strathern (1988), have argued that people in many traditional societies defined themselves through their relationships – with other people, with their ancestors, with the places they inhabited, and with the objects they made, used, gave, and received. Their identity (or personhood) can be conceived as 'dividual' rather than individual, formed of many component parts. The division of the body after death (as we see archaeologically), could give material form to these sorts of relationships. Retention of certain parts of the body might express relationships with the dead person or the community to which they belonged. These physical mementoes created and maintained bonds between one group and another in a process that John Chapman (2000) has described as 'enchainment'.

Archaeological discussions around personhood and enchainment have generally focussed on consensual relationships between communities (exceptions include Fowler 2004, 152–3). Relationships were established through kinship, the giving of gifts, exchange of marriage partners, cooperative labour, and innumerable other mechanisms. Indeed, many of the human remains deposited in Atlantic Scotland may derive from exactly these sorts of relationships. Others, however, may not, the Hornish boy being a prime candidate. The Hornish boy was a 'dividual' in a very literal sense. The physical division of his body and the placement of distinct parts in distinct places with various animals hint at a posthumous sorting of the physical body into certain conceptual categories from which his identity was constructed – a physical rehearsal of certain metaphysical relationships.

When it comes to headhunting, concepts of personhood and enchainment are potentially highly relevant, and yet the relationships expressed are the opposite of consensual. Headhunting underscores the difference between in-group and out-group. Enemies are dehumanized and quite literally objectified. The transformation of dangerous outsiders into cultural objects that can be ritually cleansed, contained, and controlled within the domestic sphere is a powerful metaphor for the subjugation of the external world. The trophy head remains as a physical token and as a reassurance of the ability of the community to impose itself on the world and achieve some level of control over forces that might otherwise overwhelm it.

The focus on the head, which seems so dominant in southern France, and which has so often been transposed from there to the wider Celtic world, is simply one regional expression of a much broader set of concerns relating to the human body, personal and corporate identity, and the place of humanity in Iron Age cosmologies. Across Europe, the human body became a powerful medium through which cosmological, religious, and ideological relationships were all too frequently played out.

CHAPTER 8

Bodies of belief

> *They had taken Kauhiakama prisoner to Oahu and roasted him in an oven, and they used his skull as a filth pot.*
>
> Kamakau 1961, 232

A cult of the head?

The human head held an important place in the social and religious life of Iron Age Europe. In southern France, its overwhelmingly dominant role in indigenous iconography, as well as its physical presence on so many domestic and religious sites, suggests that it was fetishised above any other object. In this region we can trace the development over time of what we might reasonably describe as a headhunting society. Headhunting was by no means a static phenomenon, and as we follow the dynamics of social change in southern France from the end of the Bronze Age to the period of Roman annexation, we see ideas about the head shift, alter, and adapt.

In the earlier part of the period, headhunting seems to embody cosmological ideas relating to the fertility and reproduction of the community as a whole. It has no obvious links to individual heroic warriors or to wider power structures. Over time, however, beliefs and practices surrounding the head appear subtly to change. With the construction of monumental sanctuaries, as at Roquepertuse and Glanon, the human head is increasingly cast within a more formalised religious iconography. Places of long-standing religious significance seem to act increasingly as foci for larger and more integrated communities. In these sanctuaries, the human head remains central to the iconographic language with which new ideas were articulated.

With the emergence of the Saluvian confederacy during the third century BC, the symbolic role of the head takes on another dimension. The Entremont warrior statues embody a new iconographic expression where the person of the individual heroised warrior takes central stage. Although drawing deeply on the pre-existing cosmological and religious power of the head, these sculptures articulate an emerging social and political ideology associated with the establishment of new power structures. As well as glorifying the individual headhunter, we have seen how these statues embody an 'ideology of encompassment' (Hoskins 1996c, 225), symbolising the increasing depth and reach of Saluvian power. Following the destruction of indigenous autonomy by the Roman army, headhunting emerges once again in the first century BC at sites like La Cloche: by now, however, much of the more overt religious and cosmological imagery seems to have been stripped away. From an ideology of encompassment, we move to an 'ideology of debasement' in which human heads become little more than mutilated trophies, intended to induce fear and to demonstrate the annihilation of the enemy. The treatment meted out to the Hawaiian chief Kauhiakama (Kamakau 1961, 232), quoted at the beginning of this chapter, was likewise a deliberately heinous act that outraged and demanded retribution from his allies. Likewise, the human head in Roman-occupied southern France had become essentially a vehicle for brutalistic displays of power and aggression.

It was, of course, these same southern French communities that provided the ethnographic material for classical accounts of Celtic thought and behaviour. Poseidonius, insofar as he travelled among the Celts at all, travelled within a relatively short radius of Massalia through lands annexed a generation or so earlier by the Roman military. The attitudes and practices he attributed to the Gauls were in large measure culled from this limited region at this particular point in its history. The 'cult of the head', which has been extrapolated from the accounts of Poseidonius and others, may be an appropriate shorthand for the phenomenon of head curation and display in southern France; less appropriate has been its export to a whole range of other European societies whose attitudes to the human body were potentially more fluid.

Iron Age bodies

Special treatments of the human head were indeed a pan-European phenomenon; the head was everywhere close to the core of behaviour and belief. We have seen snapshots from places such as Raffin in eastern Ireland and the Sculptor's Cave in northeast Scotland, among many others, and iconographic representations from many times and places within Iron Age Europe. But in focussing our gaze exclusively on the head, we lose sight of a broader picture in which conceptualisations of the body were richer, stranger, and infinitely more diverse. Even though the head almost always seems to have been privileged to some extent as an object of display and deposition, in most areas there was a wider range of treatments applied to the human body, reflecting a more complex set of cosmological, religious, and ideological understandings.

We have seen in the preceding chapter the diversity of body manipulation in the Atlantic Scottish Iron Age; as ever, the head is the most commonly represented body part found in domestic contexts, but here it forms just one part of a much wider range of deposits from isolated long bones through articulated limbs to complete skeletons. The same applies to the human remains found in Iron Age Wessex. From the more hierarchical societies of northern France, in the Middle La Tène period, the highly structured deposits of bodies and body parts at Ribemont-sur-Ancre show a similarly complex set of attitudes to the body; the head was significant in its absence at Ribemont, but a broad range of ritualised acts concerned with the exposure, dismantling, and rearrangement of human and animal bodies dominates the surviving archaeological evidence.

The Iron Age body and its component parts had the power to mediate between the day-to-day world of the living and the supernatural world of ancestors, spirits, and deities. The sorts of practice seen across Iron Age Europe show that these realms were never conceived as wholly separate. The boundaries between them were permeable, and especially so at liminal places, such as caves and hilltops where the earth met the underworld, the sea, or the sky. Human and animal bodies, alone or in various combinations, acted as metaphors for more abstract ideas, such as fertility, luck, or health. As in more recent headhunting societies, the enaction of metaphorical relationships was perceived as having very real causative effects in the here and now. Just as the taking of heads in itself (without any mediating concept of 'life-fluid' or 'life-force') brought fertility and security to headhunting communities in nineteenth-century Southeast Asia and elsewhere, the curation and manipulation of bodies and body parts in the Iron Age could bring beneficial effects for the living.

Iron Age communities across Europe seem to have shared similar cosmological beliefs and these gave rise to recurrent forms of ritual performance. Violence against living bodies, or elaborate funerary treatments for members of the in-group drew on similar ideas about the power of ritualised action to bring benefits to the individual or the community. As the most enduringly recognisable part of the human body, the head seems time and again to have been the focus of both ritualised violence and funerary behaviour.

Yet although the head played a role in Iron Age cosmologies across the whole of Europe, it is only in southern France that we can see with any clarity its centrality to religious and ideological expression in chiefly or even state-level societies. The near-exclusive focus on the head here may reflect the particular historical circumstances of the region, on the cusp of Mediterranean and temperate Europe. The specific refinements of the head in cosmology, religion, and ideology in southern France lend peculiar emphasis to an aspect of bodily treatment that is prevalent throughout Iron Age Europe but which is contained elsewhere within a broader range of practice and belief. The peoples of southern France were exposed earlier and more directly to Mediterranean traditions of codified religion than their neighbours to the north. Their increasingly hierarchical social structures during the last few centuries BC fostered the development of more formalised religious and ideological expression, with the human head at its centre.

Headhunting in Iron Age Europe had nothing to do with any pan-Celtic belief in the power or divinity of the human head: there was no 'Celtic cult of the head'. Instead, a diverse range of communities across the continent drew, according to their own cultural and historical circumstances, on shared cosmological understandings of the world with their roots deep in prehistory.

Postscript

I began Chapter 1 with a discussion of the adult cranium buried in a pit at the Cnip wheelhouse in Lewis. During the rather protracted period over which this book was written, it became possible to date the Cnip cranium directly as part of a wider programme of AMS dating of human remains from Atlantic Scotland (Tucker and Armit 2009). We had suspected that this skull might have been curated for some time before its burial in the pit – a depositional event dated firmly to the first century AD by a sequence of radiocarbon dates (Armit 2006a, 221). And so it turned out, although the story was not quite what we had expected.

In fact, the Cnip skull did not date to the Iron Age at all. The AMS date revealed instead that the individual had died in the period 1540–1410 BC, nearly a millennium and a half before his cranium was buried in the pit. Although we can never rule it out completely, it seems inherently improbable that a skull, no matter how special or sacred, could have been successfully curated over this enormously long span of Hebridean prehistory, a period of perhaps sixty generations. There is, however, a more likely solution. A short walk from the Cnip wheelhouse, only a couple of hundred metres away across a low headland, is an unstable patch of wind-torn machair. Gradually, over the past forty years, human remains have been eroding from this area, and some of these have also been dated to the Middle Bronze Age (Dunwell et al. 1995). Machair goes through periodic cycles of erosion and redeposition, and it seems quite likely that a cycle of erosion may have affected this area during the Iron Age, leading to the emergence from the sand of some unexpected human remains.

For Iron Age people, farming around the headland, these bones would demand explanation. As we have seen, Iron Age communities in this region did not bury their dead in any straightforward sense. Their monuments were great houses like the broch towers and wheelhouses, rather than graves or temples. In common with other societies who occupy monumental, long-established houses, they would probably have possessed a considerable depth of genealogical knowledge, perhaps extending back several centuries to the foundation of these buildings. The bodies emerging from the sand would have had no place in these histories. Explanations for their presence would have had to invoke more remote pasts; the realm of mythological rather than genealogical histories (Gosden and Lock 1998).

The Cnip skull did not, therefore, belong to either an enemy or an ancestor, at least in any straightforward sense. Instead, it was a problematic relic from a very distant past. We will never know what happened to the other bones of the skeleton, and we cannot tell how long the skull may have been curated by its Iron Age finders before finally being deposited. Ultimately, however, this fragment formed the centrepiece of a significant deposit in the heart of their home, marking the foundation of a structure that altered and renewed the fabric of the house. The Cnip skull reminds us of the constant potential for past treatments of human remains to surprise us. It reminds us too of the need to extract every last piece of information that we can from these ‘detached fragments’ – any evidence for curation, every trace of trauma, each pathology, all signs of weathering, polishing, and scraping. The handling of human bones, then as now, was never routine: it was always charged with meaning.

Bibliography

Adeyinka, A. A.

1974 King Gezo of Dahomey, 1818–1858: A reassessment of a west African monarch in the nineteenth century. *African Studies Review* 17.3: 541–8.

Aelianus

1959 *Aelian, On the Characteristics of Animals*. Trans. A. E. Scholfield. Cambridge, Mass.: Harvard University Press.

Allason-Jones, L., and B. McKay

1985 *Coventina's Well*. Chollerford: Trustees of the Clayton Collection.

Almagro-Gorbea, M., and A. Lorrio

1993 La tête humaine dans l'art Celtique de la péninsule Ibérique. Pp. 219–37 in J. Briard and A. Duval (eds.), *Les représentations humaines du Néolithique à l'Âge du Fer*. Actes du 115e Congrès National des Sociétés Savantes. Paris: Éditions du C.T.H.S.

2004 War and society in the Celtiberian world. *E–Keltoi* 6: 73–112.

Anonymous

1909 Donations to and purchases for the Museum and Library, with exhibits. *Proceedings of the Society of Antiquaries of Scotland* 43: 17–18.

2004 CBSNews.com, 25 June.

Appian

1912 *Roman History*. Vol. 1. Trans. H. White. Cambridge, Mass.: Harvard University Press.

Aranegui Gascó, C.

2004 Sculptures Ibériques des Ve et IVe s. av. n. è. *Documents d'Archéologie Méridionale* 27: 105–16.

Arcelin, P.

2000 Arles protohistorique, agglomeration et structuration urbaine. Pp. 7–23 in M. Baudat (ed.), *Espace et urbanisme d'Arles des origins à nos jours*. Arles: Groupe Archéologique Arlésien.

2004a Entremont et la sculpture du second Âge du Fer en Provence. *Documents d'Archéologie Méridionale* 27: 71–84.

2004b La Tarasque, une oeuvre funéraire près de Puech de Noves (Bouches-du-Rhône). Pp. 41–8 in O. Cavalier (ed.), *La Tarasque de Noves: réflexions sur un motif iconographique et sa postérité*. Avignon: Musée Calvet.

2004c La Tarasque de Noves, une expression de croyances Celtiques dans la provence Romanisée. Pp. 49–60 in O. Cavalier (ed.), *La Tarasque de Noves: réflexions sur un motif iconographique et sa postérité*. Avignon: Musée Calvet.

2006 Avant Aquae Sextiae: l'oppidum d'Entremont. Pp. 125–68 in F. Mocci and N. Nin, *Aix-en-Provence, Pays d'Aix, Val de Durance*. Carte Archéologique de la Gaule No. 13/4. Paris: Fondation Maison des Sciences de l'Homme.

Arcelin, P., and G. Congès

2004 La sculpture protohistorique de Provence dans le Midi Gaulois. *Documents d'Archéologie Méridionale* 27: 10–12.

Arcelin, P., B. Dedet, and M. Schwaller

1992 Espaces public, espaces religieux protohistoriques en Gaule méridionale. *Documents d'Archéologie Méridionale* 15: 181–248.

Arcelin, P., and A. Rapin

2003 L'iconographie anthropomorphe de l'Âge du Fer en Gaule Méditerranéenne. Pp. 183–220 in O. Büchsenschütz, A. Bulard, M.-B. Chardenoux, and N. Ginoux (eds.), *Decors, images et signes de l'Âge du Fer Européen. XXVI Colloque de l'Association Française pour l'Étude de l'Âge du Fer, Paris et Saint-Denis, 2002*. Tours: FERACF.

Arens, W.

1979 *The Man-Eating Myth: Anthropology and Anthropophagy*. Oxford: Oxford University Press.

Arkush, E. N.

2006 Collapse, conflict, conquest: The transformation of warfare in the Late Prehispanic Andean Highlands. Pp. 286–335 in E. N. Arkush and M. W. Allen (eds.), *The Archaeology of Warfare: Prehistories of Raiding and Conquest*. Gainesville: University Press of Florida.

Armit, I.

2003 *Towers in the North: The Brochs of Scotland*. Stroud: Tempus.

2005 *Celtic Scotland: Iron Age Scotland in Its European Context*. 2nd ed. London: Batsford.

2006a *Anatomy of an Iron Age Roundhouse: The Cnip Wheelhouse Excavations, Lewis*. Edinburgh: Society of Antiquaries of Scotland.

2006b Inside Kurtz's compound: Headhunting and the human body in prehistoric Europe. Pp. 1–14 in M. Bonogofsky (ed.), *Skull Collection, Modification and Decoration*. British Archaeological Reports International Series. Oxford: Archaeopress.

2010a Janus in furs? Opposed human heads in the art of the European Iron Age. Pp. 279–86 in G. Cooney, K. Becker, J. Coles, M. Ryan, and S. Sievers (eds.), *Relics of Old Decency: Archaeological Studies in Later Prehistory*. Dublin: Wordwell.

2010b Porticos, pillars and severed heads: The display and curation of human remains in the southern French Iron Age. Pp. 90–100 in K. Rebay-Salisbury, M.-L. Stig Sørensen, and J. Hughes (eds.), *Body Parts and Bodies Whole*. Oxford: Oxbow.

Armit, I., C. Gaffney, and A. Hayes

2012 Space and movement in a southern French oppidum: Integrating geophysical and topographic survey at Entremont, Provence. *Antiquity* 86.

Armit, I., and V. Ginn

2007 Beyond the grave: Human remains from domestic contexts in Atlantic Scotland. *Proceedings of the Prehistoric Society* 73: 115–36.

Armit, I., T. Horsley, and F. Marty
2008 Geophysical survey on a southern French oppidum. *PAST* 58: 3–5.

Armit, I., and J. T. McKenzie
in prep. *An Inherited Place: Broxmouth Hillfort and the Southern Scottish Iron Age*. Edinburgh: Society of Antiquaries of Scotland.

Armit, I., and R. Schulting
2007 An Iron Age decapitation from the Sculptor's Cave, Covesea, NE Scotland. *PAST* 55: 1–2.

Armit, I., R. Schulting, C. Knüsel, and I. A. G. Shepherd
2011 Death, decapitation and display: Bronze and Iron Age human remains from the Sculptor's Cave, Covesea, NE Scotland. *Proceedings of the Prehistoric Society* 77: 251–78.

Arnold, B.
1999 'Drinking the Feast': Alcohol and the legitimation of power in Celtic Europe. *Cambridge Archaeological Journal* 9.1: 71–93.

Arnold, D. Y., and C. A. Hastorf
2008 *Heads of State: Icons, Power and Politics in the Ancient and Modern Andes*. Walnut Creek: Left Coast Press.

Aswani, S.
2000a (ed.), 'Essays on Head-Hunting in the Western Solomon Islands'. Special Issue of the *Journal of the Polynesian Society* 109. Auckland: Polynesian Society.
2000b Preface: On head-hunting in the Western Solomons. Pp. 4–9 in 'Essays on Head-Hunting in the Western Solomon Islands'. *Special Issue of the Journal of the Polynesian Society* 109. Auckland: Polynesian Society.
2000c Changing identities: The ethnohistory of Roviana predatory head-hunting. Pp. 39–70 in 'Essays on Head-Hunting in the Western Solomon Islands'. *Special Issue of the Journal of the Polynesian Society* 109. Auckland: Polynesian Society.

Atkinson, R. J. C., and S. Piggott
1955 The Torrs Chamfrein. *Archaeologia* 96: 197–235.

Ballin Smith, B. (ed.)
1994 *Howe: Four Millennia of Orkney Prehistory*. Monograph Series No. 9. Edinburgh: Society of Antiquaries of Scotland.

Banks, J. W.
1967 Human and animal bones. Pp. 302–5 in D. M. E. Avery, J. E. G. Sutton, and J. W. Banks, Rainsborough, Northants, England: Excavations 1961–5. *Proceedings of the Prehistoric Society* 33: 207–306.

Barber, J.
2003 Bronze Age farms and Iron Age farm mounds of the Outer Hebrides. *Scottish Archaeological Internet Reports* 3, internet publication. http://www.sair.org.uk/sair3/index.html.

Barber, J., P. Halstead, H. James, and F. Lee
1989 An unusual Iron Age burial at Hornish Point, South Uist. *Antiquity* 63: 773–8.

Barbet, A.
1987 La peinture murale pré-romaine existe-t-elle en Gaule? Pp. 45–62 in C. Bemont, B. Delplace, K. Fischer, K. Gruel, K. Peyre, and J.-C. Richard (eds.), *Mélanges offerts au Dr J.-B. Colbert de Beaulieu*. Paris: Le Léopard d'Or.

1991 Roquepertuse et la polychromie en Gaule méridionale. *Documents d'Archéologie Méridionale* 14: 53–81.

1992 Polychromie des nouvelles sculptures préromaines de Nîmes (Gard). *Documents d'Archéologie Méridionale* 15: 96–102.

Barruol, G., U. Gibert, and G. Rancoule

1961 Le défunt heroisé de Bouriège (Aude). *Revue d'Études Ligures* 27: 45–60.

Barth, F.

1987 *Cosmologies in the Making*. Cambridge: Cambridge University Press.

Bats, M.

1990 Olbia. Pp. 206–10 in Musées de Marseille, *Voyage en Massalie: 100 ans d'archéologie en Gaule du Sud*. Marseille: Edisud.

2003 Ligyens et Salyens d'Hécatée à Strabo. Pp. 147–66 in M. Bats, B. Dedet, P. Garmy, T. Janin, C. Raynaud, and M. Schwaller (eds.), *Peuples et territoires en Gaule méditerranéenne. Hommage à Guy Barruol*. Revue Archéologique de Narbonnaise Supplément 35. Montpellier.

Beirâo, C. M., C. T. Silva, J. Soares, M. V. Gomes, and R. V. Gomes

1987 Um depósito votivo da II Idade do Ferro, no sul de Portugal, e as suas relações com as culturas da Meseta. Pp. 207–21 in J. Gorrochategui Churruca, J. L. Melena, and J. F. García Santos (eds.), *Studia Paleohispanica. Actas del IV Coloquio Sobre Lenguas y Culturas Paleohispánicas*. Vitoriaco Vasconum: Instituto de Ciencias de la Antigüedad, Universidad del País Vasco.

Belarte, M. C., E. Gailledrat, and J. Principal

2011 The functional and symbolic uses of space in western Mediterranean protohistory: The Pech Maho example (Sigean, western Languedoc, France). *Oxford Journal of Archaeology* 30: 57–83.

Benoît, F.

1955a *L'art primitif méditerranéen de la Vallée du Rhône*. Gap: Ophrys.

1955b Le sanctuaire aux 'esprits' d'Entremont. *Cahiers Ligures de Préhistoire et d'Archéologie* 4: 38–69.

1957 *Entremont: capitale celto-ligure des Salyens de Provence*. Aix-en-Provence: La Pensée Universitaire.

1964 Les 'têtes sans bouches' d'Entremont. *Cahiers Ligures de Préhistoire et d'Archéologie* 8: 68–81.

1965 *Recherches sur l'Hellénisation du Midi de la Gaule*. Aix-en-Provence: Orphrys.

1975 The Celtic oppidum of Entremont. Pp. 227–59 in R. Bruce-Mitford (ed.), *Recent Archaeological Excavations in Europe*. London: Routledge.

Benton, S.

1931 The excavation of the Sculptor's Cave, Covesea, Morayshire. *Proceedings of the Society of Antiquaries of Scotland* 65: 177–216.

Bergquist, A., and T. Taylor

1987 The origin of the Gundestrup cauldron. *Antiquity* 61: 10–24.

Bernard, L.

2000 L'habitat préromain du Verduron. Pp. 158–60 in J. Chausserie-Laprée (ed.), *Les Temps des Gaulois en Provence*. Martigues: Musée Ziem.

Bessac, J.-C.

1991 Roquepertuse. Techniques du travail de la pierre. *Documents d'Archéologie Méridionale* 14: 43–51.

1992 Pierres et techniques utilisées pour les sculptures et inscriptions tardo-hellénistiques de 'Villa-Roma' à Nîmes. *Documents d'Archéologie Méridionale* 15: 103–11.

Bhootra, B. L., and E. Weiss
2006 Muti killings: A case report. *Medicine, Science and the Law* 46.3: 255–9.

Biel, J.
1981 The late Hallstatt chieftain's grave at Hochdorf. *Antiquity* 55: 16–18.

Billingsley, J.
1998 *Stony Gaze: Investigating Celtic and Other Stone Heads*. Chieveley: Capall Bann.

Blázquez, J. M.
1990 La religión de los pueblos de la hispania preromana. *Zephyrus* 43: 223–33.

Boissinot, P.
2003 Notice 7: Velaux (Bouches-du-Rhône). Pp. 238–41 in P. Arcelin and J.-L. Brunaux (eds.), Cultes et sanctuaires en France à l'Âge du Fer. *Gallia* 60: 1–268.
2004 Usage et circulation des éléments lapidaire de Roquepertuse. *Documents d'Archéologie Méridionale* 27: 49–62.

Boissinot, P., and L.-F. Gantès
2000 La chronologie de Roquepertuse. Propositions préliminaires à l'issue des campagnes 1994–1999. *Documents d'Archéologie Méridionale* 27: 249–71.

Boissinot, P., and B. Lescure
1998 Nouvelles recherches sur le 'sanctuaire' de Roquepertuse à Velaux (IIIe s.). Premier résultats. *Documents d'Archéologie Méridionale* 21: 84–9.

Bonfante, L.
1984 Human sacrifice on an Etruscan funerary urn. *American Journal of Archaeology* 88.4: 531–9.

Bonnabel, L., and B. Boulestin
2008 L'homme sans visage: la sépulture 44 de Reims 'La Neuvillette' (Marne). *Bulletin de l'Association Française de l'Âge du Fer* 26: 15–17.

Boon, G. C.
1976 The shrine of the head, Caerwent. Pp. 163–75 in G. C. Boon and J. M. Lewis (eds.), *Welsh Antiquity*. Cardiff: National Museum of Wales.

Borić, D., and J. Robb (eds.)
2008 *Past Bodies: Body-Centred Research in Archaeology*. Oxford: Oxbow.

Bouloumié, B.
1992 *Saint-Blaise (fouilles Henri Rolland). L'habitat protohistorique, les ceramiques grecques*. Travaux de Centre Camille-Jullian, 13. Aix-en-Provence: Université de Provence.

Bowers, A. W.
1950 *Mandan Social and Ceremonial Organisation*. Chicago: University of Chicago Press.

Boyd-Dawkins, W.
1917 The inhabitants of the Lake Village. Pp. 673–80 in A. Bulleid and H. St G. Gray, *The Glastonbury Lake Village*, vol. 2. Glastonbury: Glastonbury Antiquarian Society.

Bradley, R.
1995 The head of the river. *Antiquity* 69: 169–70.
2005 *Ritual and Domestic Life in Prehistoric Europe*. London: Routledge.

Brears, P.
1989 *North Country Folk Art*. Edinburgh: John Donald.

Bringer, G., and A. Dumont-Castells
2000 Le bloc aux aux têtes coupées de Badasset. P. 105 in J. Chausserie-Laprée (ed.), *Les temps des Gaulois en Provence*. Martigues: Musée Ziem.

Britton, D., and I. H. Longworth

1969 Late Bronze Age finds from Heathery Burn Cave, County Durham. *Inventaria Archaeologica*, 9th ser., GB 55M. London: British Museum.

Brück, J.

1995 A place for the dead: The role of human remains in Late Bronze Age Britain. *Proceedings of the Prehistoric Society* 61: 245–77.

Brun, J.-P.

1999 *Le Var*. Carte Archéologique de la Gaule No. 83/1. Paris: Fondation Maison des Sciences de l'Homme.

Brunaux, J.-L.

1988 *The Celtic Gauls: Gods, Rites and Sanctuaries*. London: Seaby.

2003a Notice 14: Ribemont-sur-Ancre (Somme). Pp. 64–8 in P. Arcelin and J.-L. Brunaux (eds.), Cultes et sanctuaires en France à l'Âge du Fer. *Gallia* 60: 1–268.

2003b Notice 9: Gournay-sur-Aronde (Oise). Pp. 58–9 in P. Arcelin and J.-L. Brunaux (eds.), Cultes et sanctuaires en France à l'Âge du Fer. *Gallia* 60: 1–268.

2006 Religion et sanctuaires. Pp. 95–116 in C. Goudineau (ed.), *Religion et société en Gaule*. Paris: Errance.

Brunaux, J.-L., and C. Malagoli

2003 La France du nord. Pp. 9–46 in P. Arcelin and J.-L. Brunaux (eds.), Cultes et sanctuaires en France à l'Âge du Fer. *Gallia* 60: 1–268.

Bulleid, A., and H. St G. Gray

1917 *The Glastonbury Lake Village*. Vol. 2. Glastonbury: Glastonbury Antiquarian Society.

Butler, J.

1847 *A Sketch of Assam, with Some Account of the Hill Tribes, by an Officer*. London: Smith, Elder.

Cadoux, J.-L.

1984 L'ossuaire gaulois de Ribemont-sur-Ancre (Somme): premières observations, premières questions. *Gallia* 42: 53–78.

Caesar

1951 *Caesar: The Conquest of Gaul*. Trans. S. A. Handford. London: Penguin.

Campbell, D.

1991 *Greek Lyric III: Stesichorus, Ibycus, Simonides and Others*. Cambridge, Mass.: Harvard University Press.

Campbell, E.

1991 Excavations of a wheelhouse and other structures at Sollas, North Uist, by R. J. C. Atkinson in 1956. *Proceedings of the Society of Antiquaries of Scotland* 121: 117–73.

Canet, H., and J.-L. Roudil

1978 Le village chalcolithique de Cambous à Viols-en-Laval (Hérault). *Gallia Préhistoire* 21: 143–81.

Carneiro, R.

1990 Chiefdom-level warfare as exemplified in Fiji and the Cauca Valley. Pp. 190–211 in J. Haas (ed.), *The Anthropology of War*. Cambridge: Cambridge University Press.

Carr, G., and C. Knüsel

1997 The ritual framework of excarnation by exposure as the mortuary practice of the Early and Middle Iron Ages of central southern Britain. Pp. 167–73 in A. Gwilt and C. Haselgrove (eds.), *Reconstructing Iron Age Societies*. Oxbow Monograph 71. Oxford.

Chabot, L.
2004 *L'oppidum de la Cloche: Les Pennes-Mirabeau, Bouches-du-Rhône*. Protohistoire Européenne 7. Montagnac: Mergoil.

Chacon, R. J., and D. H. Dye
2007 Introduction to human trophy taking: An ancient and widespread practice. Pp. 5–31 in R. J. Chacon and D. H. Dye, *The Taking and Displaying of Human Body Parts as Trophies by Amerindians*. New York: Springer.

Chadwick, N.
1927 *An Early Irish Reader*. Cambridge: Cambridge University Press.
1970 *The Celts*. Harmondsworth: Penguin.

Chamberlain, A.
1999 The human skeletal remains. Pp. 288–9 in M. Parker Pearson and N. Sharples (eds.), *Excavations at Dun Vulan, South Uist*. Sheffield: Sheffield Academic Press.

Chapman, J. C.
2000 *Fragmentation in Archaeology: People, Places and Broken Objects in the Prehistory of South-Eastern Europe*. London: Routledge.

Chatwin, B.
1987 *The Songlines*. London: Picador.
1982 (1980) *The Viceroy of Ouidah*. London: Picador.

Chaume, B.
2001 *Vix et son territoire à l'Âge du Fer*. Montagnac: Editions Monique Mergoil.

Chausserie-Laprée, C.
2005 *Martigues, terre gauloise entre Celtique et Méditerranée*. Paris: Errance.

Coignard, O., and R. Coignard
2004 La Tarasque de Noves: interventions et recherches. Pp. 61–82 in O. Cavalier (ed.), *La Tarasque de Noves: réflexions sur un motif iconographique et sa postérité*. Avignon: Musée Calvet.

Coignard, O., R. Coignard, N. Marcadal, and Y. Marcadal
1998 Nouveau regard sur le sanctuaire et les gravures de l'Âge du Fer de l'oppidum des Caisses (Mouriès, Bouches-du-Rhône). *Documents d'Archéologie Méridionale* 21: 67–83.

Coignard, R., and O. Coignard
1991 L'assemble lapidaire de Roquepertuse: nouvelle approche. *Documents d'Archéologie Méridionale* 14: 27–42.

Coles, J., and S. Minnitt
1995 *'Industrious and Fairly Civilized': The Glastonbury Lake Village*. Exeter: Somerset County Council Museums Service.

Collis, J. R.
1984 *The European Iron Age*. London: Batsford.
2003 *The Celts: Origins, Myths and Inventions*. Stroud: Tempus.

Congès, G.
2000 Une nouvelle stèle aux têtes coupées à Entremont. Pp. 104 in J. Chausserie-Laprée (ed.), *Les Temps des Gaulois en Provence*. Martigues: Musée Ziem.
2004 La statuaire d'Entremont: réflexions sur d'anciennes découvertes. *Documents d'Archéologie Méridionale* 27: 63–70.

Courtaud, P., H. Duday, and E. Rousseau
2010 Les vestiges humains de l'oppidum d'Entremont (B.-du-Rh.). Pp. 318–40 in P. Arcelin and G. Congès (eds.), *Entremont: document final de synthèse sur le PCR 2010–2012. Rapport intermédiaire 2010*. Aix: Service Régional de l'Archéologie de Provence-Alpes-Côte-d'Azur.

Craig, R., C. J. Knüsel, and G. Carr
2005 Fragmentation, mutilation and dismemberment: An interpretation of human remains on Iron Age sites. Pp. 165–80 in M. Parker Pearson and N. Thorpe (eds.), *Warfare, Violence and Slavery in Prehistory*. British Archaeological Reports International Series 1374. Oxford.

Creighton, J.
2000 *Coins and Power in Late Iron Age Britain*. Cambridge: Cambridge University Press.

Cunliffe, B. W.
1995 *Danebury: An Iron Age Hillfort in Hampshire*. Vol. 6, *A Hillfort Community in Perspective*. Research Report 102. London: Council for British Archaeology.
1997 *The Ancient Celts*. Oxford: Oxford University Press.
2005 *Iron Age Communities in Britain*. 4th ed. London: Routledge.

Cunnington, M. E.
1919 Lidbury Camp. *Wiltshire Archaeological and Natural History Magazine* 40: 12–36.
1923 *The Early Iron Age Inhabited Site at All Cannings Cross Farm, Wiltshire*. Devizes: Simpson.

Dandoy, J. R., P. Selinsky, and M. Voigt
2002 Celtic sacrifice. *Archaeology* 55.1: 44–9.

D'Anna, A.
1977 *Les statues-menhirs et stèles anthropomorphes du Midi Méditerranéen*. Paris: Éditions du CNRS.

Dasent, G. W.
1894 *The Orkneyingers Saga*. In *Icelandic Sagas*, vol. 3. London: HMSO.

Davidson, J.
1984 *The Prehistory of New Zealand*. Auckland: Longman Paul.

Davis, A.
1981 (1891) Naga tribes. Pp. 237–51 in *Assam Census Report, 1891*, vol. 1. Shillong: Government Press.

Déchelette, J.
1914 *Manuel d'archéologie préhistorique, celtique et gallo-romaine*. Tome II-3, *Second Âge du Fer ou Époque de La Tène*. Paris: Picard.

Dedet, B.
1992 *Rites funéraires protohistoriques dans les garrigues languedociennes*. Revue Archéologique de Narbonnaise Supplémente 24. Paris: CNRS.

Dedet, B., and M. Py
1985 *L'occupation des rivages de l'étang de Mauguio (Hérault) au Bronze Final et au Premier Âge du Fer*. Tome III, *Synthèses et annexes*. Association pour la Recherche Archéologique en Languedoc Oriental Cahier No. 13. Caveirac.

Dedet, B., M. Py, and H. Savay-Guerraz
1985 *L'occupation des rivages de l'étang de Mauguio (Hérault) au Bronze Final et au Premier Âge du Fer*. Tome II, *Sodages et sauvetages programmes*. Association pour la Recherche Archéologique en Languedoc Oriental Cahier No. 12. Caveirac.

Dedet, B., and M. Schwaller
1990 Pratiques cultuelles et funéraires en milieu domestique sur les oppidums languedociens. *Documents d'Archéologie Méridionale* 13: 137–62.

Delamare, F., and B. Guineau
1991 Roquepertuse: analyse physico-chimique des couches picturales. *Documents d'Archéologie Méridionale* 14: 83–6.

Delaney, M.
1977 Report on the human skull fragments. Pp. 59–61 in C. J. Lynn, Trial excavations at the King's Stables, Tray Townland, Co. Armagh. *Ulster Journal of Archaeology* 40: 42–62.

Deyts, S.
1983 *Les bois sculptes des sources de la Seine*. Gallia Supplémente 42. Paris: Éditions du Centre National de la Recherche Scientifique.

Dietler, M.
1995 Early 'Celtic' socio-political relations: Ideological representation and social competition in dynamic comparative perspective. Pp.64–71 in B. Arnold and D. B. Gibson (eds.), *Celtic Chiefdom, Celtic State*. Cambridge: Cambridge University Press.
1997 The Iron Age in Mediterranean France: Colonial encounters, entanglements and transformations. *Journal of World Prehistory* 11.3: 269–358.
2005 *Consumption and Colonial Encounters in the Rhône Basin of France: A Study of Early Iron Age Political Economy*. Monographies d'Archéologie Méditerranéenne 21. Lattes: CNRS.

Dietler, M., and M. Py
2003 The Warrior of Lattes: An Iron Age stone statue discovered in Mediterranean France. *Antiquity* 77: 780–95.

Diodorus Siculus
1954 *Diodorus of Sicily*. Vol. 6. Trans. C. H. Oldfather. Cambridge, Mass.: Harvard University Press.

Donald, L.
1997 *Aboriginal Slavery on the Northwest Coast of North America*. Berkeley and Los Angeles: University of California Press.

Duceppe-Lamarre, A.
2004 Les bêtes dévorantes dans l'art laténien. Pp. 28–40 in O. Cavalier (ed.), *La Tarasque de Noves: réflexions sur un motif iconographique et sa postérité*. Avignon: Musée Calvet.

Duday, H.
1984 Les restes humains de l'oppidum de La Liquière. Pp. 351–4 in M. Py, F. Py, P. Suazet, and C. Tendille, *La Liquière, village du 1er Âge du Fer en Languedoc oriental*. Revue Archéologique Narbonnaise Supplément 11. Paris: Editions de CNRS.
1985 Les restes humains découverts dans les gisements lagunaires des rives de l'étang de Mauguio. Pp. 117–22 in B. Dedet and M. Py, *L'occupation des rivages de l'étang de Mauguio (Hérault) au Bronze Final et au Premier Âge du Fer*. Tome III, *Synthèses et annexes*. Association pour la Recherche Archéologique en Languedoc Oriental Cahier No. 13. Caveirac.
1995 Études des restes humains trouvés dans les décombres de la porte incendiée. Screens 32–8 in CD-Rom accompanying D. Garcia and L. Bernard, Un témoignage de la chute de la Confédération Salyenne? L'oppidum de Buffe Arnaud (Saint-Martin-de-Brômes, Alpes-de-Haute-Provence). *Documents d'Archéologie Méridionale* 18: 113–42.

Duncan, J.
1847 *Travels in West Africa in 1845 and 1846*. Vol. 2. London: Bentley.

Dunwell, A., T. Neighbour, and T. G. Cowie
1995 A cist burial adjacent to the Bronze Age burial at Cnip, Uig, Isle of Lewis. *Proceedings of the Society of Antiquaries of Scotland* 125: 279–88.

Durán, F. D.
1951 (1579–81) *Historia de las Indias de Nueva-Espana y Islas de Tierra Firme. Notas é ilustraciones de José F. Ramírez*. Mexico: Editoria Nacional.

Duval, A., and D. Heude
1983 *L'art celtique en Gaule*. Paris: Direction des Musées de la France.

Earle, T. K.
1997 *How Chiefs Come to Power: Political Economy in Prehistory*. Stanford, Calif.: Stanford University Press.

Ebel, C.
1976 *Transalpine Gaul: The Emergence of a Roman Province*. Leiden: Brill.

Elwin, V.
1961 *Nagaland*. Shillong: Research Department, NEFA Administration.

Enright, M. J.
2006 *The Sutton Hoo Sceptre and the Roots of Celtic Kingship Theory*. Dublin: Four Courts.

Evans, C.
2003 *Power and Island Communities: Excavations at the Wardy Hill Ringwork, Coveney, Ely*. East Anglian Archaeology 103. Cambridge: Cambridge Archaeological Unit.

Fercoq du Leslay, G.
1996 Chronologie et analyse spatiale à Ribemont-sur-Ancre. *Revue Archéologique de Picardie* 1996.3–4: 189–208.
2000 L'apport des fossés de Ribemont-sur-Ancre (Somme) à la chronologie et à l'interprétation du site. *Revue Archéologique de Picardie* 2000.1–2: 113–46.

Ferris, I.
2000 *Enemies of Rome: Barbarians through Roman Eyes*. Stroud: Sutton.

Fitts, R. J., C. C. Haselgrove, P. C. Lowther, and S. H. Willis
1999 Melsonby revisited: Survey and excavation 1992–95 at the site of the discovery of the 'Stanwick', North Yorkshire hoard of 1843. *Durham Archaeological Journal* 14–15: 1–52.

Fitzpatrick, A. P.
2007 Dancing with dragons: Fantastic animals in the earlier Celtic art of Iron Age Britain. Pp. 339–57 in C. Haselgrove and T. Moore (eds.), *The Later Iron Age in Britain and Beyond*. Oxford: Oxbow.

Florus
1929 *Florus: Epitome of Roman History*. Trans. E. S. Forster. Cambridge, Mass.: Harvard University Press.

Fowler, C.
2004 *The Archaeology of Personhood: An Anthropological Approach*. London: Routledge.

Freeman, D.
1979 Severed heads that germinate. Pp. 233–46 in R. H. Hook (ed.), *Fantasy and Symbol: Studies in Anthropological Interpretation*. New York: Academic Press.

Freeman, P.
1994 The Earliest Classical Sources on the Celts: A Linguistic and Historical Study. PhD diss., Harvard University.

Frietag, B.
2004 *Sheela-na-Gigs: Unravelling an Enigma*. London: Routledge.

Fulford, T., D. Lee, and P. J. Kitson
2004 *Literature, Science and Exploration in the Romantic Era*. Cambridge: Cambridge University Press.

Gailledrat, E., and J.-C. Bessac
2000 Découverte récente d'un ensemble sculpté du premier Âge du Fer à Mailhac (Aude). Pp. 291–303 in T. Janin (ed.), *Mailhac et le Premier Âge du Fer en Europe Occidentale: hommages à Odette et Jean Taffanel. Actes du Colloque International de Carcassonne* 17–20 *Septembre* 1997. Lattes: CNRS.

Gantès, L. F.
1990 Roquepertuse: le site protohistorique, in Musées de Marseille. Pp. 162–4 in Musées de Marseille, *Voyage en Massalie: 100 ans d'archéologie en Gaule du Sud*. Marseille: Edisud.

Gantz, J. (trans.)
1976 *The Mabinogion*. London: Penguin.

Garcia, D.
1992 Les stèles de La Ramasse, à Clermont-Hérault. *Documents d'Archéologie Méridionale* 15: 158–65.
1993 *Entre Ibères et Ligures: Lodévois et Moyenne Vallée d'Hérault Protohistoriques*. Revue Archéologique de Narbonnaise Supplément 26. Paris: CNRS Éditions.
2003 Espaces sacrés et genèse urbaine chez les Gaulois du Midi. Pp. 223–32 in M. Bats, B. Dedet, P. Garmy, T. Janin, C. Raynaud, and M. Schwaller (eds.), *Peuples et territoires en Gaule méditerranéenne. Hommage à Guy Barruol*. Revue Archéologique de Narbonnaise Supplément 35. Montpellier.
2004 *La Celtique méditerranéenne*. Paris: Errance.

Garcia, D., and L. Bernard
1995 Un témoignage de la chute de la Confédération Salyenne? L'oppidum de Buffe Arnaud (Saint-Martin-de-Brômes, Alpes-de-Haute-Provence). *Documents d'Archéologie Méridionale* 18: 113–42.

Gardner, W., and H. N. Savory
1964 *Dinorben: A Hill-Fort Occupied in Early Iron Age and Roman Times*. Cardiff: National Museum of Wales.

Gell, A.
1992 The technology of enchantment and the enchantment of technology. Pp. 40–65 in J. Coote and A. Shelton (eds.), *Anthropology, Art and Aesthetics*. Oxford: Clarendon.

Gérin-Ricard, H. de
1927 *Le sanctuaire préromain de Roquepertuse à Velaux (Bouches-du-Rhône). Son trophée – ses peintures – ses sculptures. Étude sur l'art gaulois avant le temps classique*. Marseille: Société de Statistique, d'Histoire, et d'Archéologie de Marseille et de Provence.
1929 *Le sanctuaire préromain de Roquepertuse (fouilles de 1927): étude sur l'art gaulois avant le temps classiques (supplément)*. Marseille: Provincia.

Gilles, I.
1873 *Les Saliens avant la conquête romaine. Velaux (Saint-Eutropie et les deux statues de la Roque-Pertuse)*. Paris: Thorin.

González-Ruibal, A.
2004 Artistic expression and material culture in Celtic Gallaecia. *E–Keltoi* 6: 113–66.

González Rul, F.
1963 Un tzompantli en Tlatelolco. *Boletin del Instituto Nacional de Anthropologia e Historia* 13: 3–5.

Gosden, C., and G. Lock
1998 Prehistoric histories. *World Archaeology* 30: 2–12.

Goudineau, C.
1983 Marseilles, Rome and Gaul from the third to the first century BC. Pp. 76–86 in P. Garnsey, K. Hopkins, and C. R. Whittaker (eds.), *Trade in the Ancient Economy*. London: Chatto and Windus.

Green, M.
1986 *The Gods of the Celts*. Stroud: Sutton.
1997 *Exploring the World of the Druids*. London: Thames and Hudson.

Grenier, M. A.
1922 *Les Gaulois*. Paris: Payot.

Gruat, P.
2004 Contribution à un réexamen de la statuaire protohistorique du territoire des Rutènes. *Documents d'Archéologie Méridionale* 27: 85–97.

Gruel, K.
1993 Le visage humain au droit des monnaies gauloises armoricaines. Pp. 301–10 in J. Briard and A. Duval (eds.), *Les représentations humaines du Néolithique à l'Âge du Fer*. Actes du 115e Congrès National des Sociétés Savantes. Paris: Éditions du C.T.H.S.

Haddon, A. C.
1932 *Head-hunters: Black, White and Brown*. London: Watts.

Hall, M.
1997 Abernethy Stone Quadricephalous. Unpublished typescript, Perth Museum.

Halstead P.
2003 Animal bones from Baleshare and Hornish Point. Pp. 142–8 in J. Barber, Bronze Age farms and Iron Age farm mounds of the Outer Hebrides. *Scottish Archaeological Internet Reports* 3, internet publication. http://www.sair.org.uk/sair3/index.html.

Harding, D. W.
2007 *The Archaeology of Celtic Art*. London: Routledge.

Harner, M. J.
1973 *The Jivaro: People of the Sacred Waterfalls*. London: Robert Hale.
1977 The ecological basis for Aztec sacrifice. *American Ethnologist* 4: 117–35.

Harrison, R. J.
2004 *Symbols and Warriors: Images of the European Bronze Age*. Bristol: Western Academic and Specialist Press.

Harrison, S.
2006 Skull trophies of the Pacific War: Transgressive objects of remembrance. *Journal of the Royal Anthropological Institute* 12.4: 817–36.

Hartl, P.
2005 *Zur Besonderen Bedeutung und Behandlung des Menschlichen Kopfes Innerhalb de Latènekultur Mittel- und Westeuropas*. Bonn: Habelt.

Hatt, J.-J.
1970 *Celts and Gallo-Romans*. London: Barrie and Jenkins.

Hawkes, C.
1958 The hill-forts of northern France. *Antiquity* 32: 154–62.

Heald, A., J. Barber, and J. Henderson
2007 Whitegate. *Discovery and Excavation in Scotland* 8: 126–7.

Heimann J. M.
1999 *The Most Offending Soul Alive: Tom Harrisson and His Remarkable Life*. London: Aurum.

Hencken, T. C.
1938 The excavation of the Iron Age camp on Bredon Hill, Gloucestershire, 1935–7. *Archaeological Journal* 95: 1–111.

Hermary, A.
2003 Grecs et barbares cloueurs de têtes: compléments au témoignage de Poseidonios. Pp. 525–39 in M. Bats, B. Dedet, P. Garmy, T. Janin, C. Raynaud, and M. Schwaller (eds.), *Peuples et territoires en Gaule méditerranéenne. Hommage à Guy Barruol*. Revue Archéologique de Narbonnaise Supplément 35. Montpellier.

Hermary, A., A. Hesnard, and H. Tréziny
1999 *Marseille grecque: la cité phocéenne (600–49 av. J.-C.)*. Paris: Errance.

Herodotus
1954 *Herodotus: The Histories*. Trans. A. de Sélincourt. London: Penguin.

Hickey, H.
1985 *Images of Stone*. Belfast: Blackstaff.

Hill, J. D.
1995 *Ritual and Rubbish in the Iron Age of Wessex*. British Archaeological Reports, British Series 242. Oxford: Tempus Reparatum.

Hocart, A. M.
1931 Warfare in Eddystone of the Solomon Islands. *Journal of the Royal Anthropological Institute* 61: 301–24.

Hodge, A. T.
1999 *Ancient Greek France*. Philadelphia: University of Pennsylvania Press.

Hodos, T.
2006 *Local Responses to Colonization in the Iron Age Mediterranean*. London: Routledge.

Hodson, F. R., and R. M. Rowlett
1973 From 600 BC to the Roman conquest. Pp. 157–91 in S. Piggott, G. Daniel, and C. McBurney (eds.), *France before the Romans*. London: Thames and Hudson.

Hose, C., and W. McDougall
1912 *The Pagan Tribes of Borneo: A Description of Their Physical, Moral and Intellectual Condition with Some Discussion of Their Ethnic Relations*. 2 vols. London: Macmillan.

Hoskins, J.
1987 The headhunter as hero: Local traditions and their reinterpretation in national history. *American Ethnologist* 14.4: 605–22.
1996a (ed.), *Headhunting and the Social Imagination in Southeast Asia*. Stanford, Calif.: Stanford University Press.
1996b Introduction: Headhunting as practice and as trope. Pp. 1–49 in J. Hoskins (ed.), *Headhunting and the Social Imagination in Southeast Asia*. Stanford, Calif.: Stanford University Press.

1996c The heritage of headhunting: History, ideology and violence on Sumba, 1890–1990. Pp. 216–48 in J. Hoskins (ed.), *Headhunting and the Social Imagination in Southeast Asia*. Stanford, Calif.: Stanford University Press.

Houston, S., D. Stuart, and K. Taube
2006 *The Memory of Bones: Body, Being and Experience among the Classic Maya*. Austin: University of Texas Press.

Hunter, F.
2003 *Excavations at Birnie, Moray, 2002*. Edinburgh: Department of Archaeology, National Museums of Scotland.
2007 *Beyond the Edge of the Empire – Caledonians, Picts and Romans*. Rosemarkie: Groam House Museum.

Hunter, F., and M. Collard
1997 The Cramond lioness. *Current Archaeology* 13.11: 404–7.

Hutton, J. H.
1938 *A Primitive Philosophy of Life (Frazer Lecture)*. Oxford: Oxford University Press.

Hutton, R.
1991 *The Pagan Religions of the Ancient British Isles*. Oxford: Blackwell.

Insoll, T.
2004 *Archaeology, Ritual, Religion*. London: Routledge.

Isoardi, D.
2007 *Le Castellar de Cadenet (84): un agglomération et son territoire. Rapport de fouille – 2007*. Aix-en-Provence: Centre Camille Jullian, Université de Provence.

Jackson, K.
1964 *The Oldest Irish Tradition: A Window on the Iron Age*. Cambridge: Cambridge University Press.

Jackson, S.
1973 *Celtic and Other Stone Heads*. Bradford: Sidney Jackson.

Jacobi, K. P.
2007 Disabling the dead: Human trophy taking in the prehistoric southeast. Pp. 299–338 in R. J. Chacon and D. H. Dye, *The Taking and Displaying of Human Body Parts as Trophies by Amerindians*. New York: Springer.

Jacobs, J.
1990 *The Nagas: Society, Culture and the Colonial Encounter*. London: Thames and Hudson.

Jacobsthal, P.
1944 *Early Celtic Art*. 2 vols. Oxford: Oxford University Press.

James, H. F., and R. P. J. McCullagh
2003 Excavations at Hornish Point. Pp. 72–103 in J. Barber, Bronze Age farms and Iron Age farm mounds of the Outer Hebrides. *Scottish Archaeological Internet Reports* 3, internet publication. http://www.sair.org.uk/sair3/index.html.

James, S.
1993 *Exploring the World of the Celts*. London: Thames and Hudson.
1999 *The Atlantic Celts: Ancient Peoples or Modern Invention*. London: British Museum Press.

Jandial, R., S. A. Hughes, H. E. Aryen, L. F. Marshall, and M. L. Levy
2004 The science of shrinking human heads: Tribal warfare and revenge among the South American Jivaro-Shuar. *Neurosurgery* 55.5: 1215–21.

Jewitt, J. R.
1985 The headhunters of Nootka. Pp. 216–55 in F. Drimmer (ed.), *Captured by the Indians: 15 Firsthand Accounts, 1750–1870*. New York: Dover.

Joachim, H.-E.
1993 Une reconstitution de la colonne de Pfalzfeld. Pp. 265–70 in J. Briard and A. Duval (eds.), *Les représentations humaines du Néolithique à l'Âge du Fer*. Actes du 115e Congrès National des Sociétés Savantes. Paris: Éditions du C.T.H.S.

Jope, M.
2000 *Early Celtic Art of the British Isles*. Oxford: Clarendon.

Jordan, P.
2001 The materiality of shamanism as a 'world-view': Praxis, artefacts and landscape. Pp. 87–104 in N. S. Price (ed.), *The Archaeology of Shamanism*. London: Routledge.

Jørgensen, H.
2005 Exchanging the Inalienable. The Politics and Practice of Repatriating Human Remains from Museum and Maori Tribal Perspectives. MA thesis, University of Copenhagen.

Jud, P.
2007 Les ossements humains dans le sanctuaries laténiens de la région de Trois-Lacs. Pp. 391–8 in P. Barral, A. Daubigny, C. Dunning, G. Kaenel, and M.-J. Roulière-Lambert (eds.), *L'Âge du Fer dans l'Arc jurassien et ses marges. Dépôts, lieux sacres et territorialité à l'Âge du Fer*. Besançon: Presses Universitaires de Franche-Comté.

Jullian, C.
1908 *Histoire de la Gaule*. Tome II, *La Gaule indépendante*. Paris: Hachette et Cie.

Justinus
1853 *M. Junianus Justinus: Epitoma Historiarum Philippicarum*. Trans. J. S. Watson. London: Bohn.

Kamakau, S. M.
1961 *Ruling Chiefs of Hawaii*. Honolulu: Kamehameha Schools Press.

Keeley, L. H.
1996 *War before Civilization: The Myth of the Peaceful Savage*. Oxford: Oxford University Press.

Keith, A.
1923 Notes on fragments of human skulls from All Cannings Cross. Pp. 41–3 in Maud E. Cunnington, *The Early Iron Age Inhabited Site at All Cannings Cross Farm, Wiltshire*. Devizes: Simpson.

Kelly, E. P.
2006 *Kingship and Sacrifice: Iron Age Bog Bodies and Boundaries*. Archaeology Ireland Heritage Guide No. 35. Wicklow: Wordwell.

Kidd, I. G.
1988 *Posidonius*. Vol. 2, *Commentary*. Cambridge: Cambridge University Press.

Knüsel, C.
2002 More Circe than Cassandra: The princess of Vix in ritualized social context. *European Journal of Archaeology* 5.3: 275–309.

Koch, J. T.
1995 *The Celtic Heroic Age*. Malden, Mass.: Celtic Studies Publications.

Kuijt, I.

2008 The regeneration of life: Neolithic structures of symbolic remembering and forgetting. *Current Anthropology* 49.2: 171–97.

Laforet, A.

1992 *The Book of the Grand Hall*. Hull, Quebec: Musée Canadien des Civilisations.

Laing, L.

1981 *Celtic Britain*. London: Paladin.

Laing, S. (trans.)

1907 *Heimskringla: A History of the Norse Kings*. London: Norroena Society.

Lambot, B.

2006 Religion et habitat: les fouilles d'Acy-Romance. Pp. 177–90 in C. Goudineau (ed.), *Religion et société en Gaule*. Paris: Errance.

Lambrechts, P.

1954 *L'exaltation de la tête dans la pensée et dans l'art des Celtes*. Bruges: De Tempel.

Lange, G.

1983 *Die menschlichen Skelletreste aus dem Oppidum von Manching*. Wiesbaden: Franz Steiner.

Law, R.

1989 'My head belongs to the king': On the political and ritual significance of decapitation in pre-colonial Dahomey. *Journal of African History* 30.3: 399–415.

Lawson, A. J.

2000 *Potterne, 1982–5: Animal Husbandry in Later Prehistoric Wiltshire*. Salisbury: Wessex Archaeology.

Layton, R.

1991 *The Anthropology of Art*. 2nd ed. Cambridge: Cambridge University Press.

Leach, E.

1977 A view from the bridge. Pp. 161–76 in M. Spriggs (ed.), *Archaeology and Anthropology: Areas of Mutual Interest*. British Archaeological Reports 19. Oxford: British Archaeological Reports.

Lee, F.

2003 Human bone from Baleshare and Hornish Point. Pp. 140–2 in J. Barber, Bronze Age farms and Iron Age farm mounds of the Outer Hebrides. *Scottish Archaeological Internet Reports* 3, internet publication. http://www.sair.org.uk/sair3/index.html.

Lenerz-de-Wilde, M.

1993 Sculptures anthropomorphes du 2e Âge du Fer sur la péninsule Ibérique. Pp. 239–52 in J. Briard and A. Duval (eds.), *Les représentations humaines du Néolithique à l'Âge du Fer*. Actes du 115e Congrès National des Sociétés Savantes. Paris: Éditions du C.T.H.S.

Le Roux, F.

1958 La 'Branche Sanglante' du roi d'Ulster et les 'têtes coupées' des Salyens de Provence. *Ogam* 2: 139–54.

1964 Têtes coupées et religion celtique. *Ogam* 16: 451–3.

Lescure, B.

1990 Roquepertuse: collection archéologique, in Musées de Marseille. Pp. 165–71 in Musées de Marseille, *Voyage en Massalie: 100 ans d'archéologie en Gaule du Sud*. Marseille: Edisud.

2004 La statuaire de Roquepertuse et ses nouveaux indices d'interprétation à l'issue des fouilles récentes. *Documents d'Archéologie Méridionale* 27: 45–7.

Leveau, P.
1990 Constantine, in Musées de Marseille. Pp. 113–17 in Musées de Marseille, *Voyage en Massalie: 100 ans d'archéologie en Gaule du Sud*. Marseille: Edisud.

Livy
1912 *Livy, The History of Rome*. Vol. 2. Trans. C. Roberts. London: Dent.

Locker, A.
2000 Animal bone. Pp. 101–19 in A. J. Lawson, *Potterne, 1982–5: Animal Husbandry in Later Prehistoric Wiltshire*. Salisbury: Wessex Archaeology.

Lorrio, A. J., and G. Ruiz Zapatero
2005 The Celts in Iberia: An overview. *E–Keltoi* 6: 167–254.

Low, A.
1931 Comments on the human bones from the Sculptor's Cave, Covesea. P. 207 in S. Benton, The excavation of the Sculptor's Cave, Covesea, Morayshire. *Proceedings of the Society of Antiquaries of Scotland* 65: 177–216.

Lumholtz, C. S.
2004 (1920) *Through Central Borneo: An Account of Two Years Travel in the Land of the Head-Hunters between the Years 1913 and 1917*. Montana: Kessinger.

Lynn, C. J.
1977 Trial excavations at the King's Stables, Tray Townland, Co. Armagh. *Ulster Journal of Archaeology* 40: 42–62.
1997 The date of deposition of the ape cranium. P. 125 in D. Waterman, *Excavations at Navan Fort, 1961–71*. Northern Ireland Archaeological Monographs No. 3. Belfast: The Stationery Office.

MacKillop, J.
1998 *Dictionary of Celtic Mythology*. Oxford: Oxford University Press.

Mahieu, E.
1998 L'anthropologie d'Entremont. *Documents d'Archéologie Méridionale* 21: 62–5.

Mallory, J. P.
1981 The sword of the Ulster Cycle. Pp. 99–114 in B. G. Scott (ed.), *Studies on Early Ireland: Studies in Honour of M. V. Duignan*. Belfast: AYIA.
1992 The world of Cú Chulainn: The archaeology of the Táin Bó Cúailnge. Pp. 103–59 in J. P. Mallory (ed.), *Aspects of the Táin*. Belfast: December Publications.

Mallory, J. P., and T. E. McNeill
1991 *The Archaeology of Ulster: From Colonization to Plantation*. Belfast: Institute of Irish Studies.

Marcadal, Y.
2000 Habitats de plaine et de hauteur à Mouriès. Pp. 191–3 in J. Chausserie-Laprée (ed.), *Les temps des Gaulois en Provence*. Martigues: Musée Ziem.

Marcellinus
1940 *Ammianus Marcellinus: Roman History, II, Books 20–26*. Trans. J. C. Rolfe. Cambridge, Mass.: Harvard University Press.

Marco-Simón, F.
1993 Iconografía y religión celtibérica. Reflexiones sobre un vaso de Arcóbriga. Pp. 537–52 in J. Padró, M. Prevosti, M. Roca, and J. Sanmarti (eds.), *Homenatge a Miquel Tarradell*. Estudis Universitaris Catalans XXIX. Barcelona: Curial Edicions Catalanes.

2003 Signa Deorum: Comparación y contexto histórico en España y Galia. Pp. 121–36 in T. Tortosa and J. A. Santos (eds.), *Arqueología e iconografía: Indagar en las imágenes*. Monografías de la Escuela Española de Historia y Arqueología en Roma no. 26. Rome: L'Erma di Bretschneider.

Mariën, M.-E.

1970 *Le trou de l'ambre au bois de Wérimont, Eprave*. Monographies d'Archéologie Nationale, 4. Brussels: Musées Royaux d'Art et d'Histoire.

1975 Massacre et sacrifice humain: deux cas d'interprétation. Pp. 253–61 in E. Anati (ed.), *Actes du Symposium International sur les Religions de la Préhistoire (Valcamonica Symposium 1972)*. Capo di Monte: Edizioni del Centro.

Marsden, S.

1932 *Letters and Journals of Samuel Marsden, 1765–1838*. Ed. J. R. Elder. Dunedin: Somerville, Wilkie and Reed.

Marty, F.

2000 L'oppidum du Castellan. Pp. 177–9 in J. Chausserie-Laprée (ed.), *Les temps des Gaulois en Provence*. Martigues: Musée Ziem.

Marwick, H.

1927a Antiquarian notes on Stronsay. *Proceedings of the Orkney Antiquarian Society* 5: 61–83.

1927b Underground galleried building at Rennibister, Orkney. *Proceedings of the Society of Antiquaries of Scotland* 61: 296–301.

McGregor, M.

1976 *Early Celtic Art in North Britain*. Leicester: Leicester University Press.

McKinley, J.

2000 Human bone. Pp. 95–101 in A. J. Lawson, *Potterne, 1982–5: Animal Husbandry in Later Prehistoric Wiltshire*. Salisbury: Wessex Archaeology.

McKinley, R.

1976 Human and proud of it! A structural treatment of headhunting rites and the social definition of enemies. Pp. 92–126 in G. N. Appell (ed.), *Studies in Borneo Societies: Social Process and Anthropological Explanation*. Special Report no. 12. DeKalb: Northern Illinois University.

McOmish, D.

1996 East Chisenbury: Ritual and rubbish at the British Bronze Age–Iron Age transition. *Antiquity* 267: 68–76.

McSweeney, K.

2006 The human remains. Pp. 133–6 in I. Armit, *Anatomy of an Iron Age Roundhouse: The Cnip Wheelhouse Excavations, Lewis*. Monograph Series. Edinburgh: Society of Antiquaries of Scotland.

Megaw, R., and V. Megaw

1988 The stone head from Mšecké Žehrovice: A reappraisal. *Antiquity* 62: 630–41.

1989 *Celtic Art*. London: Thames and Hudson.

1993 Cumulative Celticity and the human face in Insular pre-Roman Iron Age art. Pp. 205–18 in J. Briard and A. Duval (eds.), *Les représentations humaines du Néolithique à l'Âge du Fer*. Actes du 115e Congrès National des Sociétés Savantes. Paris: Éditions du C.T.H.S.

Megaw, V.

1970 *Art of the European Iron Age: A Study of the Elusive Image*. Bath: Adams and Dart.

Melville, H.

1851 *Moby Dick*. New York: Harper.

Mendoza, R. G.
2007 The divine gourd tree: Tzompantli skull racks, decapitation rituals, and human trophies in ancient Mesoamerica. Pp. 400–43 in R. J. Chacon and D. H. Dye, *The Taking and Displaying of Human Body Parts as Trophies by Amerindians*. New York: Springer.

Mérimée, P.
1989 (1835) *Notes d'un voyage dans le Midi de la France*. Paris: Biro.

Merriman, N.
1987 Value and motivation in prehistory: The evidence for 'Celtic spirit'. Pp. 111–16 in I. Hodder (ed.), *The Archaeology of Contextual Meanings*. Cambridge University Press.

Metcalf, P.
1996 Images of head-hunting. Pp. 249–92 in J. Hoskins, *Head-Hunting and the Social Imagination in Southeast Asia*. Palo Alto: Stanford University Press.

Miller, M.
2009 Extreme makeover: How painted bodies, flattened foreheads and filed teeth made the Maya beautiful. *Archaeology* 62.1: 36–42.

Miller, V.
2007 Skeletons, skulls and bones in the art of Chichén Itzá. Pp. 165–89 in V. Tiesler and A. Cucina (eds.), *Human Sacrificial Rites among the Maya of Mayapán: A Bioarchaeological Perspective*. New York: Springer.

Mills, J. P.
1926 *The Ao Nagas*. London: Macmillan.

Momigliano, A.
1975 *Alien Wisdom: The Limits of Hellenization*. Cambridge: Cambridge University Press.

Moore, T., and Armada X. L. (eds.)
in press *Atlantic Europe in the First Millennium BC: Crossing the Divide*. Oxford: Oxford University Press.

Moser, C.
1973 *Human Decapitation in Ancient Mesoamerica*. Washington, D.C.: Dumbarton Oaks.
1974 Ritual decapitation in Moche art. *Archaeology* 27: 30–7.

Mulville, J., M. Parker Pearson, N. Sharples, H. Smith, and A. Chamberlain
2003 Quarters, arcs and squares: Human and animal remains in the Hebridean Late Iron Age. Pp. 20–34 in J. Downes and A. Ritchie (eds.), *Sea Change: Orkney and Northern Europe in the Late Iron Age, AD 300–800*. Forfar: Pinkfoot Press.

Musson, C. R.
1991 *The Breiddin Hillfort: A Later Prehistoric Settlement in the Welsh Marches*. CBA Research Report No. 76. London: Council for British Archaeology.

Napier, P. H., and P. D. Jenkins
1997 Monkey skull and mandible of the Iron Age found at Site B. Pp. 120–1 in D. Waterman, *Excavations at Navan Fort, 1961–71*. Northern Ireland Archaeological Monographs No. 3. Belfast: The Stationery Office.

Nash, D.
1976 Reconstructing Posidonios' Celtic ethnography: Some considerations. *Britannia* 7: 111–26.

Needham, R.
1976 Skulls and causality. *Man* 11.1: 71–88.

Newman, C.
1993 Sleeping in Elysium. *Archaeology Ireland* 25: 20–3.

Newman, C., M. O'Connell, M. Dillon, and K. Molloy
2007 Interpretation of charcoal and pollen data relating to an Iron Age ritual site in eastern Ireland: A holistic approach. *Vegetation History and Archaeobotany* 16.5: 349–65.

Ní Lionáin, C.
2007 Life, death and food production in Bronze Age Ireland: Recent excavations at Stramullin, Co. Meath. *Archaeology Ireland* 80: 18–21.

O'Brien, E.
1990 Iron Age burial practice in Leinster. *Emania* 7: 37–42.

Olmsted, G.
2001 *Celtic Art in Transition during the 1st Century BC*. Budapest: Archaeolingua.

Onians, R. B.
1954 *The Origins of European Thought*. 2nd ed. Cambridge: Cambridge University Press.

Orosius, Paulus
1964 *Paulus Orosius: The Seven Books of History against the Pagans*. Trans. R. Deferrari. The Fathers of the Church, vol. 50. Washington, D.C.: Catholic University of America Press.

Orschiedt, J. A.
1998 Ergebnisse einer neuen Untersuchung der spätmesolithischen Kopfbestattungen aus Süddeutschland. *Urgeschichtliche Materialhefte* 12: 147–60.
2002 Die Kopfbestattungen der Ofnet-Höhle: Ein Beleg für kriegerische Auseinandersetzungen im Mesolithikum. *Archäologische Informationen* 24: 199–207.
2005 The head burials from Ofnet cave: An example of warlike conflict in the Mesolithic. Pp. 67–73 in M. Parker Pearson and N. Thorpe (eds.), *Warfare, Violence and Slavery in Prehistory*. British Archaeological Reports International Series 1374. Oxford.

Osman Hill, W. C.
1954 Human skull showing wounds. Pp. 53–6 in M. Wheeler, *The Stanwick Fortifications: North Riding of Yorkshire*. London: Research Committee of the Society of Antiquaries.

Paillet, J.-L., and H. Tréziny
2000 Le rempart hellénistique et la porte charretière de Glanum. Pp. 189–90 in J. Chausserie-Laprée (ed.), *Les temps des Gaulois en Provence*. Martigues: Musée Ziem.

Pálsson, H., and P. Edwards (trans.)
1978 *The Orkneyinga Saga*. London: Penguin.

Parker Pearson, M.
1999 Food, sex and death: Cosmologies in the British Iron Age with particular reference to East Yorkshire. *Cambridge Archaeological Journal* 9: 43–69.

Parker Pearson, M., A. T. Chamberlain, M. J. Collins, O. E. Craig, P. Marshall, J. Mulville, H. Smith, C. Chenery, G. Cook, J. Craig, J. Evans, J. Hiller, J. Montgomery, J.-L. Schwenniger, G. Taylor, and T. Wess
2005 Evidence for mummification in Bronze Age Britain. *Antiquity* 79: 529–46.

Parker Pearson, M., and C. Richards
1994 Ordering the world: Perceptions of architecture, space and time. Pp. 1–37 in M. Parker Pearson and C. Richards (eds.), *Architecture and Order: Approaches to Social Space*. London: Routledge.

Parker Pearson, M., and N. Sharples (eds.)
1999 *Excavations at Dun Vulan, South Uist*. Sheffield: Sheffield Academic Press.

Parkinson, S.
1773 *A Journal of a Voyage to the South Seas, in His Majesty's Ship, the Endeavour*. London: Stanfield Parkinson.

Périchon, R.
1987 L'imagerie Celtique d'Aulnat (1). Pp. 677–95 in C. Bemont, B. Delplace, K. Fischer, K. Gruel, K. Peyre, and J.-C. Richard (eds.), *Mélanges offerts au Dr J.-B. Colbert de Beaulieu*. Paris: Le Léopard d'Or.

Philips, C. W.
1931 Final report on the excavations of Merlin's Cave, Symonds Yat. *Proceedings of the University of Bristol Spelealogical Society* 4.1: 11–33.

Piggott, S.
1965 *Ancient Europe: From the Beginnings of Agriculture to Classical Antiquity*. Edinburgh: Edinburgh University Press.
1968 *The Druids*. London: Pelican.

Pindar
1956 *The Odes of Pindar*. Trans. J. Sandys. Cambridge, Mass.: Harvard University Press.

Plutarch
1931 *Plutarch, Moralia, III*. Trans. F. C. Babbitt. London: Heinemann.

Pobé, M., and J. Roubier
1961 *The Art of Roman Gaul: A Thousand Years of Celtic Art and Culture*. London: Galley Press.

Polybius
1922 *Polybius. The Histories*. Trans. W. R. Paton. London: Heinemann.

Poulain, T.
1978 Le village chalcolithique de Cambous à Viols-en-Laval (Hérault): études des vestiges osseux de la Cabane 11. *Gallia Préhistoire* 21: 183–8.
1983 Les fragments osseux. Pp. 66–9 in S. Deyts, *Les bois sculptés des sources de la Seine*. Gallia Supplémente 42. Paris: Éditions du Centre National de la Recherche Scientifique.

Powell, T. G. E.
1958 *The Celts*. London: Thames and Hudson.
1961 Foreword. Pp. vii–viii in M. Pobé and J. Roubier, *The Art of Roman Gaul: A Thousand Years of Celtic Art and Culture*. London: Galley Press.

Prades, H.
1985 *L'occupation des rivages de l'étang de Mauguio (Hérault) au Bronze Final et au Premier Âge du Fer*. Tome I, *Les recherches du groupe archéologique Painlevé (1969–1976)*. Association pour la Recherche Archéologique en Languedoc Oriental Cahier No. 11. Caveirac.

Pralon, D.
1998 Les Salyens dans les textes historiques grecs. *Documents d'Archéologie Méridionale* 21: 21–6.

Proudfoot, E.
1997 Abernethy and Mugdrum: Towards reassessment. Pp. 47–63 in D. Henry (ed.), *The Worm, the Germ and the Thorn: Essays Presented to Isabel Henderson*. Balgavies: Pinkfoot Press.

Proulx, D. A.
2006 *A Sourcebook of Nasca Ceramic Iconography*. Iowa City: University of Iowa Press.

Py, M.
1984 *La Liquière, village du 1er Âge du Fer en Languedoc oriental.* Revue Archéologique Narbonnaise Supplément 11. Paris: Editions de CNRS.
1990 *Culture, économie et société dans la région nîmoise.* Vol. 2. Collection de l'École Française de Rome 131. Rome: École Française de Rome.

Py, M., and D. Lebeaupin
1994 Stratigraphie du Marduel (Saint-Bonnet-du-Gard). VI – les niveaux du Bronze final au milieu du Ve av. n. è. Sur le Chantier Central. *Documents d'Archéologie Méridionale* 17: 201–65.

Radner, J. N. (ed. and trans.)
1978 *Fragmentary Annals of Ireland.* Dublin: Dublin Institute for Advanced Studies.

Raftery, B.
1994 *Pagan Celtic Ireland.* London: Thames and Hudson.
1997 Implications of the monkey bones from Site B. Pp. 121–4 in D. Waterman, *Excavations at Navan Fort, 1961–71.* Northern Ireland Archaeological Monographs No. 3. Belfast: The Stationery Office.

Rankin, D.
1987 *Celts and the Classical World.* London: Routledge.

Rapin, A.
2003 De Roquepertuse à Entremont, la grande sculpture du Midi de la Gaule. *Madrider Mittielungen* 44: 223–46.
2004 Pour un nouvelle lecture de la sculpture préromaine de Gaule méridionale. *Documents d'Archéologie Méridionale* 27: 13–22.

Rayssiguier, G., and C. Guichard
1990 Baou de Saint Marcel. Pp. 47–53 in Musées de Marseille, *Voyage en Massalie: 100 ans d'archéologie en Gaule du Sud.* Marseille: Edisud.

Reeder, E. R.
1999 *Scythian Gold.* New York: Abrams.

Reinach, A.
1913 Les têtes coupées et les trophées en Gaule. *Revue Celtique* 34: 38–60, 253–86.

Rennell, R., and I. McHardy
2008 Sloc Sabhaidh, Baile Sear, North Uist. Unpublished Data Structure Report, SCAPE.

Roberts, C., and J. McKinley
2003 Review of trepanations in British antiquity focusing on funerary context to explain their occurrence. Pp. 55–78 in R. Amott, S. Finger, and C. U. M. Smith, *Trepanation: History – Discovery – Theory.* Lisse: Swets and Zeitlinger.

Robertson, D.
1985 The Cacaxtla Murals. Pp. 291–302 in M. G. Roberston and E. Benson (eds.), *Fourth Palenque Roundtable, 1980*, vol. 6. San Francisco: Pre-Columbian Art Research Institute.

Robinson, W.
1975 (1841) *A Descriptive Account of Asam.* Delhi: Sanskaran Prakashak.

Roche, H.
1999 Late Iron Age activity at Tara, County Meath. *Riocht na Midhe* 10: 18–30.
2002 Excavation at the Ráith na Ríg, Tara, Co. Meath, 1997. *Discovery Programme Reports* 6: 19–165.

Rolland, M. H.
1964a Saint-Blaise. *Gallia* 22: 569.

1964b Circonscription d'Aix-en-Provence (région nord). *Gallia* 22: 545–72.
1968 Nouvelles fouilles du sanctuaire des Glaniques. *Revue d'Études Ligures* 34: 7–34.

Rosaldo, M.
1977 Skulls and causality. *Man* 12.1: 168–70.

Rosaldo, R.
1980 *Ilongot Headhunting, 1883–1974: A Study in Society and History*. Stanford, Calif.: Stanford University Press.
1984 Grief and a headhunter's rage: On the cultural force of emotions. Pp. 178–95 in E. W. Bruner (ed.), *Text, Play and Story: The Construction and Reconstruction of Self and Society*. 1983 Proceedings of the American Ethnological Society. Washington, D.C.: American Ethnological Society.

Ross, A.
1958 The human head in Insular pagan Celtic religion. *Proceedings of the Society of Antiquaries of Scotland* 91: 10–43.
1962 Severed heads in wells: An aspect of the well cult. *Scottish Studies* 6: 31–48.
1967 *Pagan Celtic Britain: Studies in Iconography and Tradition*. London: Routledge.
1974 A pagan Celtic tricephalos from Netherton, Lanarkshire. *Glasgow Archaeological Journal* 3: 26–33.

Ross, A., and R. Feachem
1984 Heads baleful and benign. Pp. 338–52 in R. Miket and C. Burgess (eds.), *Between and Beyond the Walls: Essays on the Prehistory and History of North Britain in Honour of George Jobey*. Edinburgh: John Donald.

Roth Congès, A.
2004 Le contexte archéologique de la statuaire de Glanon (Saint-Rémy-de-Provence, Bouches-du-Rhône). *Documents d'Archéologie Méridionale* 27: 23–43.

Rothé, M.-P., and H. Tréziny
2005 *Marseille et ses Alentours*. Carte Archéologique de la Gaule No. 13/3. Paris: Fondation Maison des Sciences de l'Homme.

Rovira-Hortalà, M. C.
1999 Las armas-trofeo en la cultura ibérica: pautas de identificación e interpretación. *Gladius* 19: 13–32.

Rozoy, J.-G.
1965 Les tombes sans crâne à La Tène I au Mont Troté. *Bulletin de la Société Préhistorique Française* 62: 253–61.

Rubenstein, S. L.
2006 Circulation, accumulation, and the power of Shuar shrunken heads. *Cultural Anthropology* 22.3: 357–99.

Rynne, E.
1972 Celtic stone idols in Ireland. Pp. 79–98 in C. Thomas (ed.), *The Iron Age in the Irish Sea Province*. CBA Research Report 9. London: Council for British Archaeology.

Saiz Rios, P.
1992 Representaciones plásticas de la cabeza humana en la necrópolis celtibérica (Montejo de Tiermes, Soria). Pp. 603–12 in *Actas del Segundo Symposium de Arqueología Soriana*, vol. 1. Soria: Ediciones de la Excma Diputación.

Salviat, F.
1972 Circonscription de Provence. *Gallia* 1972: 511–41.

1976 Statues féminines à Entremont. *Revue Archéologique de Narbonnaise* 9: 89–104.
1993 La sculpture d'Entremont. Pp. 165–239 in D. Coutagne (ed.), *Archéologie d'Entremont au Musée Granet*. Aix-en-Provence: Association des Amis de Musée Granet.

Scheers, S.
1981 The origins and evolution of coinage in Belgic Gaul. Pp. 18–23 in B. W. Cunliffe (ed.), *Coinage and Society in Britain and Gaul: Some Current Problems*. Research Report 38. London: Council for British Archaeology.

Schlitz, V.
1994 *Les Scythes et les Nomades des Steppes, VIIIe siècle avant J.-C.–Ier siècle après J.-C.* Paris: Gallimard.

Schoppa, H.
1958 Ein Keltischer steinpfeiler von Hirzenhain, Dillkreis. *Germania* 36: 153–4.

Sébillot, P.
1911 La tête de mort. *Revue des Traditions Populaires* 26: 129–46.

Seeman, M.
1988 Ohio Hopewell trophy-skull artefacts as evidence for competition in Middle Woodland societies circa 50 BC–AD 350. *American Antiquity* 53: 565–77.

Serafin, S., and C. Peraza Lope
2007 Human sacrificial rites among the Maya of Mayapán: A bioarchaeological perspective. Pp. 232–50 in V. Tiesler and A. Cucina (eds.), *New Perspectives on Human Sacrifice and Ritual Body Treatments in Ancient Maya Society*. New York: Springer.

Serjeantson, D., and J. Morris
2011 Ravens and crows in Iron Age and Roman Britain. *Oxford Journal of Archaeology* 30: 85–107.

Sharples, N.
1991 Warfare in the Iron Age of Wessex. *Scottish Archaeological Forum* 8: 79–89.

Shepherd, I. A. G.
1995 The Sculptor's Cave, Covesea, Moray: From Bronze Age ossuary to Pictish shrine? *Proceedings of the Society of Antiquaries of Scotland* 125: 1194–5.
2007 'An awesome place': The Late Bronze Age use of the Sculptor's Cave, Covesea, Moray. Pp. 194–203 in C. Burgess, P. Topping, and F. Lynch (eds.), *Beyond Stonehenge: Essays on the Bronze Age in Honour of Colin Burgess*. Oxford: Oxbow.

Sheppard, P. J., R. Walter, and T. Nagaoka
2000 The archaeology of head-hunting in Roviana Lagoon, New Georgia. Pp. 9–37 in 'Essays on Head-Hunting in the Western Solomon Islands'. Special issue of the *Journal of the Polynesian Society* 109. Auckland: Polynesian Society.

Sievers, S.
2002 Manching revisited. *Antiquity* 76: 943–4.

Silius Italicus
1934 *Silius Italicus: Punica*. Trans. J. D. Duff. Cambridge, Mass.: Harvard University Press.

Sintès, C. (ed.)
1996 *Musée de l'Arles Antique*. Arles: Actes Sud.

Skene, W. F. (trans.)
1858 *The Four Ancient Books of Wales*. Edinburgh: Edmonston and Douglas.

Sobin, G.
1999 *Luminous Debris: Reflecting on Vestige in Provence and Languedoc*. Berkeley: University of California Press.

Solier, Y.
1979 Découverte d'inscriptions sur plombs en écriture ibérique dans un entrepôt de Pech-Maho (Sigean). *Revue Archéologiques de Narbonnaise* 12: 55–123.

Somerville, B. T.
1897 Ethnographical notes in New Georgia, Solomon Islands. *Journal of the Royal Anthropological Institute* 26: 357–412.

Sopena, G.
2005 Celtiberian ideologies and religion. *E–Keltoi* 6: 347–410.

Speal, S.
2006 Postmortem skeletal modifications of the pre-Columbian North American mid-Continent. Pp. 119–31 in M. Bonogofsky (ed.), *Skull Collection, Modification and Decoration*. British Archaeological Reports International Series. Oxford: Archaeopress.

Speidel, M.
1970 The captor of Decebalus: A new inscription from Philippi. *Journal of Roman Studies* 60: 142–53.

Spencer, C.
1982 *The Cuicátlan Cañada and Monte Albán: A Study of Primary State Formation*. New York: Academic Press.

Stead, I. M.
1985 The Linsdorf monster. *Antiquity* 59: 40–2.

Stead, I. M., and V. Rigby
1999 *The Morel Collection: Iron Age Antiquities from Champagne in the British Museum*. London: British Museum Press.

Sterckx, C.
2005 *Les mutilations des ennemis chez les Celtes prechretiens: la tête, les seins, le Graal*. Paris: Association Kubaba.

Stokes, W.
1862 *Three Irish Glossaries*. London: Williams and Northgate.
1903 The Battle of Allen. *Revue Celtique* 24: 41–70.

Strathern, M.
1988 *The Gender of the Gift: Problems with Women and Problems with Society in Melanesia*. Berkeley: University of California Press.

Stuart, G.
1992 Mural masterpieces of ancient Cacaxtla. *National Geographic* 182.3: 120–36.

Szabó, M.
1965 A Celtic double head from Badaczony-Lábdi. *Acta Archaeologia Hungaria* 17: 233–50.
1993 Eléments anthropomorphes dans les décor des fourreaux Danubiens. Pp. 271–86 in J. Briard and A. Duval (eds.), *Les représentations humaines du Néolithique à l'Âge du Fer I*. Actes du 115e Congrès National des Sociétés Savantes. Paris: Éditions du C.T.H.S.

Tacitus
1942 *The Complete Works of Tacitus*. Trans. A. J. Church and W. J. Brodribb. New York: Random House.

Taracena, B.
1943 Cabezas-trofeo en la España céltica. *Archivo Español de Arqueologia* 16: 157–71.

Tarlow, S.
2008 The extraordinary history of Oliver Cromwell's head. Pp. 69–78 in D. Borić and J. Robb (eds.), *Past Bodies: Body-Centred Research in Archaeology*. Oxford: Oxbow.

Taube, K.
1983 The classic Maya maize god: A reappraisal. Pp. 171–81 in M. G. Roberston and V. M. Fields (eds.), *Fifth Palenque Round Table, 1983*. San Francisco: Pre-Columbian Art Research Institute.

Taylor, T.
1991 Celtic Art. *Scottish Archaeological Review* 8: 129–32.
2002 *The Buried Soul*. London: 4th Estate.

Tierney, J. J.
1960 The Celtic ethnography of Poseidonius. *Proceedings of the Royal Irish Academy* 60: 189–275.

Trancik Petitpierre, V.
1996 Anthropologische Beobachtungen an Isoliert Vorliegerden Skelletresten aus der Siedlung Basel-Gasfabrik. *Archéologie Suisse* 4: 153–5.

Tremearne, A. J. N.
1912 *The Tailed Headhunters of Nigeria: An Account of an Official's Seven Years' Experiences in the Northern Nigerian Pagan Belt, and a Description of the Manners, Habits and Customs of the Native Tribes*. London: Seeley, Service.

Tress Barry, F.
nd (c.1901–8) Hillhead. Unpublished typescript, National Museums of Scotland.

Tucker, F. C.
2010 Woven into the Stuff of Other Men's Lives: The Treatment of the Dead in Iron Age Atlantic Scotland. PhD diss., University of Bradford.

Tucker, F. C., and I. Armit
2009 Human remains from Iron Age Atlantic Scotland Dating Project. *Discovery and Excavation in Scotland* 10: 214–16.

Turk, P.
2005 *Images of Life and Myth*. Ljubljana: Narodni Muzej Slovenije.

Underhill, A. P.
2006 Warfare and the development of states in China. Pp. 253–85 in E. N. Arkush and M. W. Allen (eds.), *The Archaeology of Warfare: Prehistories of Raiding and Conquest*. Gainesville: University Press of Florida.

Up de Graff, F.
1923 *Head Hunters of the Amazon: Seven Years of Exploration and Adventure*. New York: Duffield.

van der Kroef, J.
1952 Some head-hunting traditions of southern New Guinea. *American Anthropologist* 54.2.1: 221–35.

Vayda, A. P.
1960 *Maori Warfare*. Polynesian Society Maori Monographs 2. Wellington.

Verdin, F.
2003 Glanum et Nîmes: un même processus d'urbanisation, des destinées différentes. Pp. 565–73 in M. Bats, B. Dedet, P. Garmy, T. Janin, C. Raynaud, and M. Schwaller (eds.), *Peuples et*

territoires en Gaule méditerranéenne. Hommage à Guy Barruol. Revue Archéologique de Narbonnaise Supplément 35. Montpellier.

von Fürer-Haimendorf, C.
1969 *The Konyak Nagas: An Indian Frontier Tribe*. New York: Holt, Rinehart and Winston.

Waddell, J.
1998 *The Prehistoric Archaeology of Ireland*. Galway: Galway University Press.

Wait, G. A.
1985 *Ritual and Religion in Iron Age Britain*. British Archaeological Reports, British Series 149. 2 vols. Oxford: Archaeopress.

Walker, L.
1984 The deposition of human remains. Pp. 442–63 in B. Cunliffe, *Danebury: An Iron Age Hillfort in Hampshire*, vol. 2, *The Excavations, 1969–78: The Finds*. CBA Research Report 52. London: Council for British Archaeology.

Walsh, P.
1894 Maori preserved heads. *Transactions and Proceedings of the Royal Society of New Zealand* 27: 610–16.

Warmenbol, E.
1996 L'or, la mort et les hyperboréens. La bouche des enfers ou le Trou de Han à Han-sur-Lesse. Pp. 203–34 in *Archäologische Forschungen zum Kultgeschehen in der jüngeren Bronzezeit und frühen Eisenzeit Alteuropas Erbgenisse eines Kolloquiums in Regensburg 4–7 okt 1993*. Bonn: Habelt.
2007 *Les Grottes de Han-sur-Lesse*. Centre de Recherches Archéologiques. http://dev.ulb.ac.be/crea/AccueilFrancais.php?page=HanSurLesse.

Waterman, D.
1997 *Excavations at Navan Fort, 1961–71*. Northern Ireland Archaeological Monographs No. 3. Belfast: The Stationery Office.

Weber, K.
2002 *Das Ratsel der Kelten vom Glauberg*. Stuttgart: Konrad Theiss.

Webster, J.
1996 Ethnographic barbarity: Colonial discourse and 'Celtic warrior societies'. Pp. 111–23 in J. Webster and N. Cooper (eds.), *Roman Imperialism: Post-Colonial Perspectives*. University of Leicester.

West, M. L.
1999 *The East Face of Helicon: West Asiatic Elements in Greek Poetry Myth*. Oxford: Clarendon Press.

Wheeler, R. E. M.
1954 *The Stanwick Fortifications: North Riding of Yorkshire*. London: Research Committee of the Society of Antiquaries.
1958 Review of *Entremont: capitale celto-ligure des Salyens de Provence* by F. Benoît. *Journal of Roman Studies* 48: 211–12.

Whimster, R.
1981 *Burial Practices in Iron Age Britain: A Discussion and Gazetteer of the Evidence, c. 700 BC–AD 43*. British Archaeological Reports, British Series 90. Oxford: Archaeopress.

Williams, J. H. C.
2001 *Beyond the Rubicon: Romans and Gauls in Republican Italy*. Oxford: Oxford University Press.

Wilson, C. E.
1981 Burials within settlements in southern Britain during the pre-Roman Iron Age. *Bulletin of the Institute of Archaeology London* 18: 127–69.

Woodward, A.
1993 The cult of relics in prehistoric Britain. Pp. 1–8 in M. Carver (ed.), *In Search of Cult*. Woodbridge: Boydell.

Woolf, G.
1998 *Becoming Roman: The Origins of Provincial Civilisation in Gaul*. Cambridge: Cambridge University Press.

Wright, G. R. H.
1988 The severed head in earliest Neolithic times. *Journal of Prehistoric Religion* 2: 51–6.

Zegwaard, G. A.
1959 Headhunting practices of the Asmat of Netherlands New Guinea. *American Anthropologist* 61: 1020–41.

Index

For EU product safety concerns, contact us at Calle de José Abascal, 56–1°,
28003 Madrid, Spain or eugpsr@cambridge.org.

www.ingramcontent.com/pod-product-compliance
Lightning Source LLC
LaVergne TN
LVHW062121200926
848484LV00007B/1732
9780521877565